T0179301

New Directions in Civil Engineering

SERIES EDITOR: W. F. CHEN *Purdue University*

Response Spectrum Method in Seismic Analysis and Design of Structures
Ajaya Kumar Gupta *North Carolina State University*

Stability Design of Steel Frames
W. F. Chen *Purdue University*
E. M. Lui *Syracuse University*

Concrete Buildings: Analysis for Safe Construction
W. F. Chen *Purdue University*
K. H. Mosallam *Ministry of Interior, Saudi Arabia*

Unified Theory of Reinforced Concrete
Thomas T. C. Hsu *University of Houston*

Stability and Ductility of Steel Structures under Cyclic Loading
Yuhshi Fukumoto *Osaka University*
George C. Lee *State University of New York at Buffalo*

Advanced Analysis in Steel Frames
W. F. Chen *Purdue University*
S. Toma *Hokkaigakuen University*

Analysis and Software of Tubular Members
W. F. Chen *Purdue University*
S. Toma *Hokkaigakuen University*

Flexural-Torsional Buckling of Structures
N. S. Trahair *University of Sydney*
(only for sale by CRC Press in North America, the Philippines, or Mexico)

Water Treatment Processes: Simple Options
S. Vigneswaran *University of Technology, Sydney*
C. Visvanathan *Asian Institute of Technology, Bangkok*

UNIFIED THEORY OF REINFORCED CONCRETE

Thomas T. C. Hsu

Professor of Civil Engineering
University of Houston
Houston, Texas

CRC Press
Taylor & Francis Group
Boca Raton London New York

CRC Press is an imprint of the
Taylor & Francis Group, an **informa** business

CRC Press
Taylor & Francis Group
6000 Broken Sound Parkway NW, Suite 300
Boca Raton, FL 33487-2742

© 1993 by Taylor & Francis Group, LLC
CRC Press is an imprint of Taylor & Francis Group, an Informa business

First issued in paperback 2019

No claim to original U.S. Government works

ISBN 13: 978-0-367-45013-7 (pbk)
ISBN 13: 978-0-8493-8613-8 (hbk)

Visit the Taylor & Francis Web site at
http://www.taylorandfrancis.com

and the CRC Press Web site at
http://www.crcpress.com

Library of Congress Cataloging-in-Publication Data

Hsu, Thomas T. C. (Thomas Tseng Chuang), 1933-
    Unified theory of reinforced concrete/by Thomas T. C. Hsu.
        p.   cm.   —(New directions in civil engineering)
    Includes bibliographical references and index.
    ISBN 0-8493-8613-6
    1. Reinforced concrete construction. I. Title. II. Series.
IN PROCESS   92-19058
                                                                                          CIP

Foreword

Over a period of about 100 years, reinforced concrete has developed into the most widely used material for civil engineering and large building structures. Its popularity stems from its architectural flexibility, the ready availability of its constituent materials in most parts of the world, and its consequent economy. Its shortcomings, such as excessive cracking and deflection, have been overcome through the development of prestressed concrete and a better understanding of the requirements for the manufacture of durable concrete. Using these techniques in combination with high strength steel and concrete, structural concrete spans undreamed of a relatively few years ago are now commonplace.

The further development of reinforced concrete is impeded by the empirical basis of some design equations in use today, particularly in the area of shear. Because these equations are empirical and not based on a rational mechanical model for behavior, it is dangerous to use them in the design of members having properties outside the range of the variables included in the experiments upon which the empirical equations are based. This is recognized, for instance, in the 1989 edition of the ACI Building Code (ACI 318-89), which prevents recognition in design of any increase in shear strength due to increase in concrete strength beyond 10,000 psi. This is accomplished by limiting the value of the term $(f'_c)^{0.5}$ to 100 psi in all calculations of shear strength. It is necessary because the relevant equations are based on tests of members having concrete strengths generally in the range of 3000 to 8000 psi.

Initially, the design of reinforced concrete members was entirely empirical: it was based on tests of typical full scale members. This was quickly recognized as unsatisfactory and the quest commenced for a rational design theory for reinforced concrete. At first, attempts were made to develop expressions for the strength of members. However, at the turn of the century French engineers developed design equations for flexure and axial load assuming linear elastic behavior of the concrete and steel at service load and zero flexural tensile strength for the concrete. This became known as the working stress method of reinforced concrete design. Over the years it was realized that although this method of design ensured satisfactory behavior at service load, it could result in very different factors of safety for differing members designed for the same allowable stresses at service load. Consequently, the middle of this century saw a strong effort to develop theories for the strength of reinforced concrete members subject to flexure and to combined flexure and axial load, which were based on rational mechanical models of behavior and on the properties of the steel and the concrete. This was achieved in the 1950s. As a result, it became possible to express code requirements for design for flexure and for flexure combined with axial load as a set of basic assumptions from which all necessary design equations could be developed. Unfortunately, the same situation does not hold for shear and torsion, for which code design provisions contain many empirical equations.

Much research work has been carried out over the last 25 years with the objective of developing a fundamental understanding of the behavior of reinforced concrete when subjected to shear, torsion, and shear and torsion in combination with flexure and axial load, so that more rational methods of design can be developed. Professor Hsu has been in the vanguard of this work for the entire period. He is, therefore, extremely well qualified to undertake the task of developing a unified theory for the behavior of reinforced concrete when subject to any combination of flexure, axial load, shear, and torsion.

Professor Hsu first describes three widely accepted models for reinforced concrete behavior: the struts-and-ties model, the Bernoulli compatibility truss model for flexure and flexure combined with axial load, and the equilibrium (plasticity) truss model. He then goes on to develop two other models for the behavior of reinforced concrete elements subject to any combination of shear and direct stresses. These models satisfy compatibility as well as equilibrium. The first, "the Mohr compatibility truss model," assumes elastic behavior in the steel and concrete and is applicable to the calculation of service load behavior. The second, "the softened truss model," also describes behavior at higher loads. It accounts for nonlinear behavior of the concrete and steel, the influence of concrete tensile stresses between cracks on the average stress–strain relationship for reinforcing steel, and the influence of orthogonal tensile stresses on the compression stress–strain relationship for concrete. Professor Hsu then applies the second model to the problem of reinforced concrete members subject to torsion, showing how the entire torque-twist curve may be calculated. Finally, Professor Hsu performs a very useful service for practicing engineers by using the softened truss model to develop a simple expression for the thickness of the shear flow zone in a member subject to torsion. This greatly facilitates the calculation of torsional strength.

I have known Professor Hsu for the past 30 years, initially as a colleague at the Portland Cement Association Research and Development Laboratories, where I was responsible for his first becoming involved with the study of torsion in reinforced concrete. Subsequently, we have worked together on the Shear and Torsion Committees of the American Concrete Institute, spending many hours (sometimes late into the night) discussing, with our fellow committee members, problems related to these subjects and to the development of relevant clauses for the ACI Building Code. I have come to have a great respect for his ability to present his case carefully and convincingly, in support of any technical position that he takes.

Professor Hsu is to be congratulated on a very clear and detailed presentation of the very complicated topic of the behavior of reinforced concrete when subject to complex combinations of stress. His book will be of interest not only to researchers and students of reinforced concrete, but also to practicing engineers, who are increasingly faced with the design of unusual structures for which the empirical design equations found in codes of practice are inadequate.

Alan H. Mattock
Professor Emeritus of Civil Engineering
University of Washington
Seattle, Washington

The RC Unified Theory and Seismic Analysis

Professor Hsu has developed a unified theory of reinforced concrete (RC) structures that is applicable to all the four basic actions of bending, axial load, shear and torsion. Based on the principles of the mechanics of materials, i.e., the stress equilibrium, the strain compatibility, and the constitutive laws of concrete and reinforcement, this is a rational and comprehensive theory that completes the knowledge of reinforced concrete subjected to static loadings. It is an important milestone in the history of the development of reinforced concrete theory and will surely become an essential part of study in any field involving the use of reinforced concrete.

As a researcher of structural engineering I would like to point out, in particular, the significance of Professor Hsu's unified theory to the development of earthquake engineering. Seismic responses of RC structures are predicted from three component sources of information: primary curves from static loadings, hysteretic loops from cyclic loadings, and dynamic models (see diagram). Of these three components, primary curves (representing the basic nature of concrete structures) are by far the most important and fundamental. For the past 30 years the primary curves of RC

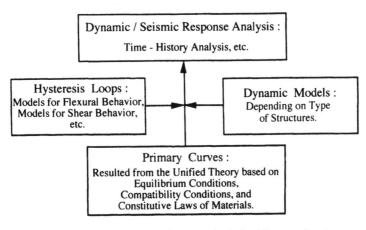

Algorithm for Dynamic / Seismic Response Analysis of Concrete Structures

structures could be rationally constructed only for bending or bending with axial load. That is, prediction of dynamic behavior, to date, has been limited to frame-type structures. Now that primary curves can be scientifically determined for shear and torsion according to Professor Hsu's unified theory, seismic response analysis can consequently be advanced to accurately predict the behavior of wall-type and shell-type structures. Because the superiority of wall-type structures (such as shear walls) in resisting earthquake loadings is gaining recognition in many countries, Professor Hsu's unified theory is increasingly being appreciated by researchers in earthquake engineering world-wide.

Finally, reading Professor Hsu's book, as taking his lectures, is a process of enlightenment. He has the unique talent to extract physical meaning out of abstract mathematics. From the complicated, he distills principles; from the seemingly random, he perceives patterns. Professor Hsu's devotion to his work and his magnanimous generosity in sharing his knowledge is an example to all of us in the teaching and research profession.

Y. L. Mo
Associate Professor of Civil Engineering
National Cheng Kung University
Tainan, Taiwan

Preface

Reinforced concrete structures are subjected to a complex variety of stresses and strains. The four basic actions are *bending, axial load, shear*, and *torsion*. Each action alone, or in combination with others, may affect structures in different ways under varying conditions. At present, there is no one comprehensive theory for reinforced concrete structural behavior that addresses all of these basic actions and their interactions. Although the theoretical bases for bending and for axial loads are better understood, the current design methods for shear and torsion are still based on empirical information. There is little consistency among countries around the world in their building codes, especially in the specifications for shear and torsion.

The purpose of this book is to integrate the collection of available information with new research data and develop one **unified theory** of reinforced concrete behavior that embraces and accounts for all of the four basic actions and their combinations. This rational, unified theory of reinforced concrete behavior is based on the three fundamental principles of the mechanics of materials, i.e., *stress equilibrium, strain compatibility*, and *constitutive laws of materials*. The theoretical methods presented in this book replace a host of empirical formulas currently used. This unified theory can serve as the foundation of a universal design code that can be adopted internationally.

The new data utilized in this book are primarily from research on shear and torsion through the development of the various types of truss model theories. These theories provide a clear concept of how reinforced concrete structures resist shear and torsional stresses, and show that the treatment of these two actions can be combined. These new studies also show that the bending and axial loads can be integrated into the truss model theories together with shear and torsion. Additionally, the effect of prestressing can be included in these theories in a logical way. Altogether, it is now possible to develop a unified truss model theory that encompasses all four of the actions, in reinforced as well as prestressed concrete.

The **unified theory** has five component models: *the struts-and-ties model, the Bernoulli compatibility truss model, the equilibrium (plasticity) truss model, the Mohr compatibility truss model*, and *the softened truss model*. The **unified theory** is presented in increasing complexity from the simplest struts-and-ties concept neglecting compatibility conditions to the most complicated action of torsion involving both Mohr and Bernoulli compatibility conditions. This sequence of presentation is desirable not only from a pedagogical point of view, but also because it corresponds closely to the historical development of the five models.

The first model, the struts-and-ties model, illustrates the powerful truss concept for reinforced concrete structures in which the compressive stresses are resisted by the concrete struts and the tensile stresses by the reinforcing ties. The second model, the Bernoulli compatibility truss model, deals with bending action alone or in combination with axial loads. This bending theory satisfies not only the equilibrium condition, but also the Bernoulli linear compatibility condition and the uniaxial constitutive laws of materials. As a result, it can predict not only the flexural strength, but also the flexural load-deformation history of a member. In the treatment of bending action the three fundamental principles of the mechanics of materials are highlighted, and strong emphasis is placed on the capability of this model to predict strains, deflections, and ductilities. Mathematical verifications of these flexural deformations are required to ensure the serviceability of reinforced concrete structures in general and to design reinforced concrete structures in earthquake regions.

The third, fourth, and fifth models deal with shear, torsion, and their interaction with axial loads and bending. The third model, the equilibrium (plasticity) truss model, is based on the equilibrium condition and the theory of plasticity. Because compatibility conditions and constitutive laws of materials are not considered, this third model is particularly clear and elegant in expressing the interactions of the four actions. The fourth and fifth models, i.e., the Mohr compatibility truss model and the softened truss model, are developed for shear action and its combination with biaxial, normal stresses in membrane elements. These models are based on the two-dimensional equilibrium condition and Mohr's circular compatibility condition. The Mohr compatibility truss model, which utilizes Hooke's linear laws for the materials, is consistent with the theory of elasticity and is applicable to structures under service loads. The softened truss model, which utilizes the softened biaxial constitutive law of concrete, can predict not only the shear strengths, but also the shear load–shear deformation history of a member. Because of its capability to predict shear deformations and shear ductility, the softened truss model can be used for the shear design of wall-type or shell-type reinforced concrete structures to resist earthquake loadings (Hsu and Mau, 1992).

Torsional action is particularly interesting, because it requires not only the Mohr compatibility to treat the two-dimensional shear action, but also the Bernoulli compatibility to consider the bending action of the concrete struts. This three-dimensional torsional action can now be solved by using a combination of the softened truss model and the Bernoulli compatibility truss model.

In this book the process of developing the integrated theory from its five components is presented in a systematic manner pedagogically and historically, emphasizing the fundamental principles of equilibrium, compatibility, and the constitutive laws of materials. In this way, the intrinsic consistency of the five component models is illustrated. Because of its rational manner, this book can well serve as a text for a graduate course in structural engineering. In fact, the manuscript of this book did serve as the text for a new 3-credit course in the Fall semester of 1990 at the University of Houston, and in the Fall quarter of 1991 at Northwestern University.

In summary, this book presents, for the first time, a comprehensive **unified theory** for reinforced concrete behavior, advancing both the science and the technology of structural engineering. Because the consumption of reinforced concrete in the U.S. well exceeds 100 billion dollars a year, this unified theory and its application infer a significant economic impact in this country and worldwide. Practicing structural engineers will need this book to update current knowledge in this field.

This book was completed at Northwestern University, Evanston, Illinois, during a sabbatical leave made possible by the Eshbach Distinguished Visiting Professorship, Robert R. McCormick School of Engineering and Applied Sciences. Special thanks are given to Professor Surendra P. Shah, Director, NSF Center for ACBM, and his wife, Dorothie, for providing an intellectual and congenial environment, conducive to the completion of this book.

Thomas T. C. Hsu
University of Houston
and Northwestern University

To

my wife, my mother,
and the memory of my father

Contents

CHAPTER 8. SOFTENED TRUSS MODEL FOR TORSION

1

Introduction

1.1   Overview

A reinforced concrete structure may be subjected to four basic types of actions: bending, axial load, shear, and torsion. All of these actions can, for the first time, be analyzed and designed by a single unified theory based on the three fundamental principles of mechanics of materials: namely, the stress equilibrium condition, the strain compatibility condition, and the constitutive laws of concrete and steel. Because the compatibility condition is taken into account, this theory can be used to reliably predict the strength of the structure, as well as its load-deformation behavior.

Extensive research of shear action in recent years resulted in the development of various types of truss model theories. The newest theories for shear can now rigorously satisfy the two-dimensional stress equilibrium, Mohr's two-dimensional circular strain compatibility, and the softened biaxial constitutive laws of concrete. In practice, this new information on shear can be used to predict the shear load–shear deformation histories of reinforced concrete structures including deep beams and low-rise shear walls. Understanding the interaction of shear and bending is essential to the design of beams, bridge girders, high-rise shear walls, etc.

The simultaneous application of shear and biaxial loads on a two-dimensional element produces the important stress state known as *membrane stresses*. The two-dimensional element, known as *membrane element*, represents the basic building block of a large variety of structures made of walls and shells. Such structures, including submerged containers, offshore platforms, and nuclear containment vessels, can be very large with walls several feet thick. The information in this book opens the way for the rational analysis and design of these wall- and shell-type structures, based on the three fundamental principles of the mechanics of materials for two-dimensional stress and strain states.

The simultaneous application of bending and axial load is also an important stress state prevalent in columns, piers, caissons, etc. The design and analysis of these essential structures are presented in a new light, emphasizing the three principles of mechanics of materials for parallel stress state, i.e., the parallel stress equilibrium, the Bernoulli linear strain compatibility, and the uniaxial constitutive laws of materials.

1

The three-dimensional stress state of a member subjected to torsion must take into account the two-dimensional shear action in the shear flow zone, as well as the bending action of the concrete struts caused by the warping of the shear flow zone. Because both the shear action and the bending action can be taken care of by the simultaneous applications of Mohr's compatibility condition and Bernoulli's compatibility condition, the torsion action becomes, for the first time, solvable in a scientific way. This book provides all of the necessary information leading up to the rational solution of the problem in torsion.

Because each of the four basic actions experienced by reinforced concrete structures has been found to adhere to the fundamental principles of the mechanics of materials, a unified theory is developed encompassing bending, axial load, shear, and torsion in reinforced as well as prestressed concrete structures. This book is devoted to the integration of all the individual theories for the various stress states. Upon synthesis, the new rational theories should replace the many empirical formulas currently in use for shear, torsion, and membrane stresses.

The unified theory is divided into five model components based on the fundamental principles employed and the degree of adherence to the rigorous principles of mechanics of materials. The five models are: (1) the struts-and-ties model, (2) the equilibrium (plasticity) truss model, (3) the Bernoulli compatibility truss model, (4) the Mohr compatibility truss model, and (5) the softened truss model. In this book the five models are presented in a systematic manner, focusing on the significance of their intrinsic consistencies and their interrelationships. Because of its inherent rationality, this unified theory of reinforced concrete can serve as the basis for the formulation of a universal and international design code.

In Section 1.2 of this chapter, the position of the unified theory in the field of structural engineering is presented. Then the five components of the unified theory are introduced and defined in Section 1.3, followed by a historical review of the five truss models and some projections of future development. The organization of this book is explained in Section 1.4 before embarking on the introduction of the first model in Section 1.5.

In the main, this book presents a systematic and rigorous study of the last four truss models in Chapters 2 to 8.

1.2   Structural Engineering

1.2.1   STRUCTURAL ANALYSIS

We will now look at the structural engineering of a typical reinforced concrete structure, and will use, for example, a typical frame-type structure for a manufacturing plant as shown in Figure 1.1. The main portal frame with high ceiling accommodates the processing work. The columns have protruding corbels to support an overhead crane. The space on the right with low ceiling serves as offices. The roof beams of the office are supported by spandrel beams that, in turn, are supported by corbels on the left and columns on the right.

The structure in Figure 1.1 is subjected to all four types of basic actions—bending M, axial load N, shear V, and torsion T. The columns are subjected to bending and axial load, while the beams are under bending and shear. The spandrel beam carries torsional moment in addition to bending moment and shear force. Torsion frequently occurs in edge beams where the loads are transferred to the beams from one side only. The magnitudes of these four actions are obtained by performing a frame analysis under specified loads. The analysis can be based on the elastic or the inelastic

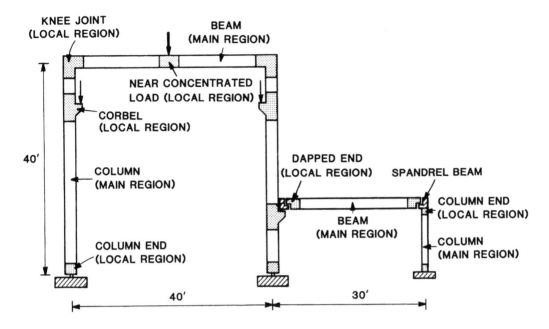

Figure 1.1 A typical frame-type reinforced concrete structure.

material laws, and the cross sections can either be uncracked or cracked. In this way, the four M, N, V, and T diagrams are obtained for the whole structure. This process is known as the *structural analysis*.

Table 1.1 illustrates a four-step general scheme in the structural engineering of a reinforced concrete structure. The process of structural analysis is the first step as indicated in row 1 of the table. Although this book will not cover structural analysis in detail, information on this topic can be found in many standard textbooks on the subject.

1.2.2 MAIN REGIONS VERSUS LOCAL REGIONS

The second step in the structural engineering of a reinforced concrete structure is to recognize the two types of regions in the structure, namely, the main regions and the local regions. The local regions are indicated by the shaded areas in Figure 1.1. They include the ends of a column or a beam, the connections between a beam and a column, the corbels, the region adjacent to a concentrated load, etc. The large unshaded areas, which include the primary portions of each member away from the local regions, are called the main regions.

From a scientific point of view, a main region is the region where the stresses and strains are distributed so regularly that they easily can be expressed mathematically. That is, the stresses and strains in the main regions are governed by simple equilibrium and compatibility conditions. For columns that are under bending and axial load, the equilibrium equations come from the parallel force equilibrium condition, whereas the compatibility equations are governed by Bernoulli's hypothesis of plane section remaining plane. In the case where beams are subjected to shear and torsion, the stresses and strains should satisfy the two-dimensional equilibrium and compatibility conditions, i.e., Mohr's stress and strain circles.

In contrast, a local region is the region where the stresses and strains are so disturbed and irregular that they are not amenable to mathematical solution. In

TABLE 1.1 Unified Theory of Reinforced Concrete Structures

(1) Structural analysis	(2) Division of regions	(3) Design actions	(4) Principles of analysis and design			
Whole structure (elastic or inelastic) Bending (M), axial load (N), shear (V), torsion (T)	Main regions	Sectional actions		Equilibrium	Uniaxial	Bernoulli compatibility
		M, N	Bernoulli compatibility truss model (uniaxial)			Nonsoftened laws of material
		V, T	Mohr compatibility truss model (biaxial)		Biaxial	Mohr compatibility
			Softened truss model			Softened laws of material
	Local regions	Boundary stresses	M, N, V, T	Equilibrium (plasticity) truss model	Struts-and-ties model	

particular, the compatibility conditions are difficult to apply. In the design of the local regions the stresses are usually determined by equilibrium condition alone; the strain conditions are neglected. Numerical analysis by computer (such as finite element method) can possibly determine the stress and strain distributions in the local regions, but it is seldom employed because of its complexity.

The local region is sometimes referred to as the D region. The prefix D indicates that the stresses and strains in the region are <u>d</u>isturbed or that the region is discontinuous. Analogously, the main region is often called the B region, noting that the strain condition in a flexural region satisfies <u>B</u>ernoulli's compatibility condition. This terminology does not take into account the strain conditions of structures subjected to shear and torsion, which should satisfy <u>M</u>ohr's compatibility condition. For this reason, the term "B region" is best avoided.

The second step of structural engineering is the division of the main regions and local regions in a structure as indicated in row 2 of Table 1.1. On the one hand, the main regions of a structure are designed directly by the four sectional actions, M, N, V, and T, according to the four sectional action diagrams obtained from structural analysis. On the other hand, the local regions are designed by stresses acting on the boundaries of the regions. These boundary stresses are calculated from the four action diagrams at the boundary sections. A local region is actually treated as an isolated free body subjected to external boundary stresses. The third step of structural engineering is the determination of the design actions for the two regions. The third step of finding the sectional actions for main regions and the boundary stresses for local regions is indicated in row 3 of Table 1.1.

1.2.3   MEMBER AND JOINT DESIGN

Once the diagrams of the four actions are determined by structural analysis and the two regions are identified, all the main regions and local regions can be designed. This fourth step of structural engineering is commonly known as the *member and joint design*. More precisely, it means the *design and analysis of the main and local regions*. By this process the size and the reinforcement of the members as well as the arrangement of reinforcement in the joints are determined.

The unified theory aims to provide this fourth and most important step with a rational method of design and analysis for all of the main and local regions in a typical reinforced concrete structure, such as the one in Figure 1.1. It serves to synthesize all the rational theories and to replace all the empirical design formulas for these regions.

The position of the unified theory in the scheme of structural engineering is shown in row 4 of Table 1.1. The five model components of the unified theory are distinguished by their adherence to the three fundamental principles of the mechanics of materials (the equilibrium condition, the compatibility condition, and the constitutive laws of materials). The five models are named to reflect the most significant principle(s) embodied in each as listed in the following section.

1.3   Five Component Models of the Unified Theory

1.3.1   PRINCIPLES AND APPLICATIONS OF THE FIVE MODELS

Some of the five models are intended for the main regions and some for the local regions. Others may be particularly suitable for the service load stage or the ultimate load stage. A summary of the five models, together with their basic principles and the scope of their applications, follows.

1 *Struts-and-Ties Model*:
        Principles: Equilibrium condition only.
        Applications: Design of local regions.

2 *Equilibrium (Plasticity) Truss Model*:
        Principles: **Equilibrium** condition and the theory of **plasticity**.
        Applications: Analysis and design of M, N, V, and T in the main regions at the ultimate load stage.

3 *Bernoulli Compatibility Truss Model*:
        Principles: Equilibrium condition, **Bernoulli compatibility** condition, and the uniaxial constitutive laws of concrete and reinforcement. The constitutive laws may be linear or nonlinear,
        Applications: Analysis and design of M and N in the main regions at both the serviceability and the ultimate load stages.

4 *Mohr Compatibility Truss Model*:
        Principles: Equilibrium condition, **Mohr compatibility** condition, and Hooke's uniaxial constitutive law for both concrete and reinforcement.
        Applications: Analysis and design of V and T in the main regions at the serviceability load stage.

5 *Softened Truss Model*:
        Principles: Equilibrium condition, Mohr's compatibility condition, and the **softened** biaxial constitutive law of concrete. The constitutive law of reinforcement may be linear or nonlinear.
        Applications: Analysis and design of V and T in the main regions at both the serviceability and the ultimate load stages.

1.3.2   HISTORICAL DEVELOPMENT OF TRUSS MODELS

A historical review of the five truss models of the unified theory is briefly presented here, with a focus on the development of the basic principles of analysis and design of reinforced concrete structures.

Concrete is a material that is very strong in compressive strength but weak in tensile strength. When concrete is used in a structure to carry loads, the tensile regions are expected to crack and, therefore, must be reinforced by materials of high tensile strength, such as steel. The concept of utilizing concrete to resist compression and reinforcement to carry tension gave rise to the **struts-and-ties model**. In this model, concrete compression struts and the steel tension ties form a truss that is capable of resisting applied loadings. Ever since Joseph Monier, a French gardener, used iron meshes to reinforce his flower pots in 1857, the struts-and-ties model has been used, intuitively, by engineers to design reinforced concrete structures.

The struts-and-ties concept was easily applied to reinforced concrete beams. For example, under bending, the compressive stress in the upper part of a simply

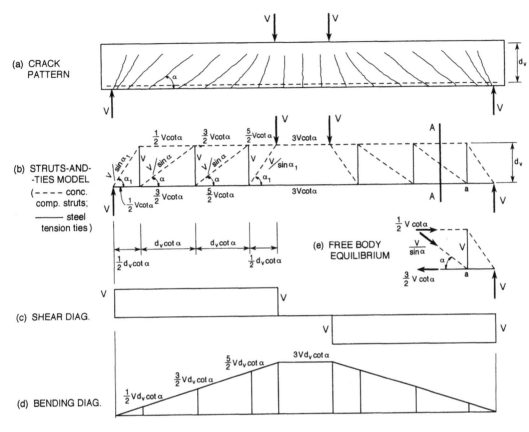

Figure 1.2 Struts-and-ties model of a concrete beam with bottom longitudinal rebars and stirrups resisting shear and bending.

supported beam is resisted by concrete in the form of a horizontal strut; the tensile stress in the lower portion is taken by the bottom steel in the form of a horizontal tie. The forces in the concrete and in the steel must be in equilibrium, and they form a couple to resist the applied bending moment. The distribution of stresses along the depth of the cross section, however, cannot be determined by equilibrium alone. The solution to the stress distribution must rely also on the strain compatibility of the beam and the constitutive relationship of the materials in order to correlate the strains and the stresses. The strain compatibility condition for bending was Bernoulli's hypothesis, which essentially states that a plane section before bending will remain a plane after bending. The simultaneous application of Bernoulli's hypothesis, the equilibrium condition, and the uniaxial constitutive relationships of materials to the analysis of reinforced concrete beams gave rise to the **Bernoulli compatibility truss model**. Although not clearly documented, this model had been used by engineers since the late 19th century, serving as the fundamental theory of reinforced concrete behavior for more than a 100 years. This theoretical model was easily expanded to address columns subjected to bending as well as axial load.

The first application of the concept of truss model to shear was proposed by Ritter (1899) and Morsch (1902) in connection with reinforced concrete beams subjected to shear and bending (Figure 1.2). In their view, a reinforced concrete beam acts like a parallel-stringer truss to resist bending and shear. Due to the bending moment, the concrete strut near the upper edge serves as the top stringer in a truss and the steel

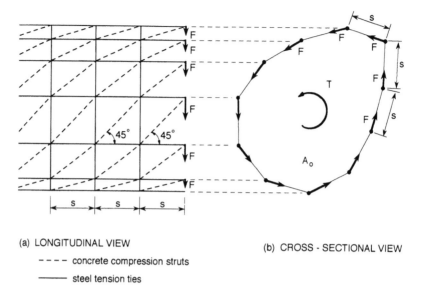

(a) LONGITUDINAL VIEW

(b) CROSS - SECTIONAL VIEW

- - - - concrete compression struts

——— steel tension ties

Figure 1.3   **Space truss model of a concrete beam with longitudinal and hoop steel resisting torsion.**

bar near the lower edge assumes the function of the bottom stringer. From shear stresses, the web region would develop diagonal cracks at an angle α inclined to the longitudinal steel. These cracks would separate the concrete into a series of diagonal concrete struts. To resist the applied shear forces after cracking, the transverse steel bars in the web would carry tensile forces and the diagonal concrete struts would resist the compressive forces. The transverse steel, therefore, serves as the tensile web members in the truss and the diagonal concrete struts become the diagonal compression web members.

The plane truss model for beams was extended by Rausch (1929) to treat members subjected to torsion. In Rausch's concept, a torsional member is idealized as a space truss formed by connecting a series of component plane trusses capable of resisting shear action (Figure 1.3). The circulatory shear stresses, developed in the cross section of the space truss, form an internal torsion moment capable of resisting the applied torsional moment.

Although Ritter, Morsch, and Rausch contributed significantly to the understanding of reinforced concrete structural behavior, their truss models could not explain some behavior of reinforced concrete, particularly regarding the so-called contribution of concrete. Researchers did not pursue this line of study until the late 1960s when Nielson (1967) and Lampert and Thurlimann (1968, 1969) derived the three fundamental equilibrium equations for shear based on the theory of plasticity. The interactive relationships of bending, shear, and torsion were further elucidated by Elfgren (1972). These theories were collectively known as the **plasticity truss model** because they were based on the yielding of steel. In the unified theory for reinforced concrete behavior, the **equilibrium (plasticity) truss model** takes into consideration the equilibrium condition alone. The role of the compatibility condition and the constitutive laws of materials need to be investigated in the future.

The next significant development was the determination, by Collins (1973), of the angle of inclination of the concrete struts by using the compatibility condition in a reinforced concrete element subjected to shear. Because this angle was assumed to coincide with the angle of inclination of the principal compressive stress and strain, this theory was known as the compression field theory. However, in this book this theory will be referred to as the **Mohr compatibility truss model** in the unified theory, because the average strain condition must satisfy Mohr's compatibility condition. The introduction of the Mohr compatibility condition into shear analysis was an important advancement. Three compatibility equations can be established from the geometric relationships in Mohr's circle.

Prior to 1972, the stress–strain curve of the concrete struts was assumed to be the same as that obtained from the uniaxial compression tests of standard concrete cylinders. This assumption led to serious overestimations of the shear and torsional strengths. In 1972, Robinson and Demorieux (1972) observed that a reinforced concrete panel subjected to compression in one direction was softened by tension in the perpendicular direction. This biaxially softening phenomenon was later quantified by Vecchio and Collins (1981, 1982), who proposed a stress–strain curve incorporating a softening coefficient.

By combining the equilibrium, the compatibility, and the softened constitutive laws of concrete, the author and his colleagues developed a theory that, with good accuracy, can predict the behavior of various types of structures subjected to shear. These structures include low-rise shear walls (Hsu and Mo, 1985d, Mau and Hsu, 1986), framed shear walls (Mau and Hsu, 1987a), shear transfer (Hsu, Mau, and Chen, 1987), and deep beams (Mau and Hsu, 1987b, 1989). By including an additional equilibrium equation and four additional compatibility equations, the theory became applicable to torsion (Hsu and Mo, 1985a, 1985b, 1985c). This theory unified shear and torsion (Hsu, 1988), and was called the **softened truss model**. The name emphasizes the importance of incorporating the softened constitutive law of concrete in the analysis of reinforced concrete structures. This theory can predict not only the shear and torsional strengths, but also the shear and torsional load-deformation behavior of a structure throughout its post-cracking loading history.

1.3.3   FUTURE DEVELOPMENTS

The unified theory of reinforced concrete behavior as presented in this work is necessarily limited to the present state of knowledge. Further developments of this rational theory are expected in the following areas.

(1) Fixed-Angle Theory and Contribution of Concrete

All of the three models for shear and torsion presented in this book (i.e., equilibrium truss model, Mohr compatibility truss model, and softened truss model) are based on a fundamental assumption that is not exactly correct. It is assumed that the direction of the cracks in a membrane element coincides with the direction of the principal stresses and strains in the concrete *after* cracking. In actuality, the direction of the first crack is determined by the direction of the principal stresses *before* cracking, which, in general, is different from the direction of post-cracking principal stresses. With increasing loading, additional cracks occur and the direction of these additional cracks tends to "rotate" toward the direction of the post-cracking principal stresses. Thus, the three models for shear and torsion can be called the *rotating-angle theory*.

The three models of the rotating-angle theory have a common weakness. They are incapable of predicting the so-called contribution of concrete. Tests have shown that the shear strength of a membrane element is made up of two terms: the "major term" attributed to the steel and the "minor term" attributed to the concrete. The existence of the minor term is apparently caused by the fact that the actual direction of cracks is, in general, different from the assumed direction of the principal stresses and strains after cracking (Pang and Hsu, 1992).

Models for shear and torsion based on the direction of the first crack (different from the direction of post-cracking principal stresses) have been developed. Such models can be categorized as the *fixed-angle theory*. The fixed-angle theory is capable of predicting the "contribution of concrete," but is considerably more complicated than the corresponding rotating-angle theory. The complexity stems from two sources. First, the fixed-angle theory must incorporate the constitutive law that relates the shear stress to the shear strain in the direction of the cracks. Extensive efforts are required to generate this constitutive law from the testing of reinforced concrete panels. Second, the equilibrium and compatibility equations in the fixed-angle theory become considerably more complicated than those in the rotating-angle theory. Efficient algorithms to solve the complex equations are needed before the "contribution of concrete" can be derived mathematically.

In actuality, the angle of inclination of the concrete struts at failure lies between those assumed by the fixed-angle theory and the rotating-angle theory. Therefore, these two theories furnish the two boundaries for the true situation. Because the more complicated fixed-angle theory is still being developed, this book includes only the simpler rotating-angle theory.

(2) Interaction of Four Actions

The interactions of the four actions—moment, axial load, shear, and torsion—are presented in Section 3.2 based on the equilibrium (plasticity) truss model. As such, these interaction relationships do not take into account the compatibility condition and the constitutive laws of materials. Future work that incorporates all of the three principles remains to be done.

(3) Local Regions

As discussed in Section 1.2.2, the local regions of a reinforced concrete structure are those areas where stresses and strains are irregularly distributed. These regions include the knee joints, corbels and brackets, dapped ends of beams, ledgers of spandrel beams, column ends, anchorage zone of prestressed beams, etc. The struts-and-ties model can be used for the design of local regions by providing a clear concept of the stress flows, following which the reinforcing bars can be arranged. However, the struts-and-ties model does not provide a unique solution, and the best solution may elude an engineer. The best solution is usually the one that best ensures the serviceability of the local region and its ultimate strength. Such service and ultimate behavior of a local region are difficult to predict because they are strongly affected by the cracking and the bond slipping between the reinforcing bars and the concrete. Fortunately, research in the analysis and design of local regions is currently being carried out in earnest around the world. Rapid development is expected in this area.

(4) Numerical Methods by Computer

The softened truss model is directly applicable to the membrane elements in wall- or shell-type structures. The question of how these membrane elements interact with each other and with the boundary elements remains to be investigated for each specific type of structure. Numerical methods, which take advantage of high-speed electronic computers, are expected to be very helpful in resolving these problems. These methods include the finite element method and the finite difference method. At present, the application of numerical methods to structural analysis is based mostly on Hooke's law, which results in a linear analysis. The softened truss model, which is based on the softened constitutive law of concrete, could serve to establish more realistic nonlinear numerical methods. In short, there is a great potential for the applications of numerical methods to the analysis of wall- and shell-type reinforced concrete structures.

(5) Fracture Mechanics

The softened constitutive laws of reinforced concrete are the most crucial component in the softened truss model. At present they are obtained directly from testing reinforced concrete panels. However, the nature of the two-dimensional softened phenomenon is unclear at present. Fracture mechanics could provide the key to the understanding of the biaxial constitutive laws of reinforced concrete.

1.4   Organization of Book Content

The unified theory is presented in increasing complexity from the simplest struts-and-ties concept (neglecting compatibility condition) to the most complicated action of torsion (involving both Mohr and Bernoulli compatibility conditions). This sequence of presentation is designed for effective teaching, and also corresponds to the historical development of the five models.

The first three chapters address the first three models, i.e., the struts-and-ties model, and Bernoulli compatibility truss model (bending theory), and the equilibrium (plasticity) truss model. Section 1.5 begins with an introduction of the concept of concrete compression struts and steel tension ties forming stable trusses. This truss concept underpins all the models in the unified theory.

The Bernoulli compatibility truss model in Chapter 2, commonly known as the bending theory, is presented in a systematic manner to highlight the three fundamental principles of the mechanics of materials for bending: the parallel stress equilibrium, the Bernoulli linear compatibility, and the uniaxial constitutive laws of materials. The content includes both the linear bending theory for checking serviceability criteria and the nonlinear bending theory for ultimate strength design. This is followed by a study of the interaction of bending with axial load.

The equilibrium (or plasticity) truss model in Chapter 3 is a concise summary of a vast amount of knowledge developed in the 1960s and 1970s on the application of the theory of plasticity to reinforced concrete structures. Emphasis is placed on the interaction of the four actions and the application of this model in the CEB-FIP code (1978).

The second half of the book, from Chapters 4 to 8, contain new information developed in the 1980s. Some are not yet in print. In Chapters 4 and 5 we will study the stresses and strains in a membrane element. The three basic equilibrium equations for reinforced concrete are derived from the transformation of two-dimensional stresses. The two-dimensional relationship of strains, of course, gives rise to three Mohr compatibility equations for solving shear problems. These two chapters provide the basic knowledge necessary for understanding the Mohr compatibility truss model and the softened truss model discussed in the last three chapters.

The Mohr compatibility truss model, discussed in Chapter 6, is the linear theory for membrane elements, and is applicable under service loads. It satisfies two-dimensional equilibrium, Mohr's circular compatibility and Hooke's uniaxial constitutive law of materials. This model is shown to be consistent with the solution by the theory of elasticity.

The softened truss model in Chapter 7 provides the nonlinear theory for membrane elements, and is applicable throughout the post-cracking loading history. This model satisfies the softened biaxial constitutive law of concrete, as well as two-dimensional equilibrium and Mohr's circular compatibility. This model has been successfully applied to structures under predominantly shear loadings.

The softened truss model is extended to solve the torsional problem in Chapter 8 by including additional equilibrium and compatibility conditions. These additional conditions include the bending of the concrete struts, which requires the use of Bernoulli compatibility. Because both the Mohr compatibility and the Bernoulli compatibility involved in the analysis are mathematically accountable, the three-dimensional torsion action is solved in a rational and rigorous manner. For practical design of torsion, however, some simple formulas are proposed.

1.5   Struts-and-Ties Model

1.5.1   GENERAL DESCRIPTION

During the 1980s, the struts-and-ties model received renewed interest because it was found to be the most powerful method for the design of local regions [see Sections 1.2.2 and 1.3.3(3)]. Much research was carried out to study the beam–column connections, corbels, dapped ends of beams, beam openings, etc. As a result, the struts-and-ties model was considerably refined. The improvements include a better understanding of the shear flow, the behavior of the "nodes" where the struts and ties intersect, and the dimensioning of the struts and the ties.

In the modern design concept, the local region is isolated as a free body and is subjected to boundary stresses obtained from the four action diagrams (see Section 1.2 and rows 1 to 3 of Table 1.1). The local region itself is imagined to be a free-form truss composed of compression struts and tension ties. The struts and ties are arranged so that the internal forces are in equilibrium with the boundary forces. In this design method the compatibility condition is not satisfied, and the serviceability criteria may not be assured. Understanding of the stress flows, the bond between the concrete and the rebars, and the steel anchorage requirement in a local region can help to improve serviceability and to prevent undesirable premature failures. A good design for a local region depends, to a large degree, on the experience of the engineer.

Proficiency in the application of this design method requires practice. An excellent treatment of the struts-and-ties model is given by Schlaich et al. (1987). This 77-page paper provides many examples to illustrate the application of this model.

For structures of special importance, the design of local regions by the struts-and-ties model may be supplemented by a numerical analysis, such as the finite element method, to satisfy both the compatibility and the equilibrium conditions. The constitutive laws of materials may be linear or nonlinear. Although numerical analysis can clarify the stress flow and improve the serviceability, it is quite tedious even for a first-order linear analysis.

1.5.2   EXAMPLES

1.5.2.1   *Beams*

A struts-and-ties model has been applied to beams to resist bending and shear as shown in Figure 1.2. Another model simulating beams to resist torsion is given in

Figure 1.3. These elegant models clearly convey how the internal forces are mobilized to resist the applied loads.

The cracking pattern of a simply supported beam reinforced with longitudinal bottom bars and vertical stirrups is shown in Figure 1.2a. The struts-and-ties model for this beam carrying two symmetrical concentrated loads V is given in Figure 1.2b. The shear and bending moment diagrams are indicated in Figure 1.2c and d. To resist the bending moment, the top and bottom stringers represent the concrete compression struts and the steel tension ties, respectively. The distance between the top and bottom stringers is designated as d_v. To resist the shear forces, the truss also has diagonal concrete compression struts and vertical steel tension ties in the web. The concrete compression struts are inclined at an angle of α, because the diagonal cracks due to shear are assumed to develop at this angle with respect to the longitudinal axis. Each cell of the truss, therefore, has a longitudinal length of $d_v \cot \alpha$, except at the local regions near the concentrated loads where the longitudinal length is $(1/2)(d_v \cot \alpha)$.

The forces in the struts and ties of this idealized truss can be calculated from the equilibrium conditions by various procedures. According to the sectional method, a cut along the section A-A at the right side of the truss will produce a free body as shown in Figure 1.2e. Equilibrium assessment of this free body shows that the top and bottom stringers are each subjected to a force of $(1/2)V \cot \alpha$ and $(3/2)V \cot \alpha$, respectively. The force in the compression strut is $V/\sin \alpha$. From the vertical equilibrium of the node point a, the force in the vertical tie is V. The results from similar calculations are recorded on the left half of the truss for all the struts and ties.

Figure 1.3 gives a much simplified struts-and-ties model for a beam to resist torsion. The longitudinal and hoop bars are assumed to have the same cross-sectional area and are both spaced at a constant spacing of s. The concrete compression struts are inclined at an angle of 45°. Because each hoop bar is treated as a series of straight ties of length s, a long plane truss is formed in the longitudinal direction between two adjacent longitudinal bars. A series of this kind of identical plane trusses is folded into a space truss with an arbitrary cross section as shown in Figure 1.3b. Because each plane truss is capable of resisting a force F, a series of F thus forms a circulatory shear flow, resulting in the torsional resistance T. It has been proven by Rausch (1929) using the equilibrium conditions at the node points that the force F is related to the torsional moment T by the formula $F = Ts/2A_0$, where A_0 is the cross-sectional area within the truss or the circulatory shear flow [see the derivation of Eq. (3-46) in Section 3.1.4, and notice the shear flow $q = F/s$].

1.5.2.2 Knee Joints

We will now use the knee joint, which connects the beam and the column as shown in Figure 1.1, to illustrate how the struts-and-ties models are utilized to help design the reinforcing bars (or "rebars") at the corner of the frame. Under gravity loads the knee joint is subjected to a closing moment as shown in Figure 1.4a. The top rebars in the beam and the outer rebars in the column are stretched by tension, while the bottom portion of the beam and the inner portion of the column are under compression. If the frame is loaded laterally, say by earthquake forces, the knee joint may be subjected to an opening moment as shown in Figure 1.4c. In this case the bottom rebars of the beam and the inner rebars of the column are stretched by tension, while the top portion of the beam and the outer portion of the column are under compression. For simplicity, the small shear stresses on the boundaries of the knee joint are neglected.

Several rebar arrangements that are commonly used are shown in Figure 1.4. Figure 1.4b gives a type of rebar arrangement frequently utilized to resist a closing

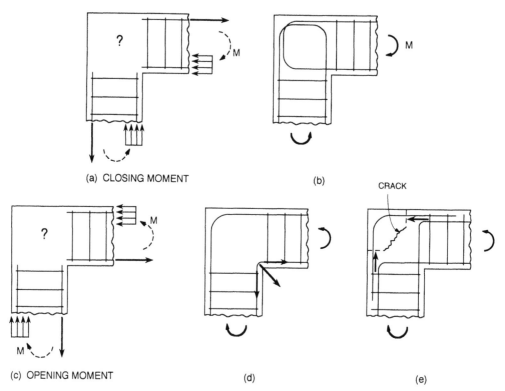

(a) CLOSING MOMENT   (b)

CRACK

(c) OPENING MOMENT   (d)   (e)

Figure 1.4 Knee joint moments and incorrect reinforcement.

moment. In this type of arrangement the top and bottom rebars of the beam are connected to form a loop in the joint region. A similar loop is formed by the outer and inner rebars of the column. This type of arrangement is very attractive because the separation of the beam steel from the column steel makes the construction easy. Unfortunately, tests (Swann, 1969) have shown that the strength of such a joint can be as low as 34% of the strength of the governing member (i.e., the beam or the column, whichever is less).

Two examples of incorrect arrangements of rebars to resist an opening moment are shown in Figure 1.4d and e. In Figure 1.4d, the bottom tension rebars of the beam are connected to the inner tension rebars of the column, whereas the top compression steel of the beam is connected to the outer compression steel of the column. This way of connecting the tension rebars of the beam and the column is obviously faulty, because the bottom tensile force of the beam and the inner tensile force of the column would produce a diagonal resultant that tends to straighten the rebar and to tear out a chunk of concrete at the inner corner. In fact, the strength of such a knee joint is only 10% of the strength of the governing member according to Swann's tests.

In Figure 1.4e, the bottom tension rebars of the beam are connected to the outer compression rebars of the column, whereas the top compression rebars of the beam are connected to the inner tension rebars of the column. Additional steel bars would be needed along the outer edge to protect the concrete core of the joint region. Such an arrangement also turns out to be flawed. The compression force at the top of the beam and the compression force at the outer portion of the column tend to push out a triangular chunk of concrete at the outer corner of the joint after the appearance of a diagonal crack. The strength of such a knee joint could be as low as 17% of the strength of the governing member (Swann, 1969).

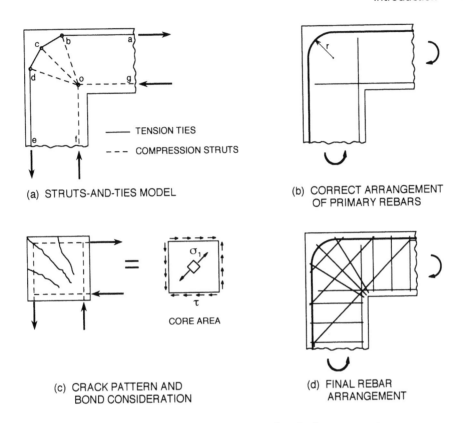

TENSION TIES

COMPRESSION STRUTS

(a) STRUTS-AND-TIES MODEL

(b) CORRECT ARRANGEMENT
OF PRIMARY REBARS

CORE AREA

(c) CRACK PATTERN AND
BOND CONSIDERATION

(d) FINAL REBAR
ARRANGEMENT

Figure 1.5 Rebar arrangement for closing moment.

(1) Knee Joint under Closing Moment

The struts-and-ties models will now be used to guide the design of rebars at the knee joint. The most important idea in the selection of the struts-and-ties assembly is to recognize the stress flow in the local region. The concrete struts should follow the compression trajectories as closely as possible, and the steel ties should trace the tension trajectories. A model that observes the stress flow pattern is expected to best satisfy the compatibility condition and the serviceability requirements.

Figure 1.5a gives the struts-and-ties model for the knee joint subjected to closing moment. First, the top tensile rebars of the beam and the outer tensile rebars of the column are represented by the ties (solid lines) *ab* and *de*, respectively. The centroidal lines of the compression zones in the beam and in the column are replaced by the struts (dotted lines) *og* and *fo*, respectively. Because the tensile stresses should flow from the top rebar of the beam to the outer rebar of the column, the top tie, *ab*, is connected to the outer tie, *de*, by two straight ties, *bc* and *cd*, along the outer corner. Because of the changes of angles at the node points, *b*, *c*, and *d*, the tensile force in the ties will produce a resultant bearing force on the concrete at each node point directed toward the point *o* at the inner corner of the joint. Following these compressive stress trajectories, three diagonal struts *bo*, *co*, and *do*, are installed. The three compressive forces acting along these diagonal struts and meeting at the node point *o* must be balanced by the two compression forces in the struts *og* and *fo*. As a whole, we have a stable struts-and-ties assembly that is in equilibrium at all the node points. The forces in all the struts and ties can be calculated.

It is clear from this struts-and-ties model that the correct arrangement of the primary tensile rebar, Figure 1.5b, should follow the tension ties *abcde* as shown in

Figure 1.5a. The radius of the bend, r, of the rebars in Figure 1.5b has a significant effect on the local crushing of concrete under the curved portion, as well as the compression failure of the diagonal struts near the point o. According to an extensive series of full-scale tests by Bai et al. (1991), the reinforcement index, $\omega = \rho f_y / f'_c$, of the tension steel at the end section of the beam should be a function of the radius ratio, r/d, and should be limited to

$$\omega = 2.2 \frac{r}{d} \tag{1-1}$$

and

$$\omega = 0.33 + 0.2 \frac{r}{d} \quad \left(\frac{r}{d} \leq 0.6 \right) \tag{1-2}$$

where d is the effective depth at the end section of the beam. The compressive strength of standard cylinder f'_c has been taken as 0.77 of the compressive strength of the 20-cm cubes used in the tests to define the reinforcement index ω. Equation (1-1) is governed by the local crushing of concrete directly under the curved portion of the tension rebars, and Eq. (1-2) is governed by the compression failure of the diagonal struts near the point o.

In addition to the primary tension rebars, the two compression rebars in Figure 1.5b should first follow the compression struts og and fo in Figure 1.5a, and then each be extended into the joint region for a length sufficient to satisfy the compression anchorage requirement.

A comparison of Figure 1.4b with Figure 1.5b reveals why the rebar arrangement in Figure 1.4b is deficient. Instead of connecting the top rebar of the beam and the outer rebar of the column by a single steel bar, these two rebars in Figure 1.4b are actually spliced together along the edge of the outer corner. This kind of splicing is notoriously weak because the splice is unconfined along the edge of the outer corner and its length is limited. If splices are desired, they should be located away from the joint region and be placed in a well confined region of the member, either inside the column or inside the beam.

In addition to the primary rebars, we must now consider the secondary rebar arrangement. Secondary rebars are provided for two purposes. First, they are designed to control cracks; second, they are added to prevent premature failure. The crack pattern of a knee joint under a closing moment is shown in Figure 1.5c. The direction of the cracks is determined by the stress state in the joint region. To understand this stress state, we notice that the four tension and compression rebars introduce, through bonding, the shear stresses τ around the core area. The shear stress produces the principal tensile stress, σ_1, which determines the direction of the diagonal cracks as indicated. To control these diagonal cracks, a set of opposing diagonal rebars perpendicular to the diagonal cracks are added as shown in Figure 1.5d.

Figure 1.5d also includes a set of inclined closed stirrups that radiate from the inner corner toward the outer corner. This set of closed stirrups is added to prevent two possible types of premature failure. First, because the curved portion of a primary tension rebar in Figure 1.5b exerts a severe bearing pressure on the concrete, it could split the concrete directly beneath, along the plane of the frame. The outer horizontal branches of the inclined stirrups, which are perpendicular to the plane of the frame, would serve to prevent such premature failures. Second, concrete in the vicinity of

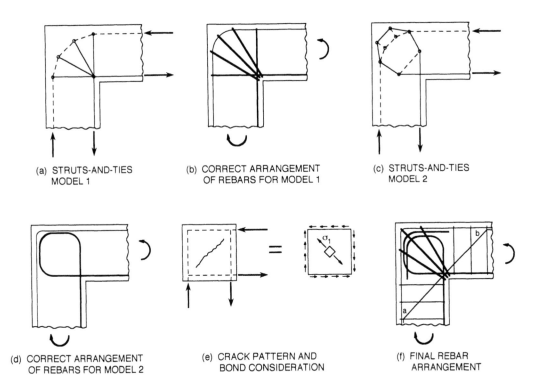

(a) STRUTS-AND-TIES MODEL 1

(b) CORRECT ARRANGEMENT OF REBARS FOR MODEL 1

(c) STRUTS-AND-TIES MODEL 2

(d) CORRECT ARRANGEMENT OF REBARS FOR MODEL 2

(e) CRACK PATTERN AND BOND CONSIDERATION

(f) FINAL REBAR ARRANGEMENT

Figure 1.6 Rebar arrangement for opening moment.

point o (Figure 1.5a) is subjected to extremely high compression stresses from all directions. The three remaining branches of the inclined closed stirrups in Figure 1.5d would serve to confine the concrete in this area.

(2) Knee Joint under Opening Moment

A struts-and-ties model (Model 1) for the knee joint subjected to an opening moment is shown in Figure 1.6a. This model is essentially a reverse case of the model for a closing moment, Figure 1.5a. This means that the ties and the struts are interchanged. Such a model should also be stable and in equilibrium. According to this model, the correct arrangement of the primary tensile rebars should follow the tension ties as shown in Figure 1.6b. It is noted that this set of radially oriented rebars should, in reality, be designed as closed stirrups because of anchorage requirements.

A second struts-and-ties model (Model 2) for an opening moment is given in Figure 1.6c. The rebars arranged according to this model are given in Figure 1.6d. This model explains why the bottom tension rebar of the beam should be connected to the inner tension rebar of the column by first forming a big loop around the core of the joint area. The connection of the rebars may be achieved by welding.

Design of rebars according to a combination of Models 1 and 2 is given in Figure 1.6f. The big loop of the tension rebar is formed by splicing rather than welding. The closed stirrups in the radial direction could also serve to control cracking. The direction of the crack is shown in Figure 1.6e. The final rebar arrangement also includes the diagonal rebars, ab, perpendicular to the diagonal line connecting the inner and outer corners. The effectiveness of such diagonal rebars is explained by the two struts-and-ties models in Figure 1.7. The first one, Figure 1.7a, is an extension of

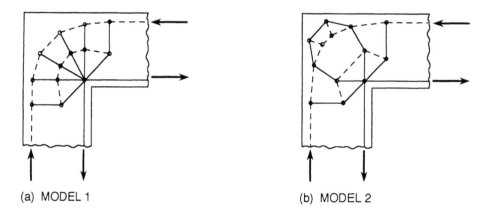

(a) MODEL 1 (b) MODEL 2

Figure 1.7 Struts-and-ties models showing effectiveness of diagonal rebars.

Model 1 in Figure 1.6a; the second one, Figure 1.7b, is a generalization of Model 2 in Figure 1.6c.

(3) Knee Joint under both Closing and Opening Moments

The rebar arrangement in a knee joint subjected to both closing and opening moments, as in earthquake loadings, is shown in Figure 1.8a. This arrangement is a combination of the two schemes designed separately for a closing moment (Figure 1.5d) and for an opening moment (Figure 1.6f).

Knee joints can also be strengthened by adding a fillet as shown in Figure 1.8b. The rebar arrangement could be simplified in such a joint because the stresses at the joint region are significantly reduced.

1.5.3   COMMENTS

Despite the clear concepts inherent in the struts-and-ties models, they are difficult to use in the actual design of the main regions of beams because of three reasons.

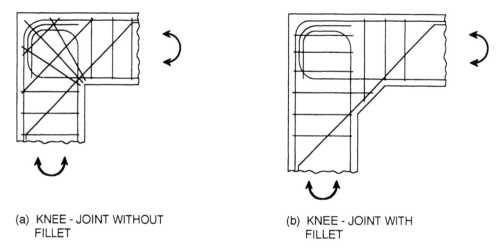

(a) KNEE - JOINT WITHOUT
    FILLET

(b) KNEE - JOINT WITH
    FILLET

Figure 1.8 Knee-joint rebar arrangement for both closing and opening moments (earthquake loading).

First, the design can be very tedious because the forces in every strut and tie must be calculated and proportioned. Instead of treating a beam as a whole to simplify the design, the struts-and-ties model is heading in the opposite direction by treating a beam as an assembly of a large number of struts and ties. Second, there are no definite objective criteria for the selection and the proportioning of the struts and ties, even though recommendations have been made. Third, and probably most important, there are indeed better and more sophisticated models for the design of members that can take into account the compatibility condition and the constitutive laws of materials. As mentioned in Section 1.2.2, Bernoulli's linear compatibility can be used for bending and axial load in the main regions, and Mohr's circular compatibility can be used for shear and torsion. Detailed study of these better models is the main objective of this book.

Nevertheless, struts-and-ties models are well suited to guide the design of local regions, such as the knee joints, because of two reasons: First, the strain compatibility conditions in the local regions are usually too complicated to be employed. We have no choice but to rely on the stress equilibrium condition. Second, the local regions are of limited lengths as compared to the main regions; see Figure 1.1. Because the number of struts and ties in a local region is limited, the effort required to design the struts and the ties is relatively easy.

2

Bernoulli Compatibility Truss Model

2.1   Bending of Reinforced Concrete Members

2.1.1   GENERAL BENDING THEORY

The analysis of a prismatic member subjected to bending, or flexure, is illustrated in Figure 2.1. On a rectangular cross section with width b and height h (Figure 2.1a), a bending moment M creates compression stresses in the top part of the cross section and tensile stresses in the bottom part. Because concrete is weak in tension, the bottom part of the cross section will crack and the tensile stresses will be picked up by the reinforcing bars (rebar, for short) indicated by the area A_s. The basic concept of reinforced concrete is to utilize the high compressive strength of concrete to resist the compression at the top, and utilize the high tensile strength of reinforcement, such as steel, synthetic fibers, carbon fibers, etc., to resist tension at the bottom. This action is similar to a truss, where the top and bottom chords resist the compressive and tensile forces, respectively.

The bending moment, M, will induce a curvature, ϕ, which is defined as the bending rotation per unit length of the member (Figure 2.1b). The fundamental assumption relating M and ϕ is the well-known Bernoulli hypothesis, which states that a plane section before bending will remain plane after bending. In other words, the strains along the depth of the cross section will be distributed linearly as shown in Figure 2.1b. The maximum compressive strain of concrete at the top surface is ε_c. The tensile strain ε_s at the centroid of the rebars is located at a distance d from the top surface. The neutral axis, where the strain is zero, is located at a distance c from the top surface.

The distribution of stresses along the depth of the cross section is given in Figure 2.1c. Above the neutral axis is a compression stress block sketched by a curve. The stresses in the compression stress block are related to the strain in Figure 2.1b through the stress–strain relationship of the concrete. The stress–strain relationship of concrete is assumed to be identical to the stress–strain curve obtained from the compression test of a standard concrete cylinder, shown in Figure 2.1d. For a typical

21

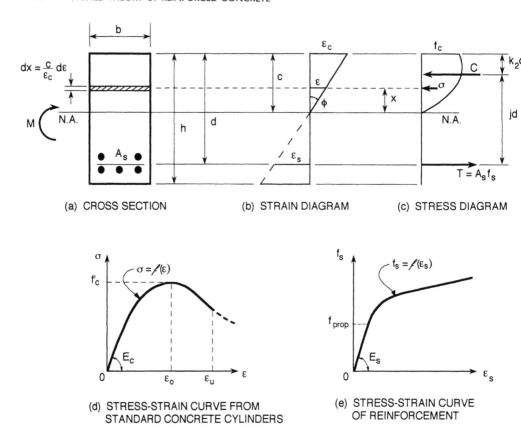

(a) CROSS SECTION   (b) STRAIN DIAGRAM   (c) STRESS DIAGRAM

(d) STRESS-STRAIN CURVE FROM STANDARD CONCRETE CYLINDERS

(e) STRESS-STRAIN CURVE OF REINFORCEMENT

Figure 2.1 Bending analysis of reinforced concrete members.

stress–strain curve of concrete, the concrete stress σ first increases linearly with a slope of E_c up to about $0.3f'_c$, then increases nonlinearly to a maximum stress of f'_c at a strain of ε_0. The strain ε_0 is usually taken as 0.002. Beyond this strain the concrete stress decreases until an ultimate strain ε_u is reached. The ultimate strain ε_u is specified by the codes for practical application.

Below the neutral axis in Figure 2.1c, the small tensile stress of concrete adjacent to the neutral axis is neglected. All the tensile stresses in the rebars are assumed to be concentrated at the centroid of the reinforcing bars with a total area of A_s. The stress in the rebar f_s is assumed to be related to the strain ε_s in Figure 2.1b through the stress–strain curve obtained from the tension tests of bare reinforcing bars (Figure 2.1e). For most types of reinforcing bars, the stress–strain curve is first linear, with a slope of E_s, up to the proportional limit f_{prop}, then forms a knee and follows roughly a straight line with a much smaller slope.

Now that the stress and strain diagrams and the stress–strain relationships of concrete and reinforcing bar are explained, we can apply the basic principles of mechanics of materials to analyze the bending action. Let x be the distance from the neutral axis to a level where the strain is ε and the stress is σ (Figure 2.1b and c). From a similar triangle of the strain diagram, $x = (c/\varepsilon_c)\varepsilon$. Differentiating x gives $dx = (c/\varepsilon_c)d\varepsilon$ and the resultant C can be expressed by integrating the compression stress block:

$$C = b\int_0^c \sigma \, dx = \frac{bc}{\varepsilon_c}\int_0^{\varepsilon_c} \sigma \, d\varepsilon \tag{2-1}$$

The integration can be performed if the concrete stress–strain relationship $\sigma = f(\varepsilon)$ in Figure 2.1d is specified analytically and a maximum strain ε_c is selected. The resultant C is then only a function of the position of the neutral axis c.

The location of the resultant C can be defined by $k_2 c$, measured from the resultant C to the extreme compression fiber at the top surface (Figure 2.1c). The coefficient k_2 is determined by taking the moment about the neutral axis:

$$C(c - k_2 c) = b \int_0^c \sigma x \, dx = b \left(\frac{c}{\varepsilon_c} \right)^2 \int_0^{\varepsilon_c} \sigma \varepsilon \, d\varepsilon \tag{2-2}$$

Substituting C from Eq. (2-1) into Eq. (2-2) gives

$$k_2 = 1 - \frac{1}{\varepsilon_c} \frac{\int_0^{\varepsilon_c} \sigma \varepsilon \, d\varepsilon}{\int_0^{\varepsilon_c} \sigma \, d\varepsilon} \tag{2-3}$$

Equation (2-3) shows that the coefficient k_2 can also be determined if the stress–strain curve of concrete (Figure 2.1d) is specified and the maximum concrete strain ε_c is selected. The location of the resultant defined by $k_2 c$ is then only a function of c.

The magnitude and the location of the resultant C defined by Eqs. (2-1) and (2-3) can both be calculated if the depth of the compression zone c is known. The depth c can be solved using two conditions: the *equilibrium of forces* in the horizontal direction and the *compatibility of strains* over the depth of the cross section. From the first condition, the compression resultant C should be equal to the tensile force resultant of the reinforcement, $T = A_s f_s$, giving

$$C = A_s f_s \tag{2-4}$$

Substituting C from Eq. (2-1) into Eq. (2-4) the depth of the neutral axis is expressed as

$$c = \frac{A_s f_s}{(b/\varepsilon_c) \int_0^{\varepsilon_c} \sigma \, d\varepsilon} \tag{2-5}$$

The rebar stress f_s in Eq. (2-5) is related to the rebar strain ε_s by the stress–strain relationship of Figure 2.1d, which, in turn, is related to the concrete strain ε_c at the top surface of the member by the strain compatibility condition. According to the straight-line distribution of strains,

$$\varepsilon_s = \varepsilon_c \frac{d - c}{c} \tag{2-6}$$

Because the cross-sectional variables b, d, and A_s are given in this analysis type of problem, Eqs. (2-5) and (2-6) contain only four unknown variables: c, f_s, ε_s, and ε_c. Once a value of ε_c is selected for a given stress–strain curve of concrete, the three unknowns c, f_s, and ε_s can be solved by Eqs. (2-5) and (2-6), together with the stress–strain relationship of the reinforcing bar, $f_s = f(\varepsilon_s)$, shown in Figure 2.1e. At failure, ε_c becomes ε_u, which is usually taken as 0.003.

Now that the location of the neutral axis c is obtained, the internal resisting moment can be calculated by the third condition, i.e., the *moment equilibrium condition.* Taking the moment about the compression resultant gives

$$M = A_s f_s d \left(1 - k_2 \frac{c}{d}\right) \tag{2-7}$$

Define $j = (1 - k_2 c/d)$. Then Eq. (2-7) becomes

$$M = A_s f_s (jd) \tag{2-8}$$

where jd is called the lever arm. Because the external moment M is resisted by a pair of equal, opposite, and parallel forces T and C, located at a distance jd from each other, this idea of bending resistance is frequently referred to as the *internal couple concept.*

The curvature of the member can also be calculated once the depth of the neutral axis c is determined. From the linear strain distribution given in Figure 2.1b the curvature ϕ is expressed as

$$\phi = \frac{\varepsilon_c}{c} \tag{2-9}$$

By selecting a series of incremental values for ε_c and calculating the corresponding moments M and curvatures ϕ from Eqs. (2-7) and (2-9), the moment–curvature relationship can be established. This so-called M-ϕ curve represents the fundamental characteristic of a flexural member, from which the deformations, i.e., deflections and rotations, of the member are calculated.

The previously described bending theory is given in the most general terms for the stress–strain relationships of concrete and reinforcement (Figure 2.1d and e). Practical applications of this theory with various simplifications of the constitutive laws of materials will be studied in Sections 2.2 and 2.3 of this chapter.

2.1.2   FUNDAMENTAL PRINCIPLES

The bending theory described in Section 2.1.1 is rational and rigorous because it satisfies the three fundamental principles of the mechanics of deformable bodies. First, the stresses in the concrete and the reinforcement (Figure 2.1c) satisfy the equilibrium condition. Second, the linear strain distribution as shown in Figure 2.1b satisfies Bernoulli's hypothesis. A plane cross section that remains plane after the deformation means that the compatibility condition is satisfied. Third, the constitutive laws of concrete and reinforcing bar as shown in Figure 2.1d and e are obeyed. As a result, the deformations of a flexural member can be determined throughout its loading history.

A reinforced concrete member subjected to bending is expected to crack before the service load stage. The strains in the rebars at the cracked sections should be greater than those between the cracked sections. Therefore, the application of Bernoulli's hypothesis to reinforced concrete members should be based on the *average strains, or smeared strains,* of the reinforcement. Accordingly, the stresses in the reinforcement must be the *average stresses,* and the constitutive law for the reinforcement must be relating the average stresses to the average strains. For this reason, the use of a stress–strain relationship obtained from testing of bare reinforcing bars (Figure 2.1e) can only be considered as approximate. An accurate relationship between the average stress and the average strain of mild-steel reinforcing bars embedded in concrete is presented in Section 7.2.3. In this bending theory of reinforced concrete the stress and the strain of a reinforcing bar always mean the average stress and the average strain.

The bending theory has been used to analyze the behavior of flexural members since the advent of the theory for reinforced concrete more than a century ago. This well known theory will be called the *Bernoulli compatibility truss model* for three reasons. First, Bernoulli's hypothesis of plane section remaining plane is the fundamental compatibility condition in this theory, allowing the flexural deformation of a member to be determined. Second, in the internal-couple concept of bending, the concrete is taking compression as in the top chord of a truss and the reinforcement is resisting tension as in the bottom chord of a truss. These top and bottom chords of a truss can resist only axial stresses, but not shear or dowel stresses. Third, the terminology of the Bernoulli compatibility truss model for bending is created to emphasize its difference from the Mohr compatibility truss model for shear that will be presented in Chapter 6. The three fundamental principles of equilibrium, compatibility, and constitutive laws of materials will be applied to shear in Chapters 6 and 7 and to torsion in Chapter 8.

The constitutive laws of concrete and reinforcement are in general nonlinear. To study the behavior of reinforced concrete beams at ultimate load stage, a nonlinear stress–strain curve of concrete must be assumed. Bending theory based on the nonlinear stress–strain relationship of concrete will be called the *nonlinear bending theory* and will be studied in Section 2.3. However, up to the service load the stress–strain relationships of concrete and rebars can be approximated as linear. Consequently, Hooke's constitutive law is generally assumed to be applicable to both concrete and rebars up to the service load stage. Bending theory based on Hooke's law is called the *linear bending theory* and is studied in the next section, Section 2.2.

2.2   Linear Bending Theory

2.2.1   BENDING THEORY BASED ON HOOKE'S LAW

In this section the analysis and design of flexural members will be made according to the Bernoulli compatibility truss model in connection with Hooke's constitutive law. Such a method is applicable to flexural members at service load. The cross section of a reinforced concrete beam, shown in Figure 2.2a, is subjected to a bending moment M. The width and the height of the cross section are designated b and h, respectively. The reinforcement has an area of A_s and is located at an effective depth of d from the top surface of extreme compression. The thickness of the concrete cover for the rebars is always specified.

According to Bernoulli hypothesis the moment M is assumed to induce a linear strain diagram as shown in Figure 2.2b. The maximum compressive strain at the top surface is ε_c and the tensile strain in the rebar is ε_s. The neutral axis (N.A.), which indicates the level of zero strain, is located at a distance kd from the top surface. A compatibility equation can, therefore, be expressed in terms of the three variables ε_c, ε_s, and k.

The geometric relationship among ε_c, ε_s, and k involves a second assumption, namely, the strain in the rebar is identical to the strain in the concrete at the rebar level. In other words, a perfect bond is assumed to exist between the rebars and the concrete.

The stresses in the cross section are obtained from the strains through the stress–strain relationships of concrete and rebars. A third assumption is made that

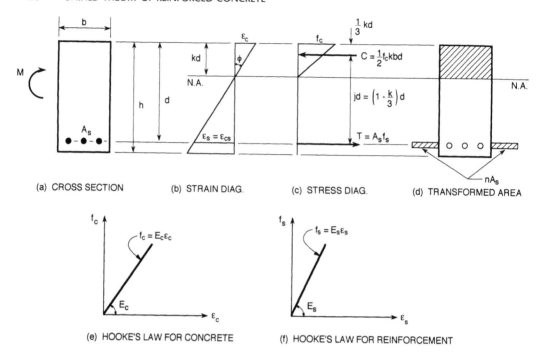

Figure 2.2 Cracked rectangular sections (singly reinforced).

both the materials obey Hooke's law (Figure 2.2e and f). Based on the linear stress–strain relationship of concrete, we can then draw the stress diagram as shown in Figure 2.2c. The stresses vary linearly from the maximum of f_c at the top surface to zero at the neutral axis. Furthermore, the resultant force C has a magnitude of $(1/2)f_c kbd$ and is located at a distance $(1/3)kd$ from the top surface.

The fourth assumption of the theory is to neglect all the tensile stresses of concrete after the cracking of the cross section. As such, no tensile stress exists below the neutral axis (Figure 2.2c) except the tensile stress in the rebars f_s. The tensile resistance is, therefore, concentrated in the rebar tensile force T, which is equal to $A_s f_s$. The rebar tension force T and the concrete compression force C constitute the internal couple. The lever arm of the couple jd is $(1 - k/3)d$.

In a singly reinforced concrete member, the bending action involves nine variables, b, d, A_s, M, f_s, f_c, ε_s, ε_c, and k as indicated in Figure 2.2a, b, and c. The stresses in the cross section (Figure 2.2c), form a parallel coplanar force system and are oriented in the longitudinal direction of the member. For this type of force system, the equilibrium condition will supply two independent equations, relating the seven variables b, d, A_s, M, f_s, f_c, and k. The three simplest forms of equilibrium equations are: (1) equilibrium of forces in the longitudinal direction, (2) equilibrium of moments about the resultant compression force of concrete, and (3) equilibrium of moments about the tensile force of rebars. The selection of two of these three equilibrium equations is strictly a matter of convenience. It must be emphasized that the equilibrium condition in bending can only be used to solve two unknowns, even if all three equilibrium equations are used.

Bernoulli's compatibility (Figure 2.2b) will give one equation, relating the strains ε_s, ε_c with k. Hooke's constitutive laws will provide two equations: one for rebar relating f_s and ε_s; one for concrete relating f_c to ε_c. Therefore, a total of five equations can be derived from the three fundamental principles of equilibrium, compatibility, and constitutive laws for the solution of a bending problem in a singly reinforced member. Because such problems involve nine unknown variables, four variables must be given before the remaining five unknown variables can be solved by the five equations.

Depending on the given variables, the problems are generally categorized into five types as shown in the following table:

Types of Problems	Given Variables				Unknown Variables				
(1) First type of analysis	b	d	A_s	M	f_s	f_c	ε_s	ε_c	k
(2) Second type of analysis	b	d	A_s	f_s (or f_c)	M	f_c (or f_s)	ε_s	ε_c	k
(3) Balanced design	b	d	f_s	f_c	M	A_s	ε_s	ε_c	k
(4) First type of design	b	d	M	f_s (or f_c)	A_s	f_c (or f_s)	ε_s	ε_c	k
(5) Second type of design	b (or d)	M	f_s	f_c	d (or b)	A_s	ε_s	ε_c	k

The first two types of problem are called analysis, because the cross-sectional properties of concrete and rebars, b, d, and A_s, are given. The last three types are called design, because at least one of these three cross-sectional properties, primarily A_s, is an unknown.

The last three types of problems were known as the allowable stress design method, prevalent prior to 1971. The allowable stress design method was made obsolete by the ultimate strength design method and, therefore, will not be treated in this book. However, the two types of analysis problems are still very relevant at present for checking the serviceability criteria, i.e., the deflections and the crack widths.

The problems in the analysis and design of flexural members boil down to finding the most efficient way to solve the five available equations for each type of problem. The methodology of the solution process is demonstrated in the next section, where the five equations are applied to the two types of analysis problems.

2.2.2 SOLUTIONS FOR ANALYSIS PROBLEMS

We will first look at the first type of analysis problem, which is posed as follows:

> Given four variables: b, d, A_s, and M.
> Find five unknown variables: f_s, f_c, ε_s, ε_c, and k.

The nine variables b, d, A_s, M, f_s, f_c, ε_s, ε_c, and k, are defined in Figure 2.2 for a singly reinforced flexural member. Because four variables are given, the remaining five unknown variables can be solved by the five available equations. These equations and their unknown variables are summarized as follows:

Type of Equations	Equations	Unknowns			
Equilibrium of forces	$A_s f_s = \frac{1}{2} f_c k b d$	f_s f_c			k (2-10)
Equilibrium of moments	$M = \frac{1}{2} A_s f_s (1 - k/3)d$	f_s			k (2-11)
Bernoulli compatibility	$\dfrac{\varepsilon_s}{\varepsilon_c} = \dfrac{1-k}{k}$		ε_s	ε_c	k (2-12)
Hooke's law for rebar	$f_s = E_s \varepsilon_s$	f_s	ε_s		(2-13)
Hooke's law for concrete	$f_c = E_c \varepsilon_c$	f_c		ε_c	(2-14)

The solution of the five unknowns (f_s, f_c, ε_s, ε_c, and k) by the five equations, Eqs. (2-10) to (2-14), is facilitated by identifying the unknown variables in each equation. These variables are shown after each equation under the column heading "Unknowns."

In the set of five equations, Eqs. (2-10) to (2-14), the first two equilibrium equations deal with unknown stresses; the third equation expresses Bernoulli's compatibility condition in terms of unknown strains. Substituting the stress–strain relationships of Eqs. (2-13) and (2-14) into Eq. (2-12) we obtain Bernoulli's compatibility equation in terms of stresses:

$$\frac{f_s}{n f_c} = \frac{1-k}{k} \tag{2-15}$$

where $n = E_s/E_c$ = modulus ratio. With this simple maneuvering, we have now reduced a set of five equations to a set of three equations, Eqs. (2-10), (2-11), and (2-15), which involve only three unknowns, f_s, f_c, and k.

Of the three equations, both Eqs. (2-10) and (2-15) have the unknown k and the unknown stress ratio f_s/f_c. Substituting the ratio f_s/f_c from Eq. (2-10) into Eq. (2-15) we obtain an equation with only one unknown k:

$$\frac{kbd}{2nA_s} = \frac{1-k}{k} \tag{2-16}$$

Defining $\rho = A_s/bd$ = percentage of rebars, Eq. (2-16) becomes

$$\frac{k}{2n\rho} = \frac{1-k}{k} \tag{2-17}$$

The unknown k can be solved by Eq. (2-17) using whatever method is convenient. Two methods are generally used. The first is the trial-and-error method. A value of k is assumed and inserted into Eq. (2-17). If the equation is satisfied, the assumed k is the solution. If the equation is not satisfied, another k value is assumed and the process is repeated. This trial-and-error method could be quite efficient if the k value could be closely estimated in the first trial. It is, therefore, convenient for engineers with experience.

The second method is to rewrite Eq. (2-17) in the form of a second order equation:

$$k^2 + (2n\rho)k - 2n\rho = 0 \tag{2-18}$$

From Eq. (2-18), k can be determined by the formula for quadratic equation:

$$k = \sqrt{(n\rho)^2 + 2n\rho} - n\rho \tag{2-19}$$

Now that the position of the neutral axis, k, is solved, the rebar stresses f_s can be obtained from the equilibrium of moments about the resultant C, Eq. (2-11):

$$f_s = \frac{M}{A_s(1 - k/3)d} \tag{2-20}$$

and the concrete stress f_c can be obtained either from the equilibrium of forces, Eq. (2-10), or more directly in this case (because M is given) from the equilibrium of moments about the tensile force T:

$$f_c = \frac{M}{\frac{1}{2}k(1 - k/3)bd^2}$$ (2-21)

Once the stresses f_s and f_c are found, the strains ε_s and ε_c can be calculated from Hooke's law by Eqs. (2-13) and (2-14).

The solution of the second type of analysis is similar to the first type described previously. The similarity can be observed by examining the five unknown variables in the five available equations. The problem posed is:

> Given four variables: b, d, A_s, and f_s (or f_c).
>
> Find five unknown variables: M, f_c (or f_s), ε_s, ε_c, and k.

The variables f_c and f_s in the parentheses should be understood as follows: If the given variable f_s is replaced by f_c in the parentheses, then the unknown variables f_c must be replaced by f_s in the parentheses.

The five equations, Eqs. (2-10) to (2-14), are still valid in this case, but the list of unknowns under the column heading "Unknowns" should be revised. The stress f_s (or f_c) becomes a known value and should be replaced by the new unknown M. The most efficient algorithm of solution still consists of the following three steps:

Step 1: Utilize the stress–strain relationships of rebar and concrete to express Bernoulli's compatibility equation in terms of stresses, thus arriving at Eq. (2-15). In this way a set of five equations is reduced to a set of three equations, Eqs. (2-10), (2-11), and (2-15), which involve only three unknowns, M, f_c (or f_s), and k.

Step 2: Of the three equations, both Eqs. (2-10) and (2-15) contain the same two unknowns, f_c (or f_s) and k. Simultaneous solution of these two equations results in Eq. (2-19) with one unknown variable k.

Step 3: Once the position of the neutral axis, k, is determined, the unknowns f_c (or f_s) can be calculated from the force equilibrium equation, and the unknown M from any one of the two moment equilibrium equations. The strains, ε_s and ε_c, can be calculated easily from the stresses f_s and f_c by Hooke's laws.

It can be seen that the solution of the second type of analysis problem is the same as the first type in Steps 1 and 2, i.e., in solving the position of the neutral axis. The only small difference is in Step 3 in the calculation of different unknowns by equilibrium equations.

2.2.3   TRANSFORMED AREA FOR REINFORCING BARS

The key problem in the flexural analysis (Section 2.2.2) is to find the location of the neutral axis, represented by k, using Bernoulli's compatibility condition, the force equilibrium condition, and Hooke's law. A short-cut method to find k will now be introduced. The method utilizes the concept of the transformed area for the rebars.

Based on the assumption that perfect bond exists between the rebar and the concrete, the rebar strain ε_s should be equal to the concrete strain at the rebar level ε_{cs} (Figure 2.2b):

$$\varepsilon_s = \varepsilon_{cs}$$ (2-22)

Applying Hooke's law, $f_s = E_s \varepsilon_s$ and $f_{cs} = E_c \varepsilon_{cs}$, to Eq. (2-22) gives

$$f_s = \frac{E_s}{E_c} f_{cs} = n f_{cs} \qquad (2\text{-}23)$$

where f_{cs} is the concrete stress at the rebar level.

Using Eq. (2-23) the tensile force T can be written as

$$T = A_s f_s = (n A_s) f_{cs} \qquad (2\text{-}24)$$

Equation (2-24) states that the tensile force T can be thought of as being supplied by a concrete area of $n A_s$ in connection with a concrete stress of f_{cs}. Physically, this means that the rebar area A_s can be transformed into a concrete area $n A_s$, as long as the rebar stress f_s is simultaneously converted to the concrete stress f_{cs}. A cross section with the transformed rebar area is shown in Figure 2.2d.

The transformation of the rebar area A_s to a concrete area $n A_s$ has a profound significance. The flexural beam, which is made up of two materials, reinforcement and concrete, can now be thought of as a homogeneous elastic material made of concrete only. For such a homogeneous elastic beam, the neutral axis coincides with the centroidal axis. So instead of solving the neutral axis from the three fundamental principles of equilibrium, compatibility, and constitutive laws, as illustrated in Section 2.2.2, we can now locate the neutral axis using the simple and well-known method of finding the centroidal axis of a homogeneous beam. In other words, the static moments of the stressed areas, shaded in Figure 2.2d, about the centroidal axis must be equal to zero. Hence,

$$\tfrac{1}{2} b (kd)^2 - n A_s (d - kd) = 0 \qquad (2\text{-}25)$$

The variable k in Eq. (2-25) can be solved by two methods. The first method is the trial-and-error procedure. A value of k is assumed and inserted into Eq. (2-25). If the equation is satisfied, the assumed k is the solution. If the equation is not satisfied, another k value is assumed and the process is repeated. This trial-and-error method is very efficient for an experienced engineer.

The second method is to solve Eq. (2-25) by the formula for quadratic equations, giving

$$k = \sqrt{(n\rho)^2 + 2n\rho} - n\rho \qquad (2\text{-}26)$$

Equation (2-26) is, of course, the same as Eq. (2-19). This method requires an engineer to remember Eq. (2-26) or to have the formula in hand.

2.2.4   BENDING RIGIDITIES OF CRACKED SECTIONS

2.2.4.1   *Singly Reinforced Rectangular Sections*

For a homogeneous elastic member, the bending rigidity EI is the product of the modulus of elasticity of the material, E, and the moment of inertia of the cross section, I. For a cracked concrete member, we will define the bending rigidity $E_c I_{cr}$ as the product of the modulus of elasticity of concrete, E_c, and a cracked moment of inertia, I_{cr}. In this section we will derive I_{cr} for singly reinforced rectangular sections.

By definition, the bending rigidity of a member is the bending moment per unit curvature, M/ϕ. Hence, we can write

$$E_c I_{cr} = \frac{M}{\phi} \qquad (2\text{-}27)$$

From the strain diagram in Figure 2.2b the curvature ϕ can be expressed by

$$\phi = \frac{\varepsilon_c}{kd} \qquad (2\text{-}28)$$

From the stress diagram in Figure 2.2.c the moment M can be obtained by taking the moment about the tensile rebar:

$$M = \tfrac{1}{2}f_c kjbd^2 \qquad (2\text{-}29)$$

Substituting ϕ and M from Eqs. (2-28) and (2-29) into Eq. (2-27) and noticing $f_c = E_c \varepsilon_c$ result in

$$I_{cr} = bd^3\left[\tfrac{1}{2}k^2 j\right] \qquad (2\text{-}30)$$

Equation (2-30) shows that the cracked moment of inertia, I_{cr}, can be calculated if the k value is determined from Eq. (2-26) and $j = (1 - k/3)$.

2.2.4.2 *Doubly Reinforced Rectangular Sections*

The transformed area concept can be used to great advantage in finding the cracked moments of inertia for doubly reinforced sections and flanged sections. These types of sections occur regularly in continuous beams as shown in Figure 2.3. In a continuous beam the cross sections are usually doubly reinforced at the column faces; those at the midspan are often designed as flanged sections.

Figure 2.4a shows a doubly reinforced rectangular section of height h and width b. The tensile rebar area A_s is located at a distance d'' from the bottom surface, and the compression rebar area A_s' at a distance d' from the top surface. The effective depth d is equal to $h - d''$.

The strains in the tensile and compressive rebars are designated ε_s and ε_s', respectively, as shown in Figure 2.4b. The neutral axis is assumed to be located at a distance kd from the top surface. The stresses in the tensile and compressive rebars are denoted as f_s and f_s', respectively, in Figure 2.4c.

The transformed area for tensile rebar is indicated by nA_s in Figure 2.4d. The transformed area for compression rebar is $mA_s' = (n - 1)A_s'$. The subtraction of an area of compression rebar A_s' in the expression mA_s' is to compensate for the same area included in the concrete area $b(kd)$. This is strictly for convenience in calculation.

The position of the neutral axis can be determined by taking the static moments of the shaded areas (Figure 2.4d) about the centroidal axis (same as neutral axis):

$$\tfrac{1}{2}b(kd)^2 + mA_s'(kd - d') = nA_s(d - kd) \qquad (2\text{-}31)$$

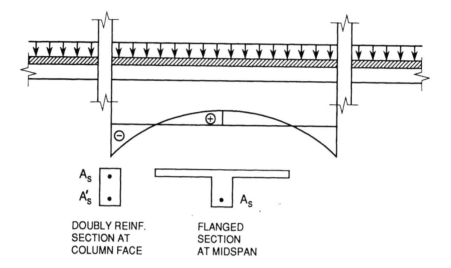

DOUBLY REINF.
SECTION AT
COLUMN FACE

FLANGED
SECTION
AT MIDSPAN

Figure 2.3 Doubly reinforced and flanged sections in continuous beams.

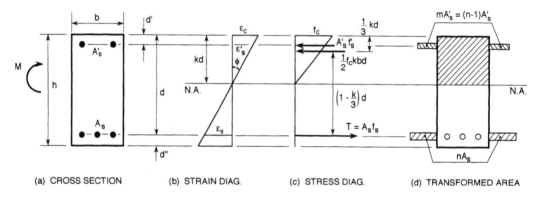

(a) CROSS SECTION   (b) STRAIN DIAG.   (c) STRESS DIAG.   (d) TRANSFORMED AREA

Figure 2.4 Cracked rectangular sections (doubly reinforced).

Dividing Eq. (2-31) by bd^2 and denoting $\rho = A_s/bd$ and $\rho' = A'_s/bd$ result in

$$k^2 + 2(n\rho + m\rho')k - 2\left(n\rho + m\rho'\frac{d'}{d}\right) = 0 \qquad (2\text{-}32)$$

Let

$$\beta_c = \frac{m\rho'}{n\rho} \qquad (2\text{-}33)$$

Solving Eq. (2-32) by the quadratic equation formula gives

$$k = \sqrt{(n\rho)^2(1 + \beta_c)^2 + 2n\rho(1 + \beta_c(d'/d))} - n\rho(1 + \beta_c) \qquad (2\text{-}34)$$

Equation (2-34) determines the location of the neutral axis for doubly reinforced cracked beams. For the special case of singly reinforced cracked beams, $\beta_c = 0$ and Eq. (2-34) degenerates into Eq. (2-26).

The moment resistance of the cross section, M, can be calculated from the stress diagram in Figure 2.4c by taking moments about the tensile force:

$$M = \frac{1}{2} f_c k \left(1 - \frac{k}{3}\right) bd^2 + A'_s f'_s (d - d') \tag{2-35}$$

From Hooke's law and Bernoulli's strain compatibility (Figure 2.4b), we can express the compressive rebar stress f'_s as

$$f'_s = E_s \varepsilon'_s = (mE_c)\left(\varepsilon_c \frac{kd - d'}{kd}\right) = m(E_c \varepsilon_c)\frac{kd - d'}{kd} \tag{2-36}$$

Substituting f'_s from Eq. (2-36), $A'_s = \rho' bd$, and $m\rho' = n\rho\beta_c$ into Eq. (2-35) gives

$$M = bd^2 (E_c \varepsilon_c)\left[\frac{1}{2}k\left(1 - \frac{k}{3}\right) + n\rho\beta_c \frac{1}{k}\left(k - \frac{d'}{d}\right)\left(1 - \frac{d'}{d}\right)\right] \tag{2-37}$$

The curvature can be written according to Bernoulli's hypothesis (Figure 2.4b), as

$$\phi = \frac{\varepsilon_c}{kd} \tag{2-38}$$

Because the bending rigidity $E_c I_{cr}$ is defined as M/ϕ, the cracked moment of inertia I_{cr} for doubly reinforced sections can be derived according to Eqs. (2-37) and (2-38):

$$I_{cr} = bd^3\left[\frac{1}{2}k^2\left(1 - \frac{k}{3}\right) + n\rho\beta_c\left(k - \frac{d'}{d}\right)\left(1 - \frac{d'}{d}\right)\right] \tag{2-39}$$

The coefficient k in Eq. (2-39) is determined from Eq. (2-34).

The cracked moment of inertia can be written simply as $I_{cr} = K_{i2} bd^3$, where K_{i2} denotes the expression in the bracket in Eq. (2-39). K_{i2} is a function of $n\rho$, β_c, and d'/d and is tabulated in ACI Handbook SP-17(73). (ACI Committee 340, 1973).

2.2.4.3 Flanged Sections

A typical flanged section—tee section (T section)—is shown in Figure 2.5a together with the transformed area in Figure 2.5b. Comparing this transformed area for T section to the transformed area for doubly rectangular reinforced section in Figure 2.4d, we observe that they are identical except for two minor differences. First, the area of the top flange is $(b - b_w)h_f$ rather than mA'_s. Second, the centroidal axis of the top flange is located at a distance $h_f/2$ from the top surface rather than at a distance d'. Therefore, the formula for cracked moment of inertia of doubly reinforced sections, Eq. (2-39), is also applicable to T sections if these two differences are taken care of as follows:

1 Redefine the original definition of the symbol β_c in Eq. (2-33) by substituting $(b - b_w)h_f/b_w d$ for $m\rho'$:

$$\beta_c = \frac{(b - b_w)h_f}{(n\rho)b_w d} \tag{2-40}$$

2 Replace d'/d by $h_f/2d$ in Eqs. (2-39) and (2-34).

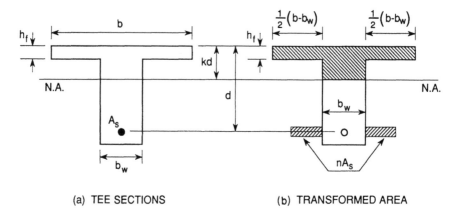

(a) TEE SECTIONS   (b) TRANSFORMED AREA

Figure 2.5 Cracked flanged sections.

In conclusion, the cracked moments of inertia, I_{cr}, for flanged sections are calculated by Eqs. (2-39) and (2-34), with the symbol β_c defined by Eq. (2-40) and the ratio d'/d replaced by $h_f/2d$.

2.2.5 BENDING RIGIDITIES OF UNCRACKED SECTIONS

2.2.5.1 *Rectangular Sections*

In a flexural member at service load, some parts of the member will be cracked and some parts will remain uncracked depending on the bending moment diagram. The deflection of the member will be a function of both the cracked and the uncracked portions. The bending rigidities of the cracked sections have been derived in Section 2.2.4.1 for singly reinforced sections and in Section 2.2.4.2 for doubly reinforced and flanged sections. The bending rigidity of uncracked rectangular sections (Figure 2.6a) will now be derived.

Before the cracking of a flexural member, the strain and the stress distributions in the cross section are shown in Figure 2.6b and c, respectively. First, the concrete stresses below the neutral axis can no longer be neglected. Second, the full height of the cross section h becomes effective, rather than the effective depth d. The transformed area of the cross section is shown in Figure 2.6d.

The internal resistance to the applied moment is contributed primarily by the concrete before cracking. Assuming that the rebars have no effect on the position of the neutral axis, then the neutral axis will lie at the middepth of the cross section. As a result, $kd = h/2$, $jd = (2/3)h$, and $C = T = (1/4)f_c bh$. Taking the moment about the neutral axis,

$$M = \tfrac{1}{6}bh^2 f_c + mA_s f_{cs}(d - h/2) \tag{2-41}$$

where $m = n - 1$, because the original rebar area A_s has been included in the first term for concrete and must be subtracted from the transformed rebar area in the second term.

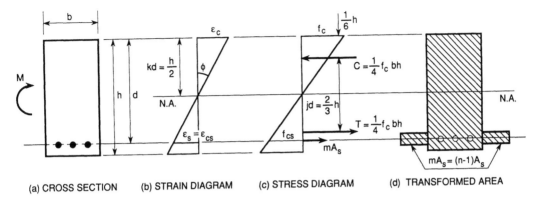

(a) CROSS SECTION   (b) STRAIN DIAGRAM   (c) STRESS DIAGRAM   (d) TRANSFORMED AREA

Figure 2.6   Uncracked rectangular sections (singly reinforced).

Observing that $A_s = \rho bd$ and $f_{cs} = f_c(d - h/2)/(h/2)$, Eq. (2-41) becomes

$$M = \frac{1}{6}bh^2 f_c\left[1 + 3m\rho\left(\frac{d}{h}\right)\left(\frac{d - h/2}{h/2}\right)^2\right] \qquad (2\text{-}42)$$

Assuming $d/h = 0.9$ and $(d - h/2) = 0.8(h/2)$, then Eq. (2-42) gives

$$M = \tfrac{1}{6}bh^2 f_c[1 + 1.7m\rho] \qquad (2\text{-}43)$$

In Eq. (2-43), $m\rho$ is the transformed area percentage of tensile rebars. The steel ratio ρ is usually in the range of 1 to 1.5% and the modulus ratio m varies from 5 to 7 for conventional concrete. Therefore, the second term contributed by rebars, $1.7\rho m$, in the bracket is roughly 10 to 15% of the first term of unity contributed by concrete.

The curvature before cracking is

$$\phi = \frac{\varepsilon_c}{0.5h} \qquad (2\text{-}44)$$

The uncracked flexural rigidity $E_c I_{uncr}$ is defined by the ratio M/ϕ according to Eqs. (2-43) and (2-44). Dividing M by ϕ and noticing that $f_c = E_c \varepsilon_c$, the uncracked moment of inertia I_{uncr} becomes

$$I_{uncr} = \tfrac{1}{12}bh^3(1 + 1.7m\rho) \qquad (2\text{-}45)$$

If the effect of the reinforcement is neglected before cracking, we have the uncracked moment of inertia for the gross section, I_g. Setting $\rho = 0$ in Eq. (2-45) gives

$$I_g = \tfrac{1}{12}bh^3 \qquad (2\text{-}46)$$

Equation (2-46) is the well-known formula for the moment of inertia of a rectangular cross section in the case of a homogeneous beam. In design practice, the effect of rebars is frequently neglected and the uncracked moment of inertia is calculated from the gross section, Eq. (2-46).

The moment of inertia of a gross section can also be obtained from the moment of inertia for a cracked section, Eq. (2-30). Substituting $k = (1/2)h/d$ and $j = (2/3)h/d$

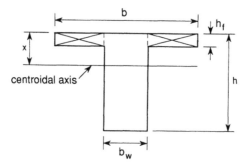

Figure 2.7 Uncracked T sections.

into Eq. (2-30), the resulting Eq. (2-30) is identical to Eq. (2-46). Hence, the transition of flexural behavior from an uncracked stage to a cracked stage can be mathematically modeled by a unified set of equations, rather than by different sets of formulas.

2.2.5.2 Flanged Sections

The calculation of the moment of inertia of uncracked T sections is quite straightforward, but rather tedious. In practice, I_g for gross T sections are obtained from graphs in handbooks, where the rebar areas are neglected. The equation for plotting the graphs is briefly derived here.

The T section shown in Figure 2.7 has a height of h and a width of b. The thickness of the flange is h_f and the width of the web is b_w. The first step is to find the position of the centroidal axis defined by the distance x measured from the top surface of the cross section. Taking the static moments of the web area, $b_w h$, and the outstanding flange area, $(b - b_w)h_f$, about an axis lying on the top surface gives

$$x = \frac{h}{2}\left[\frac{1 + (b/b_w - 1)(h_f/h)^2}{1 + (b/b_w - 1)(h_f/h)}\right] \tag{2-47}$$

The second step is to find the gross moment of inertia about the centroidal axis by the transfer axis theorem:

$$I_g = \frac{b_w h^3}{12} + b_w h\left(\frac{h}{2} - x\right)^2 + (b - b_w)h_f\left(x - \frac{h_f}{2}\right)^2 + \frac{(b - b_w)h_f^3}{12} \tag{2-48}$$

Inserting x from Eq. (2-47) into Eq. (2-48), simplifying, and grouping the terms result in

$$I_g = \frac{b_w h^3}{12}\left[1 + \left(\frac{b}{b_w} - 1\right)\left(\frac{h_f}{h}\right)^3 + \frac{3(b/b_w - 1)(h_f/h)(1 - h_f/h)^2}{1 + (b/b_w - 1)(h_f/h)}\right] \tag{2-49}$$

Let the expression within the bracket of Eq. (2-49) be denoted as K_{i4}. Then Eq. (2-49) can be written simply as $I_g = K_{i4}(b_w h^3/12)$, where K_{i4} is a function of b/b_w and h_f/h. K_{i4} is plotted as a function of b/b_w and h_f/h in a graph in the ACI Handbook SP-17 (73) (ACI Committee 340, 1973) and in other textbooks.

2.2.6   BENDING DEFLECTIONS OF REINFORCED CONCRETE MEMBERS

2.2.6.1   *Deflections of Homogeneous Elastic Members*

The bending deformations, i.e., deflections and rotations, of a member are in general, calculated from the basic moment–curvature relationships of the sections along the member. In the simple case of a homogeneous elastic member, however, the curvature ϕ is proportional to the moment M and the proportional constant is the bending rigidity EI. Because the curvature ϕ is defined as the bending rotation per unit length $(d\Phi/dx)$ the differential bending rotation $d\Phi$ is expressed for a differential length dx as

$$d\Phi = \frac{M}{EI}\,dx \tag{2-50}$$

The deflections and rotations of a member can be calculated from Eq. (2-50) by integration if the moment diagram is given analytically. Other forms of integration, such as moment area method, elastic load method, and conjugate beam method are available in any textbook on structural analysis.

In design practice, the deflections and rotations of a prismatic member are calculated from formulas available in handbooks and textbooks. A method of organizing the various formulas is given in Section 9.5.2.4 of the Commentary of the 1983 ACI Code (ACI Committee 318, 1983) (deleted in the 1989 Code Commentary). In this method the deflection Δ of a member at the midspan (simple and continuous beams) or at the support (cantilevers) is

$$\Delta = K\frac{5Ml^2}{48EI} \tag{2-51}$$

where

M = moment at midspan for simple and continuous beams or moment at support for cantilever beams

l = clear span of a beam

K = 1 for simple beam
   12/5 for cantilever beam
   $1.20 - 0.20M_0/M_c$ for continuous beam, where M_0 = simple span moment = $wl^2/8$ and M_c is the net midspan moment

For a beam with both ends fixed, $M_c = wl^2/24$ and $K = 0.60$; for a beam with one end fixed and the other end hinged, $M_c = wl^2/16$ and $K = 0.8$.

The difficulty of applying Eq. (2-51) to reinforced concrete beams with both cracked and uncracked sections is the evaluation of the bending rigidity EI. To overcome this difficulty we introduce an effective bending rigidity E_cI_e, which can be used in connection with Eq. (2-51) to calculate the deflections.

2.2.6.2 Effective Bending Rigidities

The effective bending rigidity $E_c I_e$ is the product of two quantities E_c and I_e. E_c is the modulus of elasticity of concrete defined by the ACI Code (ACI Committee 318, 1989) to be

$$E_c = 33w^{1.5} \sqrt{f_c'} \tag{2-52}$$

where E_c and f_c' are in pounds per square inch and the unit weight w is in pounds per cubic foot. In the case of normal weight concrete, w is taken as 144 lb/ft^3 and $E_c = 57{,}000\sqrt{f_c'}$.

The effective moment of inertia, I_e, will now be derived. A uniform moment M acting on a length of beam is expected to create a curvature $\phi = d\Phi/dx$ as shown in Figure 2.8a. The moment–curvature relationship (M-ϕ curve) is plotted in Figure 2.8b. Up to the cracking moment M_{cr}, the curve follows approximately a slope calculated from the bending rigidity $E_c I_g$ of the gross uncracked section. After cracking, however, the curve changes direction drastically. When the ultimate moment M_u is reached, the slope approaches the bending rigidity $E_c I_{cr}$, calculated by the cracked section.

The trend described in the preceding text can be illustrated clearly by plotting the bending rigidity EI against the moment M in Figure 2.8c. Below the cracking moment M_{cr}, the experimental curve is roughly horizontal at the level of the calculated $E_c I_g$. After cracking, however, the curve drops drastically. When the ultimate moment M_u is approached, the curve becomes asymptotic to the horizontal level of the calculated $E_c I_{cr}$. A theoretical equation to express the moment of inertia, therefore, must vary from I_g at cracking to I_{cr} at ultimate.

The trend of the experimental curve (Figure 2.8c) can be closely approximated by a theoretical curve suggested by Branson (1965):

$$I = \left(\frac{M_{cr}}{M} \right)^4 I_g + \left[1 - \left(\frac{M_{cr}}{M} \right)^4 \right] I_{cr} \le I_g \tag{2-53}$$

Equation (2-53) satisfies the two limiting cases. First, $I = I_g$ when $M_{cr}/M = 1$. Second, $I = I_{cr}$ when M_{cr}/M approaches zero. The moment of inertia I obviously has a value between I_g and I_{cr}. The power of 4 for (M_{cr}/M) in Eq. (2-53) was selected to best fit the experimental curve.

When a beam is subjected to lateral loads, the moment is no longer uniform along the length of the beam. Figure 2.9a shows a simply supported beam under uniform load. According to the parabolic bending moment diagram (Figure 2.9b), the moment in the central region of the beam exceeds the cracking moment M_{cr}, causing cracks to develop and the curvature ϕ to increase rapidly (Figure 2.9c). Near the ends, however, the regions are still uncracked because the moment is still less than the cracking moment. For such a beam with both cracked and uncracked regions, a rigorous calculation of the deflection using numerical integration would be very tedious. For practical purpose, an effective moment of inertia, I_e, was proposed by Branson (1977) and adopted by the ACI Code, Section 9.5.2 (ACI Committee 318, 1989):

$$I_e = \left(\frac{M_{cr}}{M_{max}} \right)^3 I_g + \left[1 - \left(\frac{M_{cr}}{M_{max}} \right)^3 \right] I_{cr} \le I_g \tag{2-54}$$

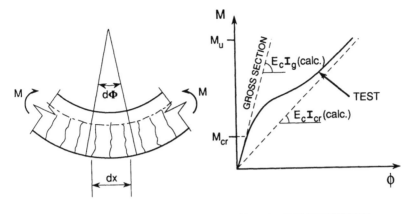

(a) CURVATURE OF BEAM
    UNDER UNIFORM BENDING

(b) MOMENT-CURVATURE
    RELATIONSHIP (M - ϕ) CURVE

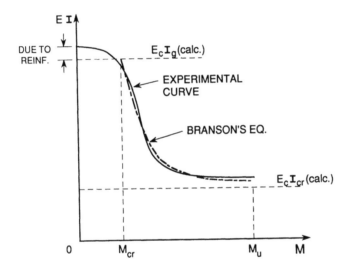

(c) BENDING RIGIDITY E**I** AS A FUNCTION OF MOMENT M

Figure 2.8 Variation of bending rigidity with moment.

where M_{max} is the maximum moment at midspan and the cracking moment is

$$M_{\text{cr}} = \frac{f_r I_g}{y_t} \tag{2-55}$$

In Eq. (2-55), y_t is the distance from the centroidal axis of gross section to the extreme fiber in tension. f_r is the modulus of rupture defined as

$$f_r = 7.5\sqrt{f_c'} \tag{2-56}$$

Equation (2-54) is a modification of Eq. (2-53). The power of 3 for $(M_{\text{cr}}/M_{\text{max}})$ in Eq. (2-54) is chosen to best fit the test results obtained for simply supported beams. Equation (2-54) is also found to be applicable to cantilever beams if M_{max} is the maximum moment at the support. For continuous beams, the ACI Code allows the

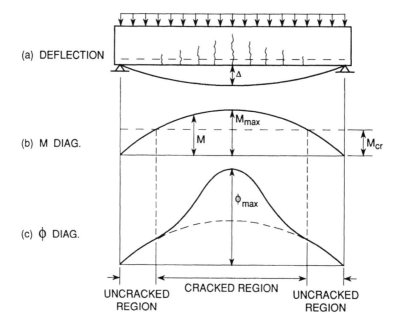

(a) DEFLECTION

(b) M DIAG.

(c) ϕ DIAG.

UNCRACKED
REGION

CRACKED REGION

UNCRACKED
REGION

Figure 2.9 Moment and curvature diagrams for simply supported beams under uniform load.

effective moment of inertia to be taken as the average of the values obtained from Eq. (2-54) for the critical positive and negative moment sections.

The effective bending rigidity $E_c I_e$, determined by Eq. (2-52) and Eqs. (2-54) to (2-56), is used to calculate the deflections in conjunction with Eq. (2-51).

2.3   Nonlinear Bending Theory

2.3.1   MILD-STEEL REINFORCED BEAMS AT ULTIMATE LOAD

A general bending theory has been briefly described in Section 2.1.1 using equilibrium, Bernoulli's compatibility, and the general nonlinear constitutive laws of concrete and reinforcement. The special case of linear bending theory has been presented in Section 2.2. This linear theory is based on equilibrium, Bernoulli's compatibility, and Hooke's constitutive laws for both concrete and reinforcement. In this section, a nonlinear bending theory will be studied which, in addition to equilibrium and compatibility, will utilize the nonlinear constitutive laws of concrete, coupled with the constitutive law of mild steel. The stress–strain relationship of mild steel is assumed to have a linear stress–strain relationship up to yielding, followed by a yield plateau. This elastic–perfectly plastic type of stress–strain relationship imparts a distinct method of analysis and design for concrete members reinforced with mild steel. The nonlinear bending theory presented in this section is, therefore, applicable only to *mild-steel* reinforced concrete members at *ultimate* load stage.

The general principle of analysis for bending action enunciated in Section 2.1.1 will be developed in conjunction with the constitutive laws of concrete and mild steel at ultimate. These principles will first be applied to singly reinforced rectangular beams in Section 2.3.2, and then to doubly reinforced rectangular beams and flanged beams in Sections 2.3.3 and 2.3.4, respectively. Finally, the moment–curvature relationship of mild-steel reinforced beams will be discussed in Section 2.3.5 using both the nonlinear bending theory in Section 2.3 and the linear bending theory in Section 2.2.

2.3.2   SINGLY REINFORCED RECTANGULAR BEAMS

A singly reinforced rectangular section (Figure 2.10a) is subjected to a bending moment M_u at the ultimate load stage. The cross section has a width of b and an effective depth of d. The mild steel reinforcing bars have a total area of A_s. The moment M_u induces a curvature ϕ_u as indicated in the strain diagram of Figure 2.10b. The linear distribution of strain based on Bernoulli's fundamental hypothesis is assumed to remain valid at ultimate. The maximum concrete strain at the top surface, ε_u, is specified by the ACI Code (ACI Committee 318, 1989) to be 0.003 at ultimate.

2.3.2.1   *Nonlinear Constitutive Laws of Concrete*

The distribution of stresses in the concrete compression stress block (Figure 2.10c) is assumed to have the same shape as the stress–strain curve of concrete (Figure 2.10e) obtained from the tests of standard 6-in. × 12-in. concrete cylinders. The stress–strain curve of concrete is frequently expressed by a parabolic equation:

$$\sigma = f_c' \left[2 \left(\frac{\varepsilon}{\varepsilon_0} \right) - \left(\frac{\varepsilon}{\varepsilon_0} \right)^2 \right] \tag{2-57}$$

where

σ = compressive stress of concrete

ε = compressive strain of concrete

f_c' = maximum compressive stress of concrete obtained from standard cylinders

ε_0 = strain at the maximum stress f_c', usually taken as 0.002

The magnitude of the compression resultant C can be expressed as

$$C = k_1 f_c' cb \tag{2-58}$$

where k_1 is defined as the ratio of the average stress of the compression stress block to the maximum stress of concrete f_c'. Substituting C from Eq. (2-1) into Eq. (2-58) and replacing the maximum strain ε_c by the ultimate strain ε_u give the expression for k_1:

$$k_1 = \frac{1}{f_c' \varepsilon_u} \int_0^{\varepsilon_u} \sigma \, d\varepsilon \tag{2-59}$$

Substituting the concrete stress σ from Eq. (2-57) into Eq. (2-59) and integrating result in

$$k_1 = \frac{\varepsilon_u}{\varepsilon_0} \left(1 - \frac{1}{3} \frac{\varepsilon_u}{\varepsilon_0} \right) \tag{2-60}$$

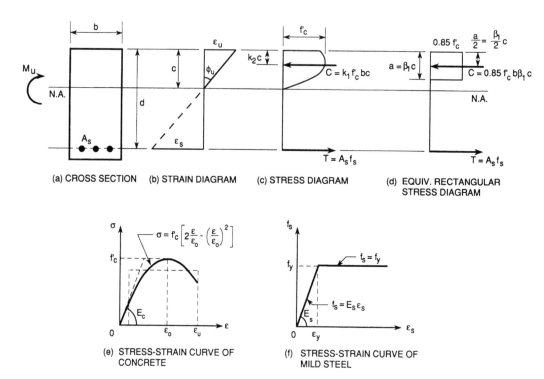

Figure 2.10   Singly reinforced rectangular sections at ultimate.

Insert the ACI value of $\varepsilon_u = 0.003$ and $\varepsilon_0 = 0.002$ into Eq. (2-60) and we have $k_1 = 0.75$.

The location of the resultant C is defined by $k_2 c$ measured from the top surface of the cross section. The coefficient k_2 has been given in Eq. (2-3) except that the maximum stress ε_c should be replaced by ε_u at ultimate:

$$k_2 = 1 - \frac{1}{\varepsilon_u} \frac{\int_0^{\varepsilon_u} \sigma\varepsilon\, d\varepsilon}{\int_0^{\varepsilon_u} \sigma\, d\varepsilon} \tag{2-61}$$

Again, substituting the concrete stress σ from Eq. (2-57) into Eq. (2-61) and integrating result in

$$k_2 = 1 - \frac{2}{3}\left[\frac{1 - \frac{3}{8}(\varepsilon_u/\varepsilon_0)}{1 - \frac{1}{3}(\varepsilon_u/\varepsilon_0)}\right] \tag{2-62}$$

Inserting the ACI value of $\varepsilon_u = 0.003$ and $\varepsilon_0 = 0.002$ into Eq. (2-62) gives $k_2 = 0.4167$.

The shape of the concrete stress–strain curve was found by tests to depend on the strength of the concrete f'_c. Apparently, the stress–strain curve becomes more linear in the ascending portion and steeper in the descending portion when the concrete strength is increased. In other words, the coefficients k_1 and k_2 should decrease with an increase of f'_c.

To simplify the analysis, ACI Code (ACI Committee 318, 1989) allows the curved concrete stress block to be replaced by an equivalent rectangular stress block as shown in Figure 2.10d. The replacement is such that the magnitude and the location of the resultant C remain unchanged. The ACI rectangular stress block has a uniform stress of $0.85f_c'$ and a depth $a = \beta_1 c$. β_1 is taken as 0.85 for $f_c' = 4000$ psi or less, and decreases by 0.05 for every increment of 1000 psi beyond $f_c' = 4000$ psi. A lower limit of $\beta_1 = 0.65$ is also specified for $f_c = 8000$ psi or greater. Based on this rectangular stress block the magnitude of the resultant $C = 0.85f_c'b\beta_1 c$, and the coefficient $k_1 = 0.85\beta_1$. The coefficient k_2 is, of course, equal to $\beta_1/2$.

The maximum value of k_2, which is equal to $\beta_1/2 = 0.85/2 = 0.425$ by the ACI Code, is very close to the value of 0.4167 calculated from Eq. (1-62). The maximum value of k_1, however, is equal to $0.85\beta_1 = 0.85(0.85) = 0.7225$ according to the ACI Code. This is somewhat less than the value of 0.75 calculated from Eq. (2-60) based on a very reasonable parabolic stress–stress curve of concrete. Apparently, this difference is not caused by the shape of the stress–strain curve, but actually takes into account the size effect, the shape effect, and the loading rate effect between the testing of beams and the testing of standard cylinders for f_c'.

Using the ACI equivalent rectangular stress block the force equilibrium equation, Eq. (2-5), and the moment equilibrium equation, Eq. (2-7), become

$$c = \frac{A_s f_s}{0.85 f_c' b \beta_1} \tag{2-63}$$

$$M_u = \varphi A_s f_s d \left(1 - \frac{\beta_1 c}{2d}\right) \tag{2-64}$$

where φ is the reduction factor for material. $\varphi = 0.9$ for bending in the ACI Code.

The steel stress f_s in Eqs. (2-63) and (2-64) has not yet been determined. The determination of f_s depends on the stress–strain relationship of mild steel.

2.3.2.2 Constitutive Law of Mild Steel

Mild steel is assumed to have an elastic–perfectly plastic stress–strain curve as shown in Figure 2.10f. If the steel is in the elastic range ($\varepsilon_s < \varepsilon_y$) when the concrete crushes at ultimate load ($\varepsilon_u = 0.003$), the beam is called overreinforced. If the steel is in the plastic range ($\varepsilon_s \geq \varepsilon_y$) when the concrete crushes, the beam is called underreinforced. The method of analysis and design for beams reinforced by mild steel depends strongly on knowing whether the beam is overreinforced or underreinforced. In order to divide these two modes of failure, we define the "balanced condition" when the steel reaches the yield point ($\varepsilon_s = \varepsilon_y$) simultaneously with the crushing of concrete ($\varepsilon_u = 0.003$).

The bending of singly reinforced beams involves nine variables, b, d, A_s, M_u, f_s, f_c', ε_s, ε_u, and c (or a), as shown in Figure 2.10a to d. The coefficients k_1 and k_2, or the coefficient β_1, are not considered variables because they are determined from the given compression stress–strain curve (Figure 2.10e) or directly from tests. A total of four equations are available: two from equilibrium, one from Bernoulli compatibility, and one from the constitutive law of mild steel (Figure 2.10f). Therefore, five variables must be given so that the remaining four unknown variables can be solved by the four equations.

2.3.2.3 Balanced Condition

The problem posed for the balanced condition is

Given: b, d, f_c', $\varepsilon_s = \varepsilon_y$, $\varepsilon_u = 0.003$

Find: A_s, M_u, f_s, and a (or c)

The four available equations and their unknowns are summarized as follows:

Type of Equations	Equations	Unknowns				
Equilibrium of forces	$A_s f_s = 0.85 f_c' ba$	A_s			a	(2-65)
Equilibrium of moments	$M_u = \varphi A_s f_s (d - a/2)$	A_s	M_u	f_s	a	(2-66)
Bernoulli compatibility	$\dfrac{a}{\beta_1 d} = \dfrac{\varepsilon_u}{\varepsilon_u + \varepsilon_y}$				a	(2-67)
Constitutive law for steel	$f_s = E_s \varepsilon_s$ for $\varepsilon_s \le \varepsilon_y$			f_s		(2-68a)
	$f_s = f_y$ for $\varepsilon_s > \varepsilon_y$			f_s		(2-68b)

The four unknown variables, A_s, M_u, f_s, and a, are indicated for each equation under the column heading "Unknowns." Because the steel strain ε_s is given to be the yield stress ε_y, the steel stress f_s should obviously be equal to f_y based on the stress–strain relationship of either Eqs. (2-68a) or (2-68b). Also, the unknown depth a of the equivalent rectangular stress block can be solved directly from the compatibility condition Eq. (2-67). Once the depth a and the steel stress f_s are determined, the steel area A_s can be obtained from the force equilibrium [Eq. (2-65)] and the ultimate moment M_u from the moment equilibrium [Eq. (2-66)]. In this particular case of balanced condition there is no need to solve simultaneous equations.

In this case of balanced condition we will now add a subscript b to the three unknown quantities: A_{sb} for balanced steel area, M_{ub} for balanced ultimate moment, and a_b for balanced depth of equivalent rectangular stress block. Substituting $\varepsilon_u = 0.003$ and $\varepsilon_y = f_y(\text{psi})/29{,}000{,}000$ into the compatibility equation [Eq. (2-67)], the balanced depth a_b is expressed in terms of yield stress f_y (psi) as

$$a_b = \beta_1 d \frac{87{,}000}{87{,}000 + f_y} \tag{2-69}$$

Substituting a_b into the force equilibrium equation [Eq. (2-65)], the balanced steel area A_{sb} is

$$A_{sb} = 0.85 \beta_1 bd \frac{f_c'}{f_y} \frac{87{,}000}{87{,}000 + f_y} \tag{2-70}$$

Dividing Eq. (2-70) by bd and defining ρ_b as the balanced percentage of steel give

$$\rho_b = 0.85 \beta_1 \frac{f_c'}{f_y} \frac{87{,}000}{87{,}000 + f_y} \tag{2-71}$$

Inserting a_b from Eq. (2-69), A_{sb} from Eq. (2-70), $f_s = f_y$, and the reduction factor φ into the moment equilibrium equation Eq. (2-66) results in

$$M_{ub} = \varphi 0.85 \beta_1 \frac{87{,}000}{87{,}000 + f_y} \left(1 - 0.5 \beta_1 \frac{87{,}000}{87{,}000 + f_y} \right) f_c' bd^2 \tag{2-72}$$

Equation (2-72) can be written simply as

$$M_{ub} = R_b bd^2 \qquad (2\text{-}73)$$

where

$$R_b = \varphi 0.85\beta_1 \frac{87{,}000}{87{,}000 + f_y}\left(1 - 0.5\beta_1 \frac{87{,}000}{87{,}000 + f_y}\right)f_c' \qquad (2\text{-}74)$$

The coefficient R_b is a function of the material properties f_c' and f_y, and has been tabulated in some textbooks.

The balanced condition expressed by a_b [Eq. (2-69)], ρ_b [Eq. (2-71)], or M_{ub} [Eqs. (2-73) and (2-74)] divides underreinforced beams from overreinforced beams. The balanced percentage of steel ρ_b is useful for the analysis type of problems when the cross sections of concrete and steel are given. The balanced ultimate moment M_{ub} is convenient for the design type of problems when the moment is given. The balanced depth of rectangular stress block, a_b, could be used either in analysis or design, but mostly in design in lieu of M_{ub}.

The application of the balanced condition discussed in the preceding text can be summarized in the following table:

Types of Problems	Underreinforced Beams	Overreinforced Beams
Analysis	$\rho < \rho_b$	$\rho > \rho_b$
Design	$M_u < M_{ub}$ or $a < a_b$	$M_u > M_{ub}$ or $a > a_b$

2.3.2.4 *Overreinforced Beams*

If the percentage of steel is greater than the balanced percentage ($\rho > \rho_b$) or the ultimate moment is greater than the balanced ultimate moment ($M_u > M_{ub}$), then the beam is overreinforced. The problem posed for the analysis of overreinforced beams is

Given: b, d, A_s, f_c', $\varepsilon_u = 0.003$
Find: M_u, f_s, $\varepsilon_s < \varepsilon_y$, and a (or c)

The four available equations and their unknowns are:

Type of Equations	Equations	Unknowns		
Equilibrium of forces	$A_s f_s = 0.85 f_c' ba$		f_s	a (2-75)
Equilibrium of moments	$M_u = \varphi A_s f_s (d - a/2)$	M_u	f_s	a (2-76)
Bernoulli compatibility	$\dfrac{a}{\beta_1 d} = \dfrac{\varepsilon_u}{\varepsilon_u + \varepsilon_s}$		ε_s	a (2-77)
Constitutive law for steel	$f_s = E_s \varepsilon_s$ for $\varepsilon_s < \varepsilon_y$		f_s ε_s	(2-78)

Based on the unknowns in each equation, the best strategy to solve the four equations is as follows:

Step 1: Substitute ε_s from the stress–strain relationship of Eq. (2-78) into Eq. (2-77) and express the compatibility equation in terms of the steel stress f_s and the depth a.

Step 2: Solve the new compatibility equation simultaneously with the force equilibrium equation [Eq. (2-75)] to determine the unknowns f_s and a.

Step 3: Insert the newly found unknowns f_s and a into the moment equilibrium equation [Eq. (2-76)] and calculate the moment M_u.

Step 4: Calculate the steel strain $\varepsilon_s = f_s/E_s$ from the stress–strain relationship of steel, if required.

Elaboration of the preceding procedures is not warranted because overreinforced beams are seldom encountered. Overreinforced beams are expected to fail brittlely with small deflections because the concrete will crush before the yielding of the steel. In fact, the design of overreinforced beams is prohibited by all building and bridge codes.

2.3.2.5 Underreinforced Beams

Underreinforced beams are useful because they fail in a ductile manner by producing large deflections. This is because the steel is expected to yield before the crushing of concrete. To ensure sufficient ductility, ACI Code (ACI Committee 318, 1989) requires the beams to satisfy the following criteria.

For analysis (when the cross sections of concrete and steel are given):

$$\rho \leq \rho_{max} = 0.75\rho_b \tag{2-79}$$

For design (when the moment is given):

$$a \leq a_{max} = 0.75a_b \tag{2-80}$$

or

$$M_u \leq M_{max} = R_u bd^2 \tag{2-81}$$

where ρ_{max}, a_{max} and M_{max} are maximum percentage of steel, maximum depth of rectangular stress block and maximum moment, respectively,

and
$$R_u = (0.75)\,\varphi 0.85\beta_1 \frac{87,000}{87,000 + f_y}\left(1 - (0.75)0.5\beta_1 \frac{87,000}{87,000 + f_y}\right)f_c' \tag{2-82}$$

Equation (2-82) for R_u is the same as Eq. (2-74) for R_b except that a constant (0.75) is included both outside and inside the parentheses. R_u is a function of f_c' and f_y, and is tabulated in Table 2.1 together with the nondimensional ratio a_{max}/d and ρ_{max}.

The stress and strain diagrams for an underreinforced beam are shown in Figure 2.11c and b, respectively. The analysis and design of underreinforced beams are considerably simplified for two reasons. First, the tensile steel stress is known to yield, $f_s = f_y$, and the stress–strain curve of steel is not required in the solution. Second, the tensile steel strain is expected to lie within the plastic range $\varepsilon_s \geq \varepsilon_y$. Bernoulli's compatibility condition, which relates the steel strain ε_s to the maximum concrete strain ε_u at the top surface, becomes irrelevant to the solution of other stress-type

TABLE 2.1 a_{max}, ρ_{max}, and R_u as functions of f_c' and f_y

f_c'	$f_y = 40,000$			$f_y = 50,000$			$f_y = 60,000$			$f_y = 75,000$		
	$\dfrac{a_{max}}{d}$	$100\,\rho_{max}$	R_u	$\dfrac{a_{max}}{d}$	$100\,\rho_{max}$	R_u	$\dfrac{a_{max}}{d}$	$100\,\rho_{max}$	R_u	$\dfrac{a_{max}}{d}$	$100\,\rho_{max}$	R_u
3000	0.436	2.78	783	0.405	2.06	740	0.378	1.61	705	0.343	1.16	653
4000	0.436	3.72	1047	0.405	2.75	987	0.378	2.14	937	0.343	1.55	870
5000	0.411	4.36	1244	0.381	3.24	1181	0.355	2.54	1119	0.322	1.82	1032
6000	0.386	4.90	1424	0.357	3.64	1346	0.333	2.83	1276	0.302	2.05	1175
7000	0.360	5.34	1580	0.333	3.96	1483	0.311	3.08	1402	0.282	2.24	1295

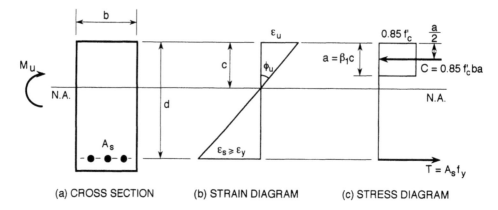

(a) CROSS SECTION   (b) STRAIN DIAGRAM   (c) STRESS DIAGRAM

Figure 2.11 Underreinforced sections at ultimate.

variables. Because of these two simplifications, the analysis and design of underreinforced beams involve only eight variables, b, d, A_s, M_u, f_y, f_c', ε_u, and a (or c). The available equations are now down to two: the only two from the equilibrium condition. Therefore, six variables must be given before the two remaining unknown variables can be solved by the two equilibrium equations.

Three types of problems are frequently encountered in the analysis and design of underreinforced beams: (1) analysis, (2) first type of design, and (3) second type of design. They are treated separately below.

(1) Analysis (find moment)

$$\text{Given: } b, d, A_s, f_y, f_c', \text{ and } \varepsilon_u$$
$$\text{Find: } M_u \text{ and } a$$

As mentioned previously, three forms of equilibrium equations are frequently used in the parallel, coplanar force system of bending action, but only two are independent. However, the three equations are given here for the purpose of finding the most convenient solution. Only two of these three equations will be selected.

Type of Equations	Equations	Unknowns	
Equilibrium of forces	$A_s f_y = 0.85 f_c' ba$	a	(2-83)
Equil. of mom. about C	$M_u = \varphi A_s f_y (d - a/2)$	M_u a	(2-84)
Equil. of mom. about T	$M_u = \varphi 0.85 f_c' ab(d - a/2)$	M_u a	(2-85)

The solution for the analysis type of problem turns out to be quite simple when the unknowns in each equations are examined. It can be seen from Eq. (2-83) that the depth a can be determined directly from the equilibrium of forces C and T. Then the moment M_u could be calculated either from the equilibrium of moments about the compressive force C [Eq. (2-84)] or from the equilibrium of moments about the tensile force T [Eq. (2-85)].

From the equilibrium of forces,

$$a = \frac{A_s f_y}{0.85 f_c' b} \tag{2-86}$$

and from the equilibrium of moments about the compression force C,

$$M_u = \varphi A_s f_y \left(d - \frac{a}{2} \right) \tag{2-87}$$

It should again be emphasized that the depth of the neutral axis c, which is equal to a/β_1, can be determined directly from the equilibrium condition [Eq. (2-86)] without using the compatibility condition and the stress–strain relationship of steel. This is a special characteristic of the mild-steel underreinforced concrete beams.

(2) First Type of Design (find area of steel)

$$\text{Given: } b, d, M_u, f_y, f_c', \text{ and } \varepsilon_u$$
$$\text{Find: } A_s \text{ and } a$$

The three equilibrium equations and their unknowns are:

Type of Equations	Equations	Unknowns		
Equilibrium of forces	$A_s f_y = 0.85 f_c' ba$	A_s	a	(2-88)
Equil. of mom. about C	$M_u = \varphi A_s f_y (d - a/2)$	A_s	a	(2-89)
Equil. of mom. about T	$M_u = \varphi 0.85 f_c' ab(d - a/2)$		a	(2-90)

Examination of the two unknowns A_s and a in the three equations reveals two ways to arrive at the solution.

First Way

The depth a is solved from the equilibrium of moment about T [Eq. (2-90)]. Then the depth a is inserted into either Eq. (2-88) or Eq. (2-89) to solve for the steel area A_s. This way of solution is quite straightforward, but not often used because Eq. (2-90) is a quadratic equation for the depth a. The solution of a quadratic equation by hand is somewhat tedious.

Second Way

The depth a is obtained by solving Eqs. (2-88) and (2-89) simultaneously using trial-and-error method. This solution is actually quite simple for an experienced engineer because the depth a can be closely estimated in the first trial. Assuming a depth a_1 to be less than $0.75a_b$, and substituting it into Eq. (2-89) give the value of A_{s1} for the first cycle as

$$A_{s1} = \frac{M_u}{\varphi f_y (d - a_1/2)} \tag{2-91}$$

Then to calculate the depth a_2 insert A_{s1} into Eq. (2-88):

$$a_2 = \frac{A_{s1} f_y}{0.85 f_c' b} \tag{2-92}$$

If $a_2 = a_1$, a solution is found. If $a_2 \neq a_1$, then assume another a_2 value and repeat the cycle. The convergence is usually very rapid, and two or three cycles of trial-and-error are usually sufficient to give an accurate solution.

(3) Second Type of Design (find cross sections of concrete and steel)

$$\text{Given: } M_u, f_y, f_c', \text{ and } \varepsilon_u$$
$$\text{Find: } b, d, A_s, \text{ and } a$$

Notice that four unknowns are shown. Obviously, only two of these four unknowns can be determined by the two available equations. Therefore, two unknowns will have to be assumed in the process of design. It is a characteristic of design problems that the number of unknowns frequently exceeds the available number of equations.

The three equilibrium equations and their unknowns are:

Type of Equations	Equations	Unknowns			
Equilibrium of forces	$A_s f_y = 0.85 f_c' ba$	b	A_s	a	(2-93)
Equil. of mom. about C	$M_u = \varphi A_s f_y d(1 - a/2d)$	d	A_s	a	(2-94)
Equil. of mom. about T	$M_u = \varphi 0.85 f_c' abd(1 - a/2d)$	b	d	a	(2-95)

Two methods of design have been used:

First Method

Assume a suitable percentage of steel $\rho = A_s/bd$, say $\rho = 0.5\rho_b$. This means that we are specifying the ratio of the two quantities that are being designed.

From equilibrium of forces [Eq. (2-93)],

$$\frac{a}{d} = \frac{A_s f_y}{bd 0.85 f_c'} = \frac{\rho f_y}{0.85 f_c'} \tag{2-96}$$

Inserting a/d from Eq. (2-96) into the moment equilibrium equation about C [Eq. (2-94)] and noticing $A_s = \rho bd$, we can solve for a parameter bd^2 for the concrete cross section:

$$bd^2 = \frac{M_u}{\varphi \rho f_y \left(1 - 0.59(\rho f_y/f_c')\right)} \tag{2-97}$$

Once bd^2 is obtained, either b or d has to be assumed to determine the remaining concrete dimension. The steel area can then be calculated from ρbd. However, because the dimensions b and d are always adjusted to become round numbers, an accurate calculation of the steel area A_s can be obtained using the "first type of design" method in which the cross section of concrete is given.

Second Method

Assume a suitable reinforcement index $\omega = \rho f_y/f_c'$. From Eq. (2-96) it can be seen that this index ω is directly proportional to the depth ratio a/d. Substituting

$a/d = \omega/0.85$ into the moment equilibrium equation [Eq. (2-95)] gives

$$bd^2 = \frac{M_u}{\varphi f_c' \omega(1 - 0.59\omega)} \tag{2-98}$$

Equation (2-98) is tabulated and used in the ACI Handbook SP-17(73) (ACI Committee 340, 1973).

(4) Deformations (find curvature and steel strain)

In conclusion, once a mild-steel reinforced concrete beam is found to be underreinforced, it is analyzed and designed by equilibrium condition without using Bernoulli's compatibility condition and the stress–strain curve of steel. It would appear that the analysis and design of underreinforced beams are based on the equilibrium (or plasticity) truss model presented in Chapter 3. This similarity is true if we are interested only in the bending strength. If we are also interested in deformations, however, we must then rely on Bernoulli's compatibility condition (Figure 2.11b) to determine the ultimate curvature ϕ_u:

$$\phi_u = \frac{\varepsilon_u}{c} \tag{2-99}$$

where $c = a/\beta_1$ and the depth a is one of the two unknowns solved by the two equilibrium equations in either the analysis or the design problems. If the steel strain ε_s is desired at the ultimate load stage, Bernoulli's compatibility condition is also required to give

$$\varepsilon_s = \varepsilon_u \frac{d - c}{c} \tag{2-100}$$

2.3.3   DOUBLY REINFORCED RECTANGULAR BEAMS

A doubly reinforced beam has compression steel A_s' in addition to the tension steel A_s (Figure 2.12a). The compression steel could be employed for various purposes: First, to increase the moment capacity of a beam when the cross section is limited. Second, in a continuous beam (Figure 2.3), ACI code requires that a portion of the bottom positive steel in the center region of a beam must be extended into the supports. These extended bars provide the compression steel for the rectangular support sections that are subjected to negative moment. Third, compression steel could be used to reduce deflections. Fourth, the ductility of a beam could be enhanced by adding compression steel (see Section 2.3.5.2).

The additional compression steel introduces three additional variables, namely, A_s', f_s', and ε_s' for the area, stress, and strain of the compression steel, respectively (Figure 2.12a to c). Therefore, the analysis and design of doubly reinforced rectangular concrete beams involve 12 variables, namely, b, d, A_s, A_s', M_u, f_s, f_s', f_c', ε_s, ε_s', ε_u, and c (or a). At the same time, two additional equations are available: one for the compatibility of compression steel and the other for the stress–strain relationship of the compression steel.

The analysis and design of doubly reinforced rectangular sections will be limited to underreinforced beams in this section (Section 2.3.3) using the stress–strain relationships of concrete and steel shown in Figure 2.12f and g. In this type of problem the tensile steel will be in the yield range $f_s = f_y$ and the tensile steel strain ε_s is irrelevant to the solution of stress-type variables. Correspondingly, the compatibility equation for the tensile steel and the stress–strain relationship of the tensile steel are

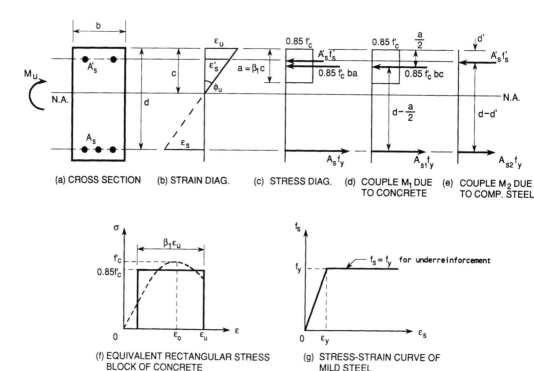

Figure 2.12 Doubly reinforced rectangular sections at ultimate.

not required. As a result, we now have 11 variables, b, d, A_s, A'_s, M_u, f_y, f'_s, f'_c, ε'_s, ε_u, and c (or a), and 4 available equations. Two of these four equations come from the equilibrium condition and the other two from the compatibility and stress–strain relationship of compression steel. If seven variables are given, the remaining four unknown variables can be solved by the four available equations.

The analysis and design of doubly reinforced beams will now be treated separately.

(1) Analysis Problems (find moment)

$$\text{Given: } b, d, A_s, A'_s, f_y, f'_c, \text{ and } \varepsilon_u$$
$$\text{Find: } M_u, f'_s, \varepsilon'_s, \text{ and } a \text{ (or } c = a/\beta_1)$$

The four available equations and their unknowns are:

Type of Equations	Equations	Unknowns		
Equil. of forces	$A_s f_y = 0.85 f'_c ba + A'_s f'_s$		f'_s	a (2-101)
Equil. mom. about T	$M_u = \varphi[0.85 f'_c ba(d - a/2)$			
	$\qquad + A'_s f'_s(d - d')]$	M_u f'_s		a (2-102)
Compat. (comp. steel)	$\dfrac{\varepsilon'_s}{\varepsilon_u} = \dfrac{c - d'}{c} \quad (c = a/\beta_1)$		ε'_s	a (2-103)
Const. law (comp. steel)	$f'_s = E_s \varepsilon'_s \quad$ for $\varepsilon'_s \le \varepsilon_y$		f'_s ε'_s	(2-104a)
	$f'_s = f_y \qquad$ for $\varepsilon'_s > \varepsilon_y$		f'_s ε'_s	(2-104b)

It should be pointed out that an approximation has been introduced in writing Eqs. (2-101) and (2-102). The compression steel force, $A'_s f'_s$, should have been written as $(A'_s f'_s - 0.85 \, A'_s f'_c)$ to include the negative force attributed to the area of concrete displaced by the compression steel. Since $0.85 \, A'_s f'_c$ is smaller than $A'_s f'_c$ by an order of magnitude, this small force. $0.85 \, A'_s f'_c$, has been neglected in the two equations.

Examination of the unknowns in the four equations indicates two methods to solve these equations:

Method 1 (Trial-and-error method)

Step 1: Assume a value of depth a $(c = a / \beta_1)$ and calculate the compression steel strain ε'_s from the compatibility condition of Eq. (2-103).

Step 2: Calculate the compression steel stress f'_s from the stress–strain relationship of compression steel, Eqs. (2-104a) or (1-104b).

Step 3: Insert f'_s into the force equilibrium equation [Eq. (2-101)] to calculate a new value of the depth a. If the new a is the same as the assumed a, a solution is obtained. If not, assume another value of a and repeat the cycle. The convergence is usually quite rapid.

Step 4: Once the depth a and the compression steel stress f'_s are solved, insert them into the moment equilibrium equation [Eq. (2-102)] to calculate the moment M_u.

Method 2 (solve quadratic equation)

Step 1: Insert the compression steel strain ε'_s from Eq. (2-104a) into Eq. (2-103) and express the compatibility of compression steel by a new equation in terms of the compression steel stress f'_s and the depth a.

Step 2: Solve the new compatibility equation simultaneously with the force equilibrium equation [Eq. (2-101)] to obtain the stress f'_s and the depth a. This is the process of solving a quadratic equation. If $f'_s \geq f_y$, then use $f'_s = f_y$ according to Eq. (2-104b) and recalculate the depth a from Eq. (2-101).

Step 3: Substitute the stress f'_s and the depth a into Eq. (2-102) to determine the moment M_u.

(2) Design Problems (find areas of tension and compression steel)

Given: b, d, M_u, f_y, f'_c, and ε_u

Find: A_s, A'_s, f'_s, ε'_s, and a (or $c = a / \beta_1$)

The problem posed above shows five unknowns, but we have only four equations. Therefore, one of the five unknowns must be assumed during the design process. It will be shown in succeeding text that the most convenient unknown to choose is the depth a.

In the design process it is convenient to take advantage of the two-internal-couple concept as shown in Figure 2.12d and e. The bending resistance of a doubly reinforced beam can be considered to consist of two internal couples. One couple, M_1, is contributed by the concrete compression stress block $0.85 f'_c ba$, and the other, M_2, by the compression steel area A'_s. The tensile steel area A_s is separated into A_{s1}

and A_{s2} for M_1 and M_2, respectively. Because the first couple is identical to a singly reinforced beam, which can be solved easily by equilibrium condition alone, the separation of the two internal couples is very convenient for the design of doubly reinforced beams.

In the separation process we have expanded the two variables A_s and M_u into four variables, A_{s1}, A_{s2}, M_1, and M_2, resulting in the addition of two unknowns. At the same time, two additional equations stating these separations are created. These two additional equations are:

Separation of M_u $$M_u = M_1 + M_2 \tag{2-105}$$

Separation of A_s $$A_s = A_{s1} + A_{s2} \tag{2-106}$$

The two groups of equations and their unknown variables are summarized as follows:

First Internal Couple M_1

Type of Equations	Equations	Unknowns		
[Equil. of forces]$_1$	$A_{s1}f_y = 0.85f_c'ba$	A_{s1}	a	(2-107)
[Equil. of mom.]$_1$	$M_1 = \varphi 0.85 f_c'ba(d - a/2)$	M_1	a	(2-108)

Second Internal Couple M_2

Type of Equations	Equations	Unknowns		
[Equil. of forces]$_2$	$A_{s2}f_y = A_s'f_s'$	$A_{s2}\ A_s'\ f_s'$		(2-109)
[Equil. of mom.]$_2$	$M_2 = \varphi A_{s2}f_y(d - d')$	$M_2\ A_{s2}$		(2-110)
Compat. (comp. steel)	$\dfrac{\varepsilon_s'}{\varepsilon_u} = \dfrac{c - d'}{c}\quad (c = a/\beta_1)$		$\varepsilon_s'\ a$	(2-111)
Const. law (comp. steel)	$f_s' = E_s\varepsilon_s'\quad$ for $\varepsilon_s' \le \varepsilon_y$		$f_s'\ \varepsilon_s'$	(2-112a)
	$f_s' = f_y\quad\ \ \ $ for $\varepsilon_s' > \varepsilon_y$		$f_s'\ \varepsilon_s'$	(2-112b)

Examination of the two groups of equations and their unknown variables shows that the only unknown variable common to the two groups is the depth a (or c). If the depth a is assumed, the unknowns M_1 and A_{s1} for the first group can be calculated directly from the two equilibrium equations [Eqs. (2-107) and (2-108)]. Then the moment M_2 for the second group is determined from Eq. (2-105) $M_2 = M_u - M_1$; the rest of the unknowns in the second groups can be solved easily.

The depth a could be used to control the ductility (see Section 2.3.5.2) of the beam because the ultimate curvature ϕ_u is inversely proportional to the depth a ($\phi_u = \varepsilon_u\beta_1/a$) according to the compatibility of strains in Figure 2.12b. The most economic way to design the beam, however, is to assume the depth $a = a_{max}$. This assumption maximizes the compression stress block in the first internal couple and, therefore, minimizes the compression steel area A_s' for the second couple. If $a = a_{max}$, the procedures for design are as follows:

Step 1: Apply $a = a_{max}$ to the first couple to find A_{s1} and M_1 directly from equilibrium equations [Eqs. (2-107) and (2-108)], respectively. Then compare M_1 to M_u. If $M_u \leq M_1$, a singly reinforced beam would suffice. If $M_u > M_1$, a doubly reinforced beam is required.

Step 2: If a doubly reinforced beam is confirmed, calculate $M_2 = M_u - M_1$ from Eq. (2-105) and insert M_2 into the moment equilibrium equation [Eq. (2-110)] of the second couple to find the steel area A_{s2}. The total steel area A_s is the sum of A_{s1} and A_{s2} [Eq. (2-106)].

Step 3: Apply $a = a_{max}$ to the second couple by inserting $c = a_{max}/\beta_1$ into the compatibility equation [Eq. (2-111)] to determine the compression steel strain ε'_s. Then find the compression steel stress f'_s from the stress–strain relationship of Eqs. (2-112a) or (2-112b).

Step 4: Substitute f'_s and A_{s2} into the force equilibrium equation [Eq. (2-109)] and calculate the compression steel area A'_s.

(3) Balanced Condition for Doubly Reinforced Beams

The balanced condition for doubly reinforced beams can easily be derived using the two-internal-couple concept as shown in Figure 2.12c to e. The only difference is that the depth a should be replaced by the balanced depth a_b in Figure 2.12c and d. Correspondingly, we can replace the symbols for steel areas A_s and A_{s1} by A_{sb} and A_{s1b}, respectively, and then write

$$A_{sb}f_y = A_{s1b}f_y + A'_s f'_s \qquad (2\text{-}113)$$

Dividing Eq. (2-113) by bd and defining $\rho_b = A_{sb}/bd$, $\rho_{1b} = A_{1b}/bd$, and $\rho' = A'_s/bd$ give

$$\rho_b = \rho_{1b} + \rho' \frac{f'_s}{f_y} \qquad (2\text{-}114)$$

The balanced percentage of steel for the first internal couple ρ_{1b} is the same as the balanced percentage of steel for singly reinforced beams and, therefore, can be calculated from Eq. (2-71). The compression steel stress f'_s for this balanced condition ($a = a_b$ and $\varepsilon_u = 0.003$) is calculated from the compatibility equation for compression steel [Eq. (2-111)] and the stress–strain relationship [Eq. (2-112)] to be

$$f'_s = 87,000\left[1 - \frac{d'}{d} \frac{87,000 + f_y}{87,000}\right] \leq f_y \qquad (2\text{-}115)$$

The maximum percentage of steel for doubly reinforced beams, ρ_{max}, is given by the ACI Code to be

$$\rho_{max} = 0.75\rho_{1b} + \rho' \frac{f'_s}{f_y} \qquad (2\text{-}116)$$

Comparing Eq. (2-116) to Eq. (2-114) shows that the reduction factor 0.75 is applied only to the first internal couple contributed by the concrete compression stress block, and not to the second internal couple contributed by the compression steel. This is because concrete is considered to be relatively brittle, whereas the compression steel is ductile.

2.3.4 FLANGED BEAMS

Flanged beams may have cross sections in the shape of a T, an L, an I, or a box. In these sections the flanges are used to enhance the compression forces of the internal couples. The flanges may be purposely added or may be available as parts of a structure, such as a floor system. In a slab-and-beam floor system the slab serves as the flange, whereas the beam serves as the web. Together they form a T beam. According to the ACI Code, the effective width of the flange of a T beam shall not exceed one quarter of the span length of the beam and the effective overhanging flange width on each side of the web shall not exceed eight times the slab thickness and one half the clear distance to the next web.

A T cross section subjected to an ultimate moment M_u is shown in Figure 2.13a. The strain and stress diagrams are given in Figure 2.13b and c. It can be seen that these strain and stress distributions are identical to those for rectangular sections (Figure 2.11b and c), except that the magnitude and the location of the compression resultant C are difficult to determine. Assuming that the average stress of $0.85f_c'$ and the coefficient β_1 are still valid for T sections, the magnitude of the resultant should include all the stresses on the shaded area in Figure 2.13a. The position of the resultant should be measured from the top surface to the centroidal axis of the shaded area. Such calculations are obviously very tedious.

A simple way to avoid the preceding difficulties is to employ the two-internal-couple concept previously used in the doubly reinforced beams. The one internal C-T couple shown in Figure 2.13c can be separated into two couples: one for the web area shown in Figure 2.13d and one for the flange area shown in Figure 2.13e. Accordingly, we denote the internal couple and the steel area in the web as M_w and A_{sw}, respectively, and the internal couple and the steel area in the flange as M_f and A_{sf}, respectively.

The analysis and design of T sections will be limited only to the underreinforced beams in this section (Section 2.3.4). As discussed previously in connection with singly and doubly reinforced rectangular beams, the tensile steel will be in the yield range, $f_s = f_y$, and the compatibility equation and the stress–strain relationship for the tensile steel become irrelevant. The only equations required for the solution come from the equilibrium condition. We will first examine the two equilibrium equations corresponding to the internal couple for the flange M_f:

Type of Equations	Equations	Unknowns	
[Equil. of forces]$_{\text{flange}}$	$A_{sf}f_y = 0.85f_c'(b - b_w)h_f$	A_{sf}	(2-117)
[Equil. of mom.]$_{\text{flange}}$	$M_f = \varphi 0.85f_c'(b - b_w)h_f(d - h_f/2)$	M_f	(2-118)

Because the area of the flange is always considered a given value, the magnitude and the position of the resultant for the flange do not change. The steel area and the moment for the flange, A_{sf} and M_f, could be determined directly from the equilibrium equations Eqs. (2-117) and (2-118).

(1) Analysis Problems (given A_s)

In the case of an analysis problem when A_s is given, we can first calculate A_{sf} from Eq. (2-117) and compare it to A_s. If $A_s \leq A_{sf}$, meaning the flange is not fully utilized, the cross section should be analyzed as a rectangular cross section with a

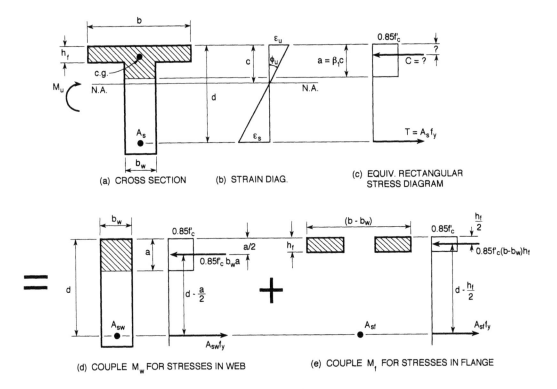

Figure 2.13 Flanged sections at ultimate.

width of b. If $A_s > A_{sf}$, meaning a T section is confirmed, then we can proceed to make an analysis of the web with a given steel area $A_{sw} = A_s - A_{sf}$.

Analysis of the web (finding moment M_w) is carried out as follows:

Given: b, d, h_f, A_{sw}, f_y, f'_c, and ε_u

Find: M_w and a (or $c = a/\beta_1$)

The equilibrium equations and their unknowns for the web are:

Type of Equations	Equations		Unknowns
[Equil. of forces]$_{\text{web}}$	$A_{sw}f_y = 0.85f'_c b_w a$	a	(2-119)
[Equil. of mom.]$_{\text{web}}$	$M_w = \varphi A_{sw}f_y(d - a/2)$	M_w a	(2-120)

Equations (2-119) and (2-120) are identical to Eqs. (2-83) and (2-84) for the analysis of a singly reinforced beam. The procedure for solution, which is very easy, is exactly the same as that given in Eqs. (2-86) to (2-87).

Once the moment for the web M_w is found, we can calculate the total moment $M_u = M_w + M_f$, where the moment for the flange M_f is obtained from Eq. (2-118).

(2) Design Problems (given M_u)

In the case of a design problem when M_u is given, we can first calculate M_f from Eq. (2-118) and compare it to M_u. If $M_u \leq M_f$, meaning the flange is not fully utilized, the cross section should be analyzed as a rectangular cross section with a width of b. If $M_u > M_f$, meaning a T section is confirmed, then we can proceed to design the web with a given moment $M_w = M_u - M_f$.

Design of the web (finding steel area A_{sw}) is carried out as follows:

$$\text{Given: } b, d, h_f, M_w, f_y, f_c', \text{ and } \varepsilon_u$$

$$\text{Find: } A_{sw} \text{ and } a \text{ (or } c = a/\beta_1)$$

The equilibrium equations and their unknowns for the web are:

Type of Equations	Equations	Unknowns		
[Equil. of forces]$_{\text{web}}$	$A_{sw}f_y = 0.85f_c'b_w a$	A_{sw}	a	(2-121)
[Equil. of mom.]$_{\text{web}}$	$M_w = \varphi A_{sw}f_y(d - a/2)$	A_{sw}	a	(2-122)

Equations (2-121) and (2-122) are exactly the same as Eqs. (2-88) and (2-89) for the design of a singly reinforced beam. The trial-and-error procedure for the solution has been described in conjunction with Eqs. (2-91) to (2-92).

Once the steel area for the web A_{sw} is found, we can calculate the total steel area $A_s = A_{sw} + A_{sf}$, where the steel area for the flange A_{sf} is obtained from Eq. (2-117).

(3) Balanced Condition of Flanged Beams

The balanced condition for flanged beams can be derived easily using the two-internal-couple concept as shown in Figure 2.13d and e. The only difference is that the depth a should be replaced by the balanced depth a_b in Figure 2.13d. Correspondingly, we can replace the symbols for steel areas A_s and A_{sw} by A_{sb} and A_{swb}, respectively, and then write

$$A_{sb}f_y = A_{swb}f_y + A_{sf}f_y \tag{2-123}$$

where

$$A_{sf} = \frac{0.85f_c'(b - b_w)h_f}{f_y} \tag{2-124}$$

Defining $\rho_b = A_{sb}/bd$, $\rho_{wb} = A_{wb}/b_w d$, and $\rho_f = A_{sf}/b_w d$ and dividing Eq. (2-123) by bd give

$$\rho_b = \frac{b_w}{b}(\rho_{wb} + \rho_f) \tag{2-125}$$

The balanced percentage of steel for the web, ρ_{wb}, is the same as the balanced percentage of steel for singly reinforced beams and, therefore, can be calculated by Eq. (2-71).

The maximum percentage of steel for flanged beams, $\rho_{\max}$, is given by the ACI Code to be

$$\rho_{\max} = 0.75\frac{b_w}{b}(\rho_{wb} + \rho_f) \tag{2-126}$$

Comparing Eq. (2-126) to Eq. (2-125) shows that the reduction factor 0.75 is applied to both the web and the flange. This is because both the web and the flange are made of relatively brittle concrete.

2.3.5 MOMENT–CURVATURE (M-ϕ) RELATIONSHIPS

2.3.5.1 *Characteristics of M-ϕ Curve*

We have now studied the general bending theory in Section 2.1, the linear bending theory in Section 2.2, and the nonlinear bending theory at ultimate load in Section 2.3. These theories will be used to understand and to calculate the fundamental moment–curvature relationships in the bending of reinforced concrete beams.

Figure 2.14a gives the moment–curvature (M-ϕ) curve for a singly reinforced beam according to the general bending theory described in Section 2.1.1. In this theory both the stress–strain curves of concrete and reinforcing bars are nonlinear as shown in Figure 2.1d and e. The curve for the rebar is roughly two straight lines connected by a curved knee. The initial straight line is very steep, whereas the final straight line is very flat. This characteristic of the stress–strain relationship is reflected in the general M-ϕ curve, which has the same shape. The knee of the M-ϕ curve is rounded. The only distinctive point on the curve is at the ultimate, point C, where the maximum strain of the concrete reaches the specified ultimate strain ε_u. The ultimate moment and the ultimate curvature at this point are designed as M_u and ϕ_u, respectively.

Mild steel has an elastic–perfectly plastic stress–strain relationship as shown in Figure 2.10f. If mild steel is used as reinforcement, the M-ϕ curve of a singly reinforced beam will have a shape as shown in Figure 2.14b. A distinct kink point, designated by B, appears. This point B reflects the yield point ($\varepsilon_s = \varepsilon_y$) of the stress–strain curve of mild steel. The moment and the curvature at this point are the yield moment M_y and the yield curvature ϕ_y, respectively. Both the curves before the yield point (OB) and after the yield point (BC) are slightly curved, reflecting the nonlinear stress–strain curve of concrete.

Point C (for ultimate) and point B (for yielding) in Figure 2.14b are calculated from cracked sections. Before cracking, however, the member is much stiffer and should be calculated according to the uncracked sections. This uncracked M-ϕ relationship is shown in Figure 2.14c at low loads and extended up to the point A, representing the cracking of concrete. The cracking moment and the cracking curvature at this distinctive point are designated as M_{cr} and ϕ_{cr}, respectively. If the four points O, A, B, and C are connected by straight lines, we have the so-called trilinear M-ϕ curve.

If the uncracked region of the M-ϕ curve is neglected and if the stress–strain curve of the concrete is assumed to be linear up to the load stage when the steel reaches the yield point, then the curve OB becomes a straight line as shown in Figure 2.14d. When points B and C are also joined by a straight line, the M-ϕ curve becomes bilinear. This bilinear M-ϕ curve captures the basic force-deformation characteristic of bending action, and is very useful in defining the ductility of reinforced concrete members as explained in the next section (Section 2.3.5.2).

The two distinctive points C (at ultimate) and B (at yield) in the bilinear M-ϕ curve can be calculated as follows:

(1) Singly Reinforced Beams

In the case of singly reinforced beams the point C in the M-ϕ curve (Figure 2.14d) can be determined according to the nonlinear bending theory of Section 2.3.2.5 for underreinforced beams. The ultimate moment M_u and the ultimate curvature ϕ_u can

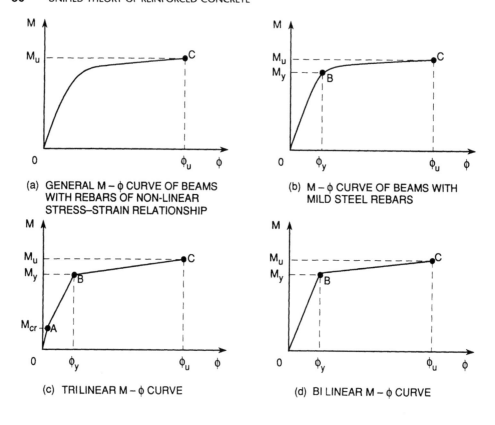

(a) GENERAL M – φ CURVE OF BEAMS
WITH REBARS OF NON-LINEAR
STRESS–STRAIN RELATIONSHIP

(b) M – φ CURVE OF BEAMS WITH
MILD STEEL REBARS

(c) TRILINEAR M – φ CURVE

(d) BILINEAR M – φ CURVE

Figure 2.14   Moment–curvature relationships.

be computed from the stress and strain diagrams of Figure 2.11c and b, respectively:

$$M_u = A_s f_y \left(d - \frac{a}{2} \right) \tag{2-127}$$

$$\phi_u = \frac{\varepsilon_u}{c} \quad (c = a/\beta_1) \tag{2-128}$$

where

$$a = \frac{A_s f_y}{0.85 f_c' b} \tag{2-129}$$

The yield moment M_y and the yield curvature ϕ_y, represented by point B in Figure 2.14d, can be calculated by the linear bending theory given in Section 2.2. Changing ε_s to ε_y in the strain diagram and f_s to f_y in the stress diagram of Figure 2.2b and c, we can derive from the equilibrium and compatibility conditions the following equations for the singly reinforced rectangular beams:

$$M_y = A_s f_y \left(d - \frac{kd}{3} \right) \tag{2-130}$$

$$\phi_y = \frac{\varepsilon_y}{d(1 - k)} \tag{2-131}$$

The coefficient k has been derived in Eq. (2-26) using the transformed area concept:

$$k = \sqrt{(n\rho)^2 + 2n\rho} - n\rho \qquad (2\text{-}132)$$

Because the lever arm $(d - kd/3)$ at yielding [Eq. (2-130)] must be less than the lever arm $(d - a/2)$ at ultimate [Eq. (2-127)], then M_y must be less than M_u. Therefore, if M_y is found to be greater than M_u, as in the cases of large percentages of tension steel, then M_y should be taken as equal to M_u.

(2) Doubly Reinforced Beams

In the case of doubly reinforced beams, the ultimate moment M_u and the ultimate curvature ϕ_u can be computed by the principles enunciated in Section 2.3.3. The moment M_u and the depth a can be obtained by solving Eqs. (2-101) to (2-104) according to Method 1 or Method 2. Once the depth a is obtained, the curvature ϕ_u is computed from Eq. (2-128).

The yield moment M_y and the yield curvature ϕ_y can be calculated by the linear bending theory explained in Section 2.2.4.2. Changing f_s to f_y in the stress diagram of Figure 2.4c, we can write from the equilibrium of moments about the concrete compression resultant the following equations:

$$M_y = A_s f_y \left(d - \frac{kd}{3} \right) + A'_s f'_s \left(\frac{kd}{3} - d' \right) \qquad (2\text{-}133)$$

Changing ε_s to ε_y in the strain diagram of Figure 2.4b, we can also obtain the following compatibility equation for the compression steel strain ε'_s:

$$\varepsilon'_s = \varepsilon_y \frac{kd - d'}{d - kd} \qquad (2\text{-}134)$$

The compression steel stress f'_s in Eq. (2-133) can now be written using Hooke's law:

$$f'_s = E_s \frac{\varepsilon_y}{1 - k} \left(k - \frac{d'}{d} \right) \leq f_y \qquad (2\text{-}135)$$

The coefficient k in Eqs. (2-133) and (2-135) is obtained from Eq. (2-34). The ratio β_c in Eq. (2-34) is defined in Eq. (2-33). Once the coefficient k is obtained, the yield curvature ϕ_y is computed from Eq. (2-131).

2.3.5.2 Bending Ductility

Because the bilinear M-ϕ curve in Figure 2.14d captures the basic force-deformation characteristic of bending action, it will be used to define the ductility of a reinforced concrete member. The bending ductility ratio μ is defined as the ratio of the ultimate curvature ϕ_u to the yield curvature ϕ_y, i.e.,

$$\mu = \frac{\phi_u}{\phi_y} \qquad (2\text{-}136)$$

The two curvatures, ϕ_u and ϕ_y, can be calculated according to the equations in the Section 2.3.5.1.

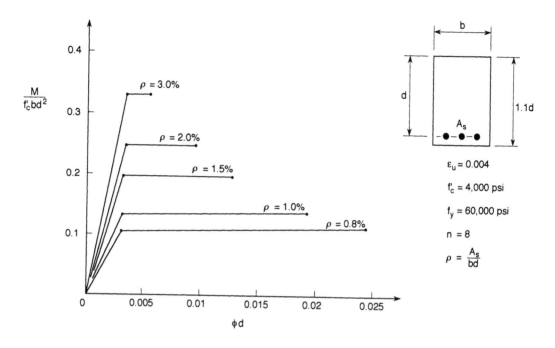

Figure 2.15 Bilinear M-ϕ curves for various percentages of tension steel.

The ultimate curvature ϕ_u, and therefore the ductility ratio μ, of a reinforced concrete member is strongly affected by the reinforcement. Five singly reinforced beams with percentages of steel varying from 0.8 to 3.0% are analyzed according to Eqs. (2-127) to (2-132) and their bilinear $M/f'_c bd^2$ vs. ϕd curves are plotted in Figure 2.15. The curvature ϕ, which has a unit of radians per unit length, has been nondimensionalized by multiplying by the effective depth d. The moment M has also been nondimensionalized by dividing by the parameter $f'_c bd^2$. In calculating these curves a more realistic ultimate strain ε_u of 0.004 is assumed rather than the conservative ACI value of 0.003. The normal material properties and geometric location of rebars are selected, namely, $f'_c = 4000$ psi, $f_y = 60,000$ psi, $n = E_s/E_c = 8$, and a concrete cover (center of rebar to surface) of 0.1d. Figure 2.15 shows clearly that the ultimate rotation $\phi_u d$ decreases rapidly with the increase of the tension steel percentage ρ. The ductility ratio μ decreases from 8.0 for $\rho = 0.8\%$ to 1.7 for $\rho = 3.0\%$.

The ductility of a bending member can be enhanced by the addition of compression steel as shown by the nondimensionalized $M/f'_c bd^2$ vs. ϕd curves in Figure 2.16. The curves for the doubly reinforced beams in this figure are calculated by the equations discussed in Section 2.3.5.1(2). For the three beams with tensile steel percentage ρ of 3%, the compression steel percentages ρ' of 0, 1, and 2% will give ultimate rotations ϕd of 0.0064, 0.0096, and 0.0193, respectively. The corresponding ductility ratio μ increases in the sequence of 1.7, 2.5, and 5.4. For the two beams with tension steel percentages ρ of 1%, an increase of compression steel percentage ρ' from 0 to 1% increases the ductility ratio μ from 6.2 to 10.8. It is clear that the compression steel is indeed very effective in increasing the ductility of flexural beams.

In short, the effects of the tension and compression reinforcement on the ductility of concrete structures are very profound. Such information is crucial in the design of reinforced concrete structures to resist earthquake.

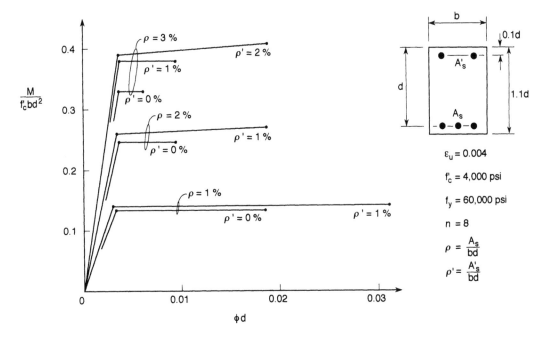

Figure 2.16 Bilinear M-ϕ curves showing effect of compression steel.

2.4   Combined Bending and Axial Load

In Sections 2.1, 2.2, and 2.3 we have studied the bending action of reinforced concrete members. The bending analysis and design of such members have been based on the three fundamental principles of parallel stress equilibrium, Bernoulli compatibility, and uniaxial constitutive laws of materials. In this section we will apply these same principles to reinforced concrete members subjected to combined bending and axial load. To limit the scope of this presentation, however, only the nonlinear theory dealing with *mild-steel* reinforced concrete members at *ultimate* load stage will be included. This section is, in fact, an extension of Section 2.3 to cover combined bending and axial load.

2.4.1   PLASTIC CENTROID AND ECCENTRIC LOADING

A member subjected to axial compression, with or without bending, is called a column. A concentric load on a column is defined as a load that produces a uniform stress in the column section and induces no bending moment. In the case of a symmetrically reinforced rectangular cross section, a concentric load is located at the geometric centroid of the concrete section. However, in the general case of an unsymmetrically reinforced rectangular section as shown in Figure 2.17a, a concentric load is said to be located at the *plastic centroid* of the cross section.

A concentric load N_n at the plastic centroid should, by definition, produce a uniform stress in the concrete as shown in Figure 2.17b. At maximum load this uniform stress is assumed to be $0.85f_c'$. At the same time, both the top and bottom steel are assumed to reach yielding, $f_s = f_s' = f_y$. The location of the plastic centroid is defined by the distance d_p measured from the concentric load to the bottom steel. The load N_n and the distance d_p can be determined from the two equilibrium

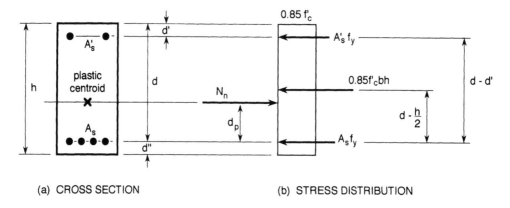

(a) CROSS SECTION   (b) STRESS DISTRIBUTION

Figure 2.17   Plastic centroid for unsymmetrically reinforced sections.

equations. Equilibrium of forces gives

$$N_n = 0.85f'_c bh + A_s f_y + A'_s f_y \qquad (2\text{-}137)$$

Note that in writing Eq. (2-137) the forces of the concrete areas displaced by the compression steel, $0.85A'_s f'_c$, and by tension steel, $0.85A_s f'_c$, have been neglected. These simplifications will be made for all the equilibrium equations in combined bending and axial load. Taking moments of all the forces about the bottom steel we have

$$N_n d_p = 0.85f'_c bh\left(d - \frac{h}{2}\right) + A'_s f_y(d - d') \qquad (2\text{-}138)$$

Inserting N_n from Eq. (2-137) into Eq. (2-138) results in

$$d_p = \frac{0.85f'_c bh(d - h/2) + A'_s f_y(d - d')}{0.85f'_c bh + A_s f_y + A'_s f_y} \qquad (2\text{-}139)$$

Figure 2.18a illustrates an unsymmetrically reinforced column section subjected to a load N_n and a bending moment M_n. The load N_n is located at the plastic centroid, and the moment M_n is taken about an axis through the plastic centroid. This combined bending and axial loading is statically equivalent to a load N_n acting at an eccentricity, $e = M_n/N_n$, measured from the plastic centroid (Figure 2.18b).

An eccentric load N_n is acting on an unsymmetrically reinforced column section as shown in Figure 2.19a. The strain and stress distributions due to the eccentric load are illustrated in Figure 2.19b and c at the ultimate load stage. The compression stress block has been converted to the ACI equivalent rectangular stress block (Figure 2.19d). The load N_n is located at an eccentricity e measured from the plastic centroid or at an eccentricity of e' measured from the centroid of the tensile steel. The latter is frequently more convenient for the analysis. The difference between e and e' is the distance d_p given by Eq. (2-139).

The mild steel has an elastic–perfectly plastic stress–strain curve as shown in Figure 2.19e. In a mild-steel reinforced column, therefore, the tensile steel may or may not yield at ultimate. If the tension steel is in the plastic range ($\varepsilon_s \geq \varepsilon_y$) when the concrete crushes at ultimate load ($\varepsilon_u = 0.003$), the column is said to fail in

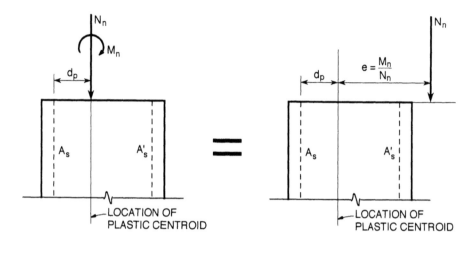

(a) COMBINED BENDING AND AXIAL LOAD

(b) ECCENTRIC LOADING

Figure 2.18 Equivalence between eccentric loading and combined bending and axial load.

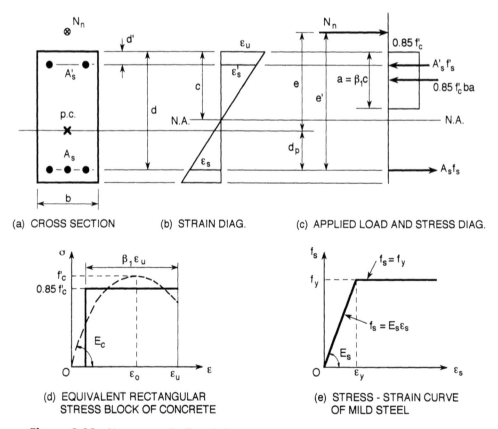

(a) CROSS SECTION   (b) STRAIN DIAG.   (c) APPLIED LOAD AND STRESS DIAG.

(d) EQUIVALENT RECTANGULAR
STRESS BLOCK OF CONCRETE

(e) STRESS - STRAIN CURVE
OF MILD STEEL

Figure 2.19 Unsymmetrically reinforced rectangular column sections at ultimate.

tension. If the tension steel is in the elastic range $(\varepsilon_s < \varepsilon_y)$ when the concrete crushes, the column is said to fail in compression. In order to divide these two types of columns, we define the "balanced condition" when the steel reaches the yield point $(\varepsilon_s = \varepsilon_y)$ simultaneously with the crushing of concrete $(\varepsilon_u = 0.003)$.

The eccentric loading of an unsymmetrically reinforced rectangular column involves 13 variables, b, d, A_s, A'_s, N_n, e', f_s, f'_s, f'_c, ε_s, ε'_s, ε_u, and a (or c), as shown in Figure 2.19a to c. The coefficient β_1 is not considered a variable because it has been determined independently from tests. A total of six equations are available: two from equilibrium (Figure 2.19c), two from Bernoulli compatibility (Figure 2.19b), and two from the constitutive law of mild steel (Figure 2.19e). Therefore, seven variables must be given before the remaining six unknown variables can be solved by the six equations.

2.4.2   BALANCED CONDITION

The problem posed for the balanced condition is

Given: b, d, A_s, A'_s, f'_c, $\varepsilon_s = \varepsilon_y$, and $\varepsilon_u = 0.003$

Find: N_n, e', f_s, f'_s, ε'_s, and a (or $c = a/\beta_1$)

The six available equations and their unknowns are:

Type of Equations	Equations	Unknowns				
Equil. of forces	$N_n = 0.85f'_c ba - A_s f_s$ $+A'_s f'_s$	N_n	f_s f'_s		a	(2-140)
Equil. of mom. about T	$N_n e' = 0.85 f'_c ba(d - \dfrac{a}{2})$ $+A'_s f'_s(d - d')$	N_n e'	f'_s		a	(2-141)
Compat. (tens. steel)	$\dfrac{a}{\beta_1 d} = \dfrac{\varepsilon_u}{\varepsilon_u + \varepsilon_y}$				a	(2-142)
Compat. (comp. steel)	$\dfrac{\varepsilon'_s}{\varepsilon_u} = \dfrac{c - d'}{c}$ $(c = a/\beta_1)$			ε'_s	a	(2-143)
Const. law (tens. steel)	$f_s = E_s \varepsilon_s$ for $\varepsilon_s \le \varepsilon_y$		f_s			(2-144a)
	$f_s = f_y$ for $\varepsilon_s > \varepsilon_y$		f_s			(2-144b)
Const. law (comp. steel)	$f'_s = E_s \varepsilon'_s$ for $\varepsilon'_s \le \varepsilon_y$		f'_s ε'_s			(2-145a)
	$f'_s = f_y$ for $\varepsilon'_s > \varepsilon_y$		f'_s ε'_s			(2-145b)

Notice that no material reduction factor φ is shown in the equilibrium equations, because φ is included in the symbol N_n. For convenience, the nominal load N_n is defined as N_u/φ.

The six unknown variables, N_n, e', f_s, f'_s, ε'_s, and a (or c) for the six equations are indicated after the equations under the heading "Unknowns." Because the tension steel strain ε_s is given as the yield strain ε_y, the tension steel stress f_s should obviously be equal to the yield stress f_y based on the stress–strain relationship of either Eqs. (2-144a) or (2-144b). The unknown depth a of the equivalent rectangular stress block can be solved directly from the compatibility condition for tension steel [Eq. (2-142)]. Once the depth a (and c) is determined, the compression steel strain ε'_s can be obtained from the compatibility equation [Eq. (2-143)] and the compression steel stress f'_s from the stress–strain equation for compression steel [Eq. (2-145)]. Substituting a, f_s, and f'_s into the force equilibrium equation [Eq. (2-140)] gives the force N_n; inserting a, f'_s, and N_n into the moment equilibrium equation [Eq. (2-141)] results in the eccentricity e'. For this particular case of balanced condition there is no need to solve simultaneous equations.

In this case of balanced condition we will now add a subscript b to the three unknown quantities: N_{nb} for balanced force, e'_b for balanced eccentricity, and a_b for balanced depth of equivalent rectangular stress block. The balanced depth a_b is expressed from Eq. (2-142) as

$$a_b = \beta_1 d \frac{\varepsilon_u}{\varepsilon_u + \varepsilon_y} \tag{2-146}$$

Equation (2-146) is, of course, identical to Eq. (2-69), derived for pure bending without axial load, except that in Eq. (2-69) the ultimate strain ε_u has been taken as 0.003 and $E_s = 29,000,000$ psi.

From Eqs. (2-140) and (2-141) the balanced force N_{nb} and the balanced eccentricity e'_b are:

$$N_{nb} = 0.85 f'_c b a_b - A_s f_y + A'_s f'_s \tag{2-147}$$

$$e'_b = \frac{1}{N_{nb}} \left[0.85 f'_c b a_b \left(d - \frac{a_b}{2} \right) + A'_s f'_s (d - d') \right] \tag{2-148}$$

The balanced depth a_b in Eqs. (2-147) and (2-148) is determined from Eq. (2-146) and the compression steel stress f'_s is obtained from Eqs. (2-143) and (2-145).

The balanced N_{nb} and e'_b can be used to divide the tension failure and the compression failure of the column as follows:

Types of Problems	Tension Failure	Compression Failure
N_n is given	$N_n < N_{nb}$	$N_n > N_{nb}$
e' is given	$e' > e'_b$	$e' < e'_b$

2.4.3   TENSION FAILURE

The analysis of column sections failing in tension is considerably simplified for the same reasons as in underreinforced beams (Section 2.3.2.5). First, the tensile steel stress is known to yield, $f_s = f_y$, and the stress–strain curve of tension steel is not required in the solution. Second, the tensile steel strain is expected to lie within the plastic range, $\varepsilon_s \geq \varepsilon_y$. Bernoulli's compatibility condition, which relates the tensile steel strain ε_s to the maximum strain of concrete ε_u, becomes irrelevant to the solution of the stress-type variables. Because of these two simplifications, analysis of column sections failing in tension involves only 12 variables, b, d, A_s, A'_s, N_n, e', f'_s, ε'_s, f_y, f'_c, ε_u, and c (or a) [Figure 2.20]. The available equations are now down to four: two from the equilibrium condition, one from the compatibility condition of compression steel, and one from the stress–strain relationship of compression steel.

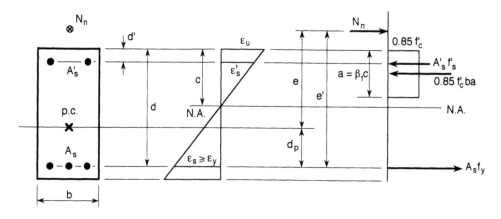

(a) CROSS SECTION   (b) STRAIN DIAG.   (c) APPLIED LOAD AND STRESS DIAG.

Figure 2.20 Tension failure in column sections.

Therefore, eight variables must be given before the four remaining unknown variables can be solved by the four available equations.

The problem posed for the analysis of columns failing in tension is

$$\text{Given: } b, d, A_s, A_s', e', f_y, f_c', \text{ and } \varepsilon_u = 0.003$$
$$\text{Find: } N_n, f_s', \varepsilon_s', \text{ and } a \text{ (or } c = a/\beta_1)$$

The four available equations and their unknowns are:

Type of Equations	Equations	Unknowns		
Equil. of forces	$N_n = 0.85f_c'ba - A_sf_y$ $+ A_s'f_s'$	N_n	f_s'	a (2-149)
Equil. of mom. about T	$N_ne' = 0.85f_c'ba(d - \dfrac{a}{2})$ $+ A_s'f_s'(d - d')$	N_n	f_s'	a (2-150)
Compat. (comp. steel)	$\dfrac{\varepsilon_s'}{\varepsilon_u} = \dfrac{c - d'}{c}$ $\quad(c = a/\beta_1)$		ε_s'	a (2-151)
Const. law (comp. steel)	$f_s' = E_s\varepsilon_s' \quad$ for $\varepsilon_s' \leq \varepsilon_y$	f_s'	ε_s'	(2-152a)
	$f_s' = f_y \quad$ for $\varepsilon_s' > \varepsilon_y$	f_s'	ε_s'	(2-152b)

Examination of the four unknowns in the four equations, Eqs. (2-149) to (2-152), suggests that the best method of solution is the generic trial-and-error procedure:

Step 1: Check $e' > e_b'$ to confirm the tension failure of the column section.

Step 2: Assume a value of the depth a, which should be less than the balanced a_b, and calculate $c = a/\beta_1$.

Step 3: Determine the compression steel strain ε_s' from the compatibility equation [Eq. (2-151)] and the compression steel stress f_s' from the stress–strain equation [Eq. (2-152)].

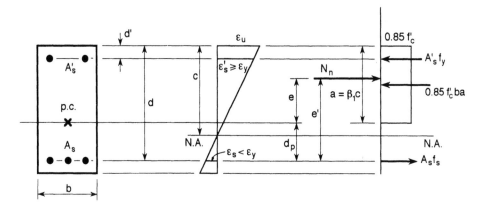

(a) CROSS SECTION   (b) STRAIN DIAG.   (c) APPLIED LOAD AND STRESS DIAG.

Figure 2.21 Compression failure ($c \leq h$) in column sections.

Step 4: Calculate the force N_n from the moment equilibrium equation [Eq. (2-150)].

Step 5: Inserting f_s' and N_n into the force equilibrium equation [Eq. (2-149)] and solve for a new value of the depth a.

Step 6: If the new depth a is equal to the assumed depth a, a solution is obtained. If not, assume another depth a and repeat the cycle. The convergence is usually quite rapid.

It would be instructive to mention here that the previously posed problem is to give the eccentricity e' and to find the force N_n. If the problem is reversed by giving N_n and finding e', the solution procedure will be simplified, because Eq. (2-149) will contain only two unknowns rather than three. This simplified solution procedure is left as an exercise for the reader.

2.4.4   COMPRESSION FAILURE

The analysis of column sections failing in compression is more complicated than sections failing in tension because two types of analysis are required to deal with two very different situations. The first situation occurs when the neutral axis lies within the concrete section ($c \leq h$) as shown in Figure 2.21. The second situation occurs when the neutral axis lies outside the concrete section ($c > h$) (Figure 2.22). These two situations will be treated separately.

When $c \leq h$ (Figure 2.21)

When the neutral axis lies within the concrete cross section, the ACI method of converting an actual compression stress block into an equivalent rectangular one remains valid. That is to say, the coefficient β_1 for the depth c and the coefficient 0.85 for the stress f_c' are still applicable. Consequently, the six equations, Eqs. (2-140) to (2-145), remain available for the analysis of columns failing in compression with $c \leq h$.

An obvious observation is available, however, to simplify the analysis of a column section failing in compression with $c \leq h$. The compression steel strain is expected to lie within the plastic range, $\varepsilon_s' \geq \varepsilon_y$, because the neutral axis is low enough such that ε_s' is close to the ultimate concrete strain ε_u of 0.003. This is true for all columns of

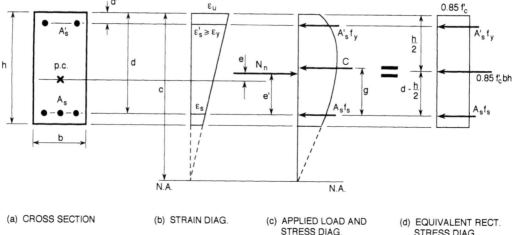

(a) CROSS SECTION   (b) STRAIN DIAG.   (c) APPLIED LOAD AND STRESS DIAG.   (d) EQUIVALENT RECT. STRESS DIAG.

Figure 2.22 Compression failure ($c > h$) in column sections.

practical sizes, except those very small ones with height h close to the lowest allowable limit of 12 in. and depth a close to the balanced a_b. Bernoulli's compatibility condition, which relates ε_s' to ε_u, becomes irrelevant to the solution of the stress-type variables. As a result, the compression steel is expected to yield, $f_s' = f_y$, and the stress–strain curve of compression steel is not required in the solution. Because of this simplification, analysis of column sections failing in compression involves only 12 variables, b, d, A_s, A_s', N_n, e', f_s, ε_s, f_y, f_c', ε_u, and c (or a) [Figure 2.21]. The number of available equations is reduced to four: two from the equilibrium condition, one from the compatibility condition of tension steel, and one from the stress–strain relationship of tension steel. Therefore, eight variables must be given before the four remaining unknown variables can be solved by the four available equations.

The problem posed for the analysis of column sections failing in compression with $c \leq h$ is

Given: b, d, A_s, A_s', e', f_y, f_c', and $\varepsilon_u = 0.003$

Find: N_n, f_s, ε_s, and a (or $c = a/\beta_1$)

The four available equations and their unknowns are:

Type of Equations	Equations	Unknowns		
Equil. of forces	$N_n = 0.85f_c'ba - A_sf_s + A_s'f_y$	N_n	f_s	a (2-153)
Equil. of mom. about T	$N_ne' = 0.85f_c'ba(d - \frac{a}{2}) + A_s'f_y(d - d')$	N_n		a (2-154)
Compat. (tens. steel)	$\frac{\varepsilon_s}{\varepsilon_u} = \frac{d-c}{c}$ $(c = a/\beta_1)$	ε_s		a (2-155)
Const. law (tens. steel)	$f_s = E_s\varepsilon_s$ for $\varepsilon_s \leq \varepsilon_y$	f_s	ε_s	(2-156)

The generic trial-and-error procedures can be used to solve these four equations:

Step 1: Check $e' < e'_b$ to confirm the compression failure of the column section.

Step 2: Assume a value of depth a, which should be greater than the balanced a_b, and calculate $c = a / \beta_1$.

Step 3: Determine the tensile steel strain ε_s from the compatibility equation [Eq. (2-155)] and the tensile steel stress f_s from the stress–strain equation [Eq. (2-156)].

Step 4: Calculate the force N_n from the moment equilibrium equation [Eq. (2-154)].

Step 5: Insert f_s and N_n into the force equilibrium equation [Eq. (2-153)] and solve for a new value of the depth a.

Step 6: If the new depth a is equal to the assumed depth a, a solution is obtained. If not, assume another depth a and repeat the cycle.

Step 7: Check the assumption for the yielding of the compression steel, $\varepsilon'_s \geq \varepsilon_y$, by Eq. (2-143). In all practical cases, this step is not necessary.

Incidentally, the posed problem here is to give the eccentricity e' and to find the force N_n. If the problem is reversed by giving N_n and finding e', the solution procedure will be simplified, because Eq. (2-153) will contain only two unknowns rather than three. The reader should now be able to work out this simplified solution procedure if required.

When $c > h$ (Figure 2.22)

When the neutral axis lies outside of the concrete cross section, the ACI rule to convert the actual compression stress block to an equivalent rectangular one is no longer valid. In other words, the coefficient β_1 derived from pure bending is not applicable. Therefore, a column section failing in compression with $c > h$ requires special analysis.

When the neutral axis leaves the bottom surface, failure begins to change from a compression failure at the extreme fiber (say $\varepsilon_u = 0.003$) to a concentric compression failure of the whole cross section ($\varepsilon_0 = 0.002$). In this range of transition no simple method is available to determine the magnitude and the location of the concrete compression stress resultant C (Figure 2.22c). Fortunately, tests have shown that the moment of C about the bottom steel is approximately equal to the moment of a rectangular stress block shown in Figure 2.22d about the same bottom steel. The rectangular stress block has a stress of $0.85f'_c$ throughout the whole cross section. Hence,

$$Cg = 0.85f'_c bh \left(d - \frac{h}{2} \right) \tag{2-157}$$

where g is the distance from the compression resultant C to the centroid of the bottom steel.

Taking all the moments about the bottom steel gives

$$N_n e' = 0.85f'_c bh \left(d - \frac{h}{2} \right) + A'_s f_y (d - d') \tag{2-158}$$

Equation (2-158) shows that $N_n e'$ is no longer a function of the depth c (or a). That is to say, the position of the neutral axis cannot be determined.

At the limiting case of concentric loading, $e = 0$ and $e' = d_p$. Inserting $e' = d_p$ from Eq. (2-139) into Eq. (2-158) gives

$$N_n = 0.85f'_c bh + A_s f_y + A'_s f_y \qquad (2\text{-}159)$$

Equation (2-159) is exactly the same as Eq. (2-137), meaning that the approximation made for the equivalent rectangular stress block is exactly correct for the limiting case of concentric loading.

We can now summarize the method of analysis for a column section failing in compression. First, assume $c \leq h$ and go through the trial-and-error procedures of Eqs. (2-153) to (2-156). If c is found to be less than h, the assumption is correct and we have a solution. Second, if c is found to be greater than h, then the load N_n can be calculated directly from Eq. (2-158) with the given eccentricity e'.

Although the approximate equation of Eq. (2-157) allows us to determine the load N_n when $c > h$, the inability to calculate the position of the neutral axis does not permit us to determine the curvature $\phi_u = \varepsilon_u/c$ in this range. Fortunately, when $c > h$, the concrete is uncracked and the curvature is very small. The effect of such curvature would be small on the overall deflection of the member. If the curvature is desired, however, the position of the neutral axis can best be obtained from numerical integration of the stress–strain curve of concrete. The method of numerical integration is beyond the scope of this book.

2.4.5 BENDING–AXIAL-LOAD INTERACTION

The methods of analysis of eccentrically loaded column sections have been presented in Sections 2.4.1 to 2.4.4. These analyses reveal an interesting trend. With an increase of the eccentricity e, the behavior of a column section changes in the following manner:

$e = 0$	axial load
$e < e_b$	compression failure
$e = e_b$	balanced condition
$e > e_b$	tension failure
$e = \infty$	pure bending

This whole range of interaction from axial load to pure bending can be visualized easily by a bending–axial-load interaction curve shown in Figure 2.23.

Figure 2.23 gives a diagram with the bending moment M as the abscissa and the load N as the ordinate. The point on the abscissa denoted as M_0 represents the pure bending, and the point on the ordinate denoted as N_0 represents the concentric load. The point B, representing the balanced condition, has a coordinate of M_{nb} and N_{nb} calculated from Eqs. (2-146) to (2-148) in Section 2.4.2. In the region between the point M_0 and the balanced point B we have the tension failure zone, which can be analyzed by the method in Section 2.4.3. The compression failure zone, which is located between the balanced point B and the point N_0, can be analyzed by the method in Section 2.4.4.

Three characteristics can be observed from the bending–axial-load interaction curve:

1 Any point on the interaction curve, such as point A, represents a pair of values M_n and N_n that will cause the column section to fail. Points inside the curve are safe and points outside the curve are unsafe.

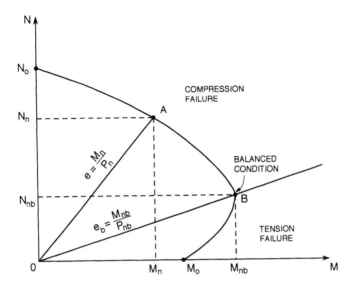

Figure 2.23   Bending–axial-load interaction curve.

2 Any radial line, such as OA, has a slope whose reciprocal represents the eccentricity $e = M_n/N_n$. The vertical slope has an eccentricity of 0 and the horizontal slope has an eccentricity of infinity.

3 In the compression zone the moment M_n *decreases* with increasing load N_n, whereas in the tension zone the moment M_n *increases* with increasing load N_n. This characteristic can be explained as follows: First, in the compression failure zone, failure is caused by the overstraining of concrete in compression ($\varepsilon_u = 0.003$). Because axial compression increases the compression strain, it decreases the capacity of the concrete to resist flexural compression. Second, in the tension failure zone, failure is caused by yielding of the steel in tension. Because axial compression decreases the tensile strain in the steel, it increases the capacity of the steel to resist flexural tension.

2.4.6   MOMENT–AXIAL-LOAD–CURVATURE (M-N-ϕ) RELATIONSHIP

The interaction of bending and axial load is shown in Figure 2.24a for a rectangular column with 2% of tension steel and 1% of compression steel. All the rebars are located at a distance of 0.1d from the surfaces. The material properties are $\varepsilon_u = 0.004$, $f_c' = 4000$ psi, $f_y = 60,000$ psi, and $n = E_s/E_c = 8$. The moment M_n has been nondimensionalized by the parameter $f_c'bd^2$ and the axial load N_n has been normalized by the capacity of the concentric load N_0.

The solid curve in this nondimensional interaction diagram represents the sectional strength of the given column under combined bending and axial load. The balanced point is calculated from Eqs. (2-146) to (2-148). The tension failure zone is computed by solving Eqs. (2-149) to (2-152), and the compression failure zone is determined either by solving Eqs. (2-153) to (2-156) or directly by Eq. (2-158), depending on the location of the neutral axis. The dotted curve from the balanced point to the point of pure bending represents the yielding of the tension steel. The method of calculation for this dotted curve, which is based on Hooke's laws for both concrete and steel, has

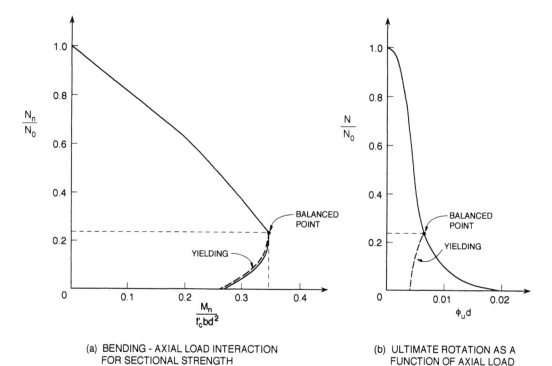

(a) BENDING - AXIAL LOAD INTERACTION
FOR SECTIONAL STRENGTH

(b) ULTIMATE ROTATION AS A
FUNCTION OF AXIAL LOAD

Figure 2.24 $M\text{-}N\text{-}\phi$ relationship for a typical column ($\rho = 2\%$, $\rho' = 1\%$, $\varepsilon_u = 0.004$, $f_c' = 4000$ psi, $f_y = 60,000$ psi, $n = 8$ and $h = 1.1d$).

not been presented. The reader should be able to derive the equations required to plot this curve using the equilibrium, Bernoulli compatibility, and Hooke's constitutive laws.

The nondimensionalized ultimate rotation $\phi_u d$ is related to the normalized axial load N/N_0 by the solid curve in Figure 2.24b. The ultimate rotation $\phi_u d$ is calculated from Bernoulli compatibility.

$$\phi_u d = \frac{\beta_1 \varepsilon_u}{(a/d)} \tag{2-160}$$

Equation (2-160) can be calculated as soon as the depth ratio, a/d, is determined in the solution process. The dotted curve in Figure 2.24b also represents the yielding of the tension steel.

Figure 2.24b shows that the ultimate rotation $\phi_u d$ decreases rapidly with increasing axial compression N_n/N_0. The ductility ratio μ drops to unity at the balanced point. In earthquake design, therefore, columns are not designed to develop flexural hinges or to dissipate energy. Flexural hinges should always be designed to occur in the beams of a frame, rather than in the columns. This so-called strong columns–weak beams concept of frame design is fundamental to the earthquake design of reinforced concrete structures.

3

Equilibrium (Plasticity) Truss Model

3.1   Basic Equilibrium Equations

3.1.1   EQUILIBRIUM IN BENDING

In Section 2.1.1 we studied a member subjected to bending. The external moment M acting on the cross section (Figure 2.1a) is resisted by an internal couple as shown in Figure 2.1c. The internal couple consists of the compression force C and the steel tensile force $T = A_s f_s$. The distance between C and T is the lever arm jd. From moment equilibrium, Eq. (2-8) was derived as

$$M = A_s f_s (jd) \qquad (3\text{-}1)$$

where $j = (1 - k_2 c/d)$. The depth of the compression zone c is determined by Bernoulli's compatibility condition, and the coefficient k_2 is a function of the shape of the stress–strain curve of concrete.

For underreinforced members, the steel yields before the concrete crushes. At the yield condition the steel stress f_s becomes f_y, where f_y is the yield strength of the tensile steel. The yield moment M_y and the tensile force at yielding, $T = A_s f_y$, are shown in Figure 3.1a and b, respectively. The compatibility condition is automatically satisfied and the analysis is reduced to finding c directly by the equilibrium of the tensile force T and the compressive force C.

The internal couple concept gives rise to the truss model concept for flexure shown in Figure 3.1c. In this model, the tensile steel bars are concentrated at the geometric centroid and constitute the tension stringer. This tension stringer is capable of resisting the tensile force $A_s f_y$. The concrete area within the compression stress block is also assumed to concentrate at the location of the resultant C. This concrete element is known as the compression stringer and is capable of resisting the resultant C. The distance between the tension and compression stringers is designated d_v, which, of course, is equal to jd.

To further simplify the analysis, ACI Code allows the replacement of the curved concrete stress block by a rectangular one, shown by the dotted line in Figure 3.1b.

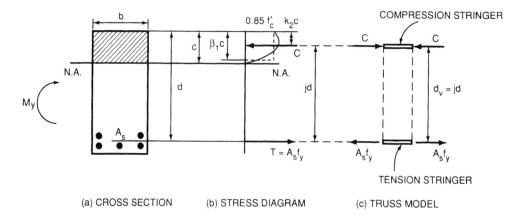

Figure 3.1 Truss model in bending.

The stress of the rectangular stress block is $0.85f'_c$ and the depth is $\beta_1 c$. The coefficient β_1 is taken as 0.85 for $f'_c = 4000$ psi or less, and decreases by 0.05 for every increment of 1000 psi beyond $f'_c = 4000$ psi. β_1 is limited to 0.65 for $f'_c > 8000$ psi. Based on this rectangular stress block the magnitude of the resultant $C = 0.85f'_c b\beta_1 c$ and the coefficient $k_2 = \beta_1/2$. Equation (3-1) becomes

$$M_y = A_s f_y d\left(1 - \frac{\beta_1 c}{2d}\right) \tag{3-2}$$

and the depth of the compression zone is obtained from $T = C$:

$$c = \frac{A_s f_y}{0.85f'_c b\beta_1} \tag{3-3}$$

In this case of underreinforced members the lever arm jd is equal to $(1 - \beta_1 c/2d)d$. For normal flexural members with steel reinforcement of 1 to 1.5%, jd or d_v is approximately $0.9d$.

3.1.2   EQUILIBRIUM IN ELEMENT SHEAR

3.1.2.1   *Element Shear Equations*

A membrane element subjected to a shear flow q is shown in Figure 3.2a. The element has a thickness of h and a square shape with a unit length in both directions. The longitudinal rebars are arranged in the l direction (horizontal axis) with a uniform spacing of s_l. The transverse rebars are arranged in the t direction (vertical axis) with a uniform spacing of s. After cracking, the concrete is separated by diagonal cracks into a series of concrete struts as shown in Figure 3.2b. The cracks are oriented at an angle α with respect to the l axis. The diagonal concrete struts, the longitudinal rebars, and the transverse rebars form a truss that is capable of resisting the shear flow q.

Equilibrium of forces on the vertical faces is shown by the force triangle on the left face of the shear element (Figure 3.2b). The shear flow q pointing upward is resolved into a longitudinal steel force n_l and a diagonal concrete force $(\sigma_d h)\cos\alpha$. The steel

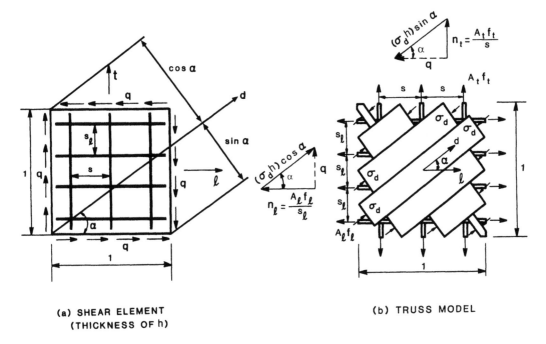

(a) SHEAR ELEMENT
(THICKNESS OF h)

(b) TRUSS MODEL

Figure 3.2 Equilibrium in element shear.

force n_l is defined as the longitudinal steel force per unit length, $A_l f_l / s_l$, where A_l is the cross-sectional area of one longitudinal rebar and f_l is the stress in the longitudinal rebars. The concrete force $(\sigma_d h)\cos\alpha$ represents the diagonal concrete stress σ_d acting on a thickness of h and a width of $\cos\alpha$. The $\cos\alpha$ relationship is shown by the geometry in Figure 3.2a. From this force triangle the shear flow q can be related to the longitudinal steel force n_l by the geometry

$$q = n_l \tan\alpha \qquad (3\text{-}4)$$

Similarly, equilibrium of forces on the horizontal faces is shown by the force triangle on the top face of the shear element (Figure 3.2b). The shear flow q pointing leftward is resolved into a transverse steel force n_t and a diagonal concrete force $(\sigma_d h)\sin\alpha$. The steel force n_t is defined as the transverse steel force per unit length, $A_t f_t / s$, where A_t is the cross-sectional area of one transverse rebar and f_t is the stress in the transverse rebars. The concrete force $(\sigma_d h)\sin\alpha$ represents the diagonal concrete stress σ_d acting on a thickness of h and a width of $\sin\alpha$. The $\sin\alpha$ relationship is also shown by the geometry in Figure 3.2a. From this force triangle the shear flow q can be related to the transverse steel force n_t by the geometry

$$q = n_t \cot\alpha \qquad (3\text{-}5)$$

The shear flow q can be related to the diagonal concrete stress σ_d using either the force triangle on the left face or the force triangle on the top face. From geometry of the triangles we obtain

$$q = (\sigma_d h)\sin\alpha \cos\alpha \qquad (3\text{-}6)$$

Assuming that yielding occurs in both longitudinal and transverse steel, then $n_l = n_{ly} = A_l f_{ly}/s_l$ and $n_t = n_{ty} = A_t f_{ty}/s$, where n_{ly} and n_{ty} are the longitudinal and transverse yield force per unit length, respectively, and f_{ly} and f_{ty} are the longitudinal and transverse yield stress, respectively. Equating Eqs. (3-4) and (3-5) to eliminate q, we have

$$\tan \alpha = \sqrt{\frac{n_{ty}}{n_{ly}}} \tag{3-7}$$

Also, multiplying Eqs. (3-4) and (3-5) to eliminate α we obtain

$$q_y = \sqrt{n_{ly} n_{ty}} \tag{3-8}$$

where q_y is the shear flow at yielding. Equation (3-7) shows that α at yielding depends on the ratio of the transverse to longitudinal steel yield forces, n_{ty}/n_{ly}. Equation (3-8) states that q_y is the square root of the product of the steel yield forces per unit length in the two directions. In other words, when both the longitudinal and transverse steel yield, the shear flow at yielding is the square-root-of-the-product average of the unit steel forces in the two directions.

   In design, the shear flow at yielding is usually given. The aim of the design is to find the yield reinforcement in both directions, n_{ly} and n_{ty}, and to check the diagonal concrete stress σ_d so that the concrete will not crush before the yielding of steel. For this purpose, Eqs. (3-4) to (3-6) can be written in the forms

$$n_{ly} = q_y \cot \alpha \tag{3-9}$$

$$n_{ty} = q_y \tan \alpha \tag{3-10}$$

$$\sigma_d = \frac{q_y}{h \sin \alpha \cos \alpha} \tag{3-11}$$

3.1.2.2   *Geometric Relationships of Equilibrium Equations*

   The equations in Section 3.1.2.1 are expressed in terms of q, n_l, and n_t, which represent forces per unit length. In order to express these equations in terms of stresses, we define the terms of smeared steel stresses as

$$\rho_l f_l = \frac{A_l f_l}{s_l h} = \frac{n_l}{h} = \text{smeared steel stress in longitudinal direction}$$

$$\rho_t f_t = \frac{A_t f_t}{sh} = \frac{n_t}{h} = \text{smeared steel stress in transverse direction}$$

where h is the thickness of the element. Note also that the shear stress $\tau_{lt} = q/h$. Then the three equilibrium equations, Eqs. (3-4) to (3-6), can be written as

$$\rho_l f_l = \tau_{lt} \cot \alpha \tag{3-12}$$

$$\rho_t f_t = \tau_{lt} \tan \alpha \tag{3-13}$$

$$\sigma_d = \tau_{lt} \frac{1}{\sin \alpha \cos \alpha} = \tau_{lt}(\tan \alpha + \cot \alpha) \tag{3-14}$$

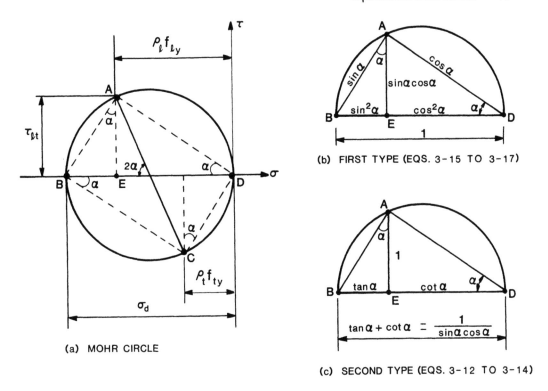

(b) FIRST TYPE (EQS. 3-15 TO 3-17)

(c) SECOND TYPE (EQS. 3-12 TO 3-14)

(a) MOHR CIRCLE

Figure 3.3 Geometric relationships in a Mohr circle.

Equations (3-12) to (3-14) can also be expressed in terms of the diagonal concrete stress σ_d by substituting τ_{lt} from Eq. (3-14) into Eqs. (3-12) and (3-13):

$$\rho_l f_l = \sigma_d \cos^2 \alpha \qquad (3\text{-}15)$$

$$\rho_t f_t = \sigma_d \sin^2 \alpha \qquad (3\text{-}16)$$

$$\tau_{lt} = \sigma_d \sin \alpha \cos \alpha \qquad (3\text{-}17)$$

These two sets of equations [Eqs. (3-12) to (3-14) and (3-15) to (3-17)] each satisfy Mohr circle as illustrated in Figure 3.3. Figure 3.3a shows a Mohr circle in a σ-τ coordinate system, where σ is the normal stress and τ is the shear stress. Point A on the circle represents the longitudinal face (the direction of a face is defined by its normal axis). On this face are acting a shear stress τ_{lt} indicated by the vertical coordinate, and a longitudinal steel stress $\rho_l f_l$, indicated by the horizontal coordinate. Point B on the σ axis of the circle represents the principal face in the d direction, on which only the diagonal concrete stress σ_d is acting. The incident angle 2α between points A and B in the Mohr circle is twice the angle α between the l axis and the d axis in the shear element. At 180° from point A we have point C, which represents the transverse face in the shear element. On this face are acting a shear stress τ_{lt} (vertical coordinate) and a transverse steel force $\rho_t f_t$ (horizontal coordinate). At 180° from point B, we have point D, which represents the face of the cracks. On this crack face the normal stress is zero. The preceding explanation of the application of the Mohr circle shows the importance of the Mohr circle in the analysis of stresses in various directions. Detailed derivation and discussion of the Mohr circle will be provided in Chapter 4.

The introduction of the Mohr circle at this point in time is intended to illustrate the geometric relationships of Eqs. (3-12) to (3-17). The geometric relationships of the half Mohr circle defined by ABD are illustrated in Figures 3.3b and c. If BD in Figure 3.3b is taken as unity, then $ED = \cos^2 \alpha$, $BE = \sin^2 \alpha$, and $AE = \sin \alpha \cos \alpha$. These three trigonometric values are actually the ratios of the three stresses $\rho_l f_l$, $\rho_t f_t$, and τ_{lt}, respectively, divided by the diagonal concrete stress σ_d, in the Mohr circle of Figure 3.3a. Hence, the set of three equilibrium equations, Eqs. (3-15) to (3-17), simply states the geometric relationships illustrated in Figure 3.3b. This set of three equations will be called the first type of expression for the Mohr circle.

Similarly, if AE in Figure 3.3c is taken as unity, then $ED = \cot \alpha$, $BE = \tan \alpha$, and $BD = \cot \alpha + \tan \alpha = 1/\sin \alpha \cos \alpha$. These three trigonometric values are actually the ratios of the three stresses $\rho_l f_l$, $\rho_t f_t$, and σ_d, respectively, divided by the shear stress τ_{lt}, in the Mohr circle of Figure 3.3a. Hence, the set of three equilibrium equations, Eqs. (3-12) to (3-14), simply states the geometric relationships illustrated in Figure 3.3c. This set of three equations will be called the second type of expression for the Mohr circle.

3.1.2.3   *Yielding of Reinforcement*

The plasticity truss model is based on the assumption that both the longitudinal and the transverse steel must yield before failure. In order to ensure this mode of failure, the shear elements are divided into two types: underreinforced and overreinforced. In an underreinforced element the yielding of steel in both directions occurs before the crushing of concrete, thus satisfying the assumption of the plasticity truss model. In an overreinforced element, however, concrete crushes before the yielding of either the longitudinal or the transverse steel, or both, thus violating the basic assumption. The state of stresses that divide the underreinforced mode of failure from the overreinforced will be called the balanced condition. Under the balanced condition, one or both of the two yielded steels just reaches yield point when the concrete crushes at an effective stress of $\zeta f_c'$. The softening coefficient ζ will be carefully studied in Chapter 7, but will be taken as 0.6 here according to the CEB-FIP Code (1978).

Balanced Condition

The criterion for the balanced condition in the plasticity truss model was derived by Nielsen and Braestrup (1975). Adding Eqs. (3-15) and (3-16) and noticing $\sin^2 \alpha + \cos^2 \alpha = 1$ give

$$\rho_l f_l + \rho_t f_t = \sigma_d \qquad (3\text{-}18)$$

At the balanced condition both steels yield ($f_l = f_{ly}$ and $f_t = f_{ty}$), whereas the concrete crushes at an effective stress of $\zeta f_c'$. Equation (3-18) then becomes

$$\rho_l f_{ly} + \rho_t f_{ty} = \zeta f_c' \qquad (3\text{-}19)$$

Define

$$\omega_l = \frac{\rho_l f_{ly}}{\zeta f_c'} = \text{longitudinal reinforcement index}$$

$$\omega_t = \frac{\rho_t f_{ty}}{\zeta f_c'} = \text{transverse reinforcement index}$$

The balanced condition, Eq. (3-19), can be written in a nondimensional form:

$$\omega_l + \omega_t = 1 \qquad (3\text{-}20)$$

Underreinforced Elements

When $\omega_l + \omega_t < 1$, we have an underreinforced element where both steels yield ($f_l = f_{ly}$ and $f_t = f_{ty}$) before the crushing of concrete. Substituting $\cos \alpha$ from Eq. (3-15) and $\sin \alpha$ from Eq. (3-16) into Eq. (3-17) results in

$$\tau_{lt} = \sqrt{(\rho_l f_{ly})(\rho_t f_{ty})} \qquad (3\text{-}21)$$

The angle α for this underreinforced element can be obtained by dividing Eq. (3-16) by Eq. (3-15):

$$\tan \alpha = \sqrt{\frac{\rho_t f_{ty}}{\rho_l f_{ly}}} \qquad (3\text{-}22)$$

Dividing both sides of Eq. (3-21) by $\zeta f_c'$, we obtain

$$\frac{\tau_{lt}}{\zeta f_c'} = \sqrt{\omega_l \omega_t} \qquad (3\text{-}23)$$

The nondimensionalized ratio $\tau_{lt}/\zeta f_c'$ will be called the shear stress ratio. Dividing both the numerator and denominator in the square root by $\zeta f_c'$, Eq. (3-22) can also be written as

$$\tan \alpha = \sqrt{\frac{\omega_t}{\omega_l}} \qquad (3\text{-}24)$$

The three pairs of equations—Eqs. (3-7) and (3-8), Eqs. (3-21) and (3-22), and Eqs. (3-23) and (3-24)—are actually the same, except that they are expressed in terms of different units. Equations (3-7) and (3-8) are in terms of force per unit length (q_y, n_{ly}, n_{ty}), Eqs. (3-21) and (3-22) are in terms of stresses ($\tau_{lt}, \rho_l f_{ly}, \rho_t f_{ty}$), and Eqs. (3-23) and (3-24) are in terms of the nondimensionalized indices ($\tau_{lt}/\zeta f_c', \omega_l, \omega_t$).

Overreinforced Elements

When $\omega_l + \omega_t > 1$, we have an overreinforced element where the concrete crushes before the yielding of steel in one or both directions. Because this failure mode violates the basic assumption of the plasticity truss model, the design of such an element is unacceptable and no further discussion is necessary.

Three Cases of Balanced Condition

The balanced condition $\omega_l + \omega_t = 1$ can be divided into three cases:

Case 1: $\omega_l = \omega_t$

When $\omega_l = \omega_t$, the yielding of both the longitudinal and the transverse steel occurs simultaneously with the crushing of concrete. The balanced condition in this case gives $\omega_l = \omega_t = 0.5$, resulting in

$$\frac{\tau_{lt}}{\zeta f_c'} = 0.5 \tag{3-25}$$

$$\alpha = 45° \tag{3-26}$$

Case 2: $\omega_t < 0.5$

In this case the transverse steel has yielded, but the concrete crushes simultaneously with the yielding of the longitudinal steel. The longitudinal reinforcement index becomes $\omega_l = 1 - \omega_t$ according to the balanced condition, and Eq. (3-23) becomes

$$\frac{\tau_{lt}}{\zeta f_c'} = \sqrt{\omega_t(1 - \omega_t)} \tag{3-27}$$

The corresponding angle α is

$$\tan \alpha = \sqrt{\frac{\omega_t}{(1 - \omega_t)}} \qquad \alpha < 45° \tag{3-28}$$

Case 3: $\omega_l < 0.5$

In this case the longitudinal steel has yielded, but the concrete crushes simultaneously with the yielding of the transverse steel. The transverse steel will then be determined by the balanced condition $\omega_t = 1 - \omega_l$, and Eq. (3-23) becomes

$$\frac{\tau_{lt}}{\zeta f_c'} = \sqrt{\omega_l(1 - \omega_l)} \tag{3-29}$$

The corresponding angle α is

$$\tan \alpha = \sqrt{\frac{(1 - \omega_l)}{\omega_l}} \qquad \alpha > 45° \tag{3-30}$$

3.1.2.4 *Design of Reinforcement*

The balanced condition can also be expressed graphically by a semicircular curve in a $\tau_{lt}/\zeta f_c'$ vs. ω_t diagram as shown in Figure 3.4. Squaring both sides of Eq. (3-27) gives

$$\left(\frac{\tau_{lt}}{\zeta f_c'}\right)^2 + \omega_t^2 - \omega_t = 0 \tag{3-31}$$

Adding 0.5^2 on both sides of Eq. (3-31) results in

$$\left(\frac{\tau_{lt}}{\zeta f_c'}\right)^2 + (\omega_t - 0.5)^2 = 0.5^2 \tag{3-32}$$

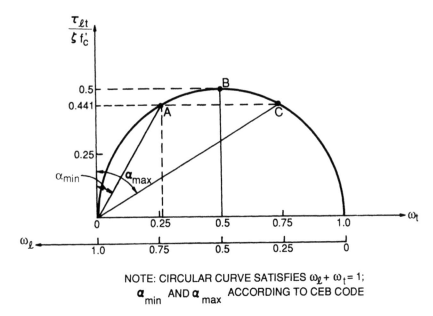

NOTE: CIRCULAR CURVE SATISFIES $\omega_l + \omega_t = 1$;
α_{min} AND α_{max} ACCORDING TO CEB CODE

Figure 3.4 Circular relationship between shear stress ratio $(\tau_{lt}/\zeta f_c')$ and reinforcement indices (ω_t, ω_l).

When $(\tau_{lt}/\zeta f_c')$ is plotted against ω_t in Figure 3.4, Eq. (3-32) represents a circle with a radius of 0.5, and the center of the circle is located on the ω_t axis at $\omega_t = 0.5$. This circle gives the nondimensionalized relationship between the shear stress τ_{lt} and the transverse steel stress $\rho_t f_{ty}$. When the longitudinal steel is chosen to satisfy the balanced condition, $\omega_l + \omega_t = 1$, an axis pointing toward the left is also drawn for ω_l.

Substituting ω_l from Eq. (3-23) into Eq. (3-24) gives

$$\tan \alpha = \frac{\omega_t}{(\tau_{lt}/\zeta f_c')} \tag{3-33}$$

Equation (3-33) shows that the angle α represents the reciprocal of the slope of a straight line through the origin in Figure 3.4. In the CEB-FIP Code (1978) the angle α is limited to $3/5 < \tan \alpha < 5/3$. These two limits, $\alpha_{min} = 31°$ and $\alpha_{max} = 59°$, are represented by the two straight lines OA and OC. At these limits the maximum shear stress ratio $\tau_{lt}/\zeta f_c'$ is 0.441.

Design of reinforcement within the semicircle will give an underreinforced element, whereas the region outside the semicircle represents overreinforcement. Design within the fan-shaped area $OABC$ further satisfies the CEB-FIP provisions that limit the range of α. On the semicircular curve that expresses the balanced condition, point B represents Case 1 of the balanced condition, whereas the arcs AB and BC give Case 2 and Case 3, respectively.

In the 1978 CEB-FIP Code, the softening coefficient ζ is specified as 0.6. Substituting $\sigma_d = 0.6f_c'$ into Eq. (3-17) provides max. $\tau_{lt} = 0.6f_c' \sin \alpha \cos \alpha$. When $\alpha = 45°$, max. $\tau_{lt} = 0.3f_c'$. At the limits of $\alpha = 31$ or $59°$, max. $\tau_{lt} = 0.265f_c'$.

The boundaries for α are supposed to serve two purposes. First, the limitations are specified to ensure the yielding of steel in both directions at failure. Second, they are specified to prevent excessive cracking. The validity of these two purposes is discussed in Sections 7.5 and 5.4, respectively, after the strain compatibility conditions are introduced.

3.1.2.5   Minimum Reinforcement

Before cracking, an element subjected to shear stress τ_{lt} will produce a principal tensile stress σ_1 that is equal in magnitude ($\tau_{lt} = \sigma_1$) and inclined at an angle of 45°. At cracking, σ_1 is assumed to reach the tensile strength of concrete f'_t. Substituting $\tau_{lt} = f'_t$ into Eqs. (3-12) and (3-13) we have

$$\rho_l f_l = f'_t \cot \alpha \qquad (3\text{-}34)$$

$$\rho_t f_t = f'_t \tan \alpha \qquad (3\text{-}35)$$

Summing Eqs. (3-34) and (3-35) gives

$$\rho_l f_l + \rho_t f_t = \frac{f'_t}{\sin \alpha \cos \alpha} \qquad (3\text{-}36)$$

A minimum amount of reinforcement to ensure that the steel will not yield immediately at cracking can be obtained by assuming the yielding of the steel, $f_l = f_t = f_y$, and by noticing that $\sin \alpha \cos \alpha$ is very close to 0.5 when α is in the vicinity of 45°. Then Eq. (3-36) becomes

$$\rho_l + \rho_t = \frac{2f'_t}{f_y} \qquad (3\text{-}37)$$

For a typical case when $f_y = 60,000$ psi and $f'_t = 300$ psi, Eq. (3-37) gives the minimum total reinforcement ($\rho_l + \rho_t$)$_{min}$ of 1%. If the total reinforcement is equally distributed in the longitudinal and transverse directions, each direction will require a minimum steel percentage of 0.5%.

In the ACI Code the minimum horizontal reinforcement in walls is specified to be in the range of 0.20 to 0.25%, and the vertical reinforcement in the range of 0.12 to 0.15%. These ACI requirements provides only about 1/4 to 1/2 the minimum reinforcement required by Eq. (3-37). Because most walls in use are designed with steel percentages close to the ACI minimum requirement, it is obvious that these walls will fail brittlely upon cracking. The common notion that shear failure of walls is brittle stems primarily from this practice of supplying an insufficient amount of steel in the walls. If a minimum steel requirement was provided according to Eq. (3-37), then the wall is expected to behave in a ductile manner.

3.1.3   EQUILIBRIUM IN BEAM SHEAR

A beam subjected to a concentrated load $2V$ at midspan is shown in Figure 3.5a. Because the reaction is V, the shear force is a constant V throughout one half of the beam, and the moment diagram is a straight line. When a beam element of length d_v is isolated and the moment on the left face is defined as M, then the moment on the right face is $M + Vd_v$. The shear forces on both the left and right faces are, of course, equal to V.

A model of the isolated beam element is shown in Figure 3.5b. The top and bottom stringers are separated from the main body of the beam element, so that the two different mechanisms to resist bending and shear can be illustrated clearly. The

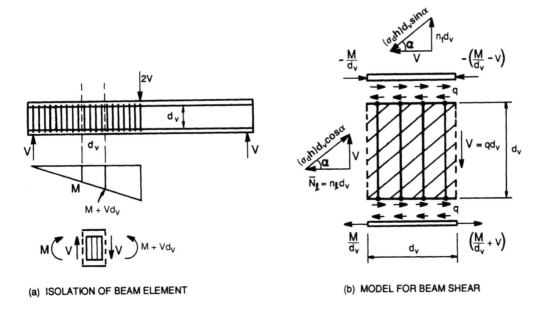

(a) ISOLATION OF BEAM ELEMENT   (b) MODEL FOR BEAM SHEAR

Figure 3.5 Equilibrium in beam shear.

stringers are resisting the bending moment and the main body is carrying the shear force. Two assumptions are made in the establishment of the model:

1 The shear flow q in the main body is distributed uniformly over the depth (i.e., in the transverse direction). Because q is a constant over the depth, $V = \int q \, d(d_v)$ $= qd_v$.

2 The shear flow q in the main body is also distributed uniformly along the length of the main body (i.e., in the longitudinal direction). Hence, the transverse steel stresses (f_t) and the stresses in the diagonal concrete struts (σ_d) vary uniformly along their lengths.

Based on these two assumptions the main body can now be treated as a large shear element as discussed in Section 3.1.2. The only difference between the shear element in Figure 3.2 and the main body of the beam in Figure 3.5b is that the longitudinal steel is uniformly distributed in the former, but is concentrated at the top and bottom stringers in the latter. This difference, however, does not alter the equilibrium condition as shown by the two force triangles on the left and top faces of the main body. To take care of the nonuniform distribution of the longitudinal steel we simply define $\bar{N}_l = n_l d_v$, which is the total force in the longitudinal steel to resist shear.

Equilibrium of forces on the main body gives three equilibrium equations for beam shear

$$V = \bar{N}_l \tan \alpha \qquad\qquad (3\text{-}38)$$

$$V = n_t d_v \cot \alpha \qquad\qquad (3\text{-}39)$$

$$V = (\sigma_d h) d_v \sin \alpha \cos \alpha \qquad\qquad (3\text{-}40)$$

This set of three equations [Eqs. (3-38) to (3-40)] is identical to the set of three equations [Eqs. (3-4) to (3-6)] for element shear if the latter three equations are multiplied by the length d_v.

Assuming the yielding of the longitudinal and transverse steel, then $\overline{N}_l = \overline{N}_{ly}$ and $n_t = n_{ty}$. Dividing Eqs. (3-38) by (3-39) to eliminate V we have

$$\tan \alpha = \sqrt{\frac{n_{ty}}{\left(\overline{N}_{ly}/d_v\right)}} \tag{3-41}$$

Multiplying Eqs. (3-38) and (3-39) to eliminate α gives

$$V_y = d_v \sqrt{\left(\overline{N}_{ly}/d_v\right) n_{ty}} \tag{3-42}$$

In design, the total longitudinal steel force $\overline{N}_l$ is divided equally between the top and bottom stringers. For the design of the bottom rebars, this bottom tensile force of $\overline{N}_l/2$ due to shear is added to the longitudinal tensile force M/d_v due to bending. The design of the top rebars, however, is less certain. In theory, the longitudinal compressive force M/d_v due to bending could be subtracted from the top tensile force of $\overline{N}_l/2$ due to shear. Such measure would not be conservative. Perhaps a better approach is to select the larger of the two forces, M/d_v or $\overline{N}_l/2$, for design purposes.

In the design of beams, it is also cost effective to select an α value less than 45°, because the transverse steel, in the form of stirrups, is more costly than the longitudinal rebars.

3.1.4 EQUILIBRIUM IN TORSION

3.1.4.1 *Bredt's Formula Relating T and q*

A hollow prismatic member of arbitrary bulky cross section and variable thickness is subjected to torsion as shown in Figure 3.6a. According to St. Venant's theory the twisting deformation will have two characteristics: First, the cross-sectional shape will remain unchanged after the twisting; second, the warping deformation perpendicular to the cross section will be identical throughout the length of the member. Such deformations imply that the in-plane normal stresses in the wall of the tube member should vanish. The only stress component in the wall is the in-plane shear stress, which exhibits as a circulating shear flow q on the cross section. The shear flow q is the resultant of the shear stresses in the wall thickness and is located on the dotted loop shown in Figure 3.6a. This dotted loop is defined as the center line of the shear flow, which may or may not lie in the middepth of the wall thickness.

A membrane wall element $ABCD$ is isolated and shown in Figure 3.6b. It is subjected to pure shear on all four faces. Let us denote the shear stress on face AD as τ_1 and that on face BC as τ_2. The thicknesses at faces AD and BC are designated h_1 and h_2, respectively. Taking equilibrium of forces on the element in the longitudinal l direction we have

$$\tau_1 h_1 = \tau_2 h_2 \tag{3-43}$$

Because shear stresses on mutually perpendicular planes must be equal, the shear stresses on face AB must be τ_1 at point A and τ_2 at point B. Equation (3-43), therefore, means that τh on face AB must be equal at points A and B. Because we

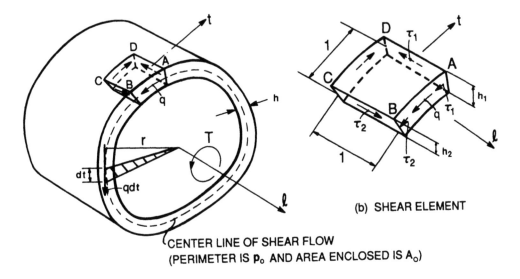

(b) SHEAR ELEMENT

CENTER LINE OF SHEAR FLOW
(PERIMETER IS p_o AND AREA ENCLOSED IS A_o)

(a) TUBE IN TORSION

Figure 3.6 Equilibrium in torsion.

define $q = \tau h$ as the shear flow, q must be equal at points A and B. Notice also that the two faces AD and BC of the element can be selected at an arbitrary distance apart without violating the equilibrium condition in the longitudinal direction. It follows that the shear flow q must be constant throughout the cross section.

The relationship between T and q can be derived directly from the equilibrium of moments about the l axis. As shown in Figure 3.6a, the shear force along a length of wall element dt is $q\,dt$. The contribution of this element to the torsional resistance is $q\,dt(r)$, where r is the distance from the center of twist (l axis) to the shear force $q\,dt$. Because q is a constant, integration along the whole loop of the center line of shear flow gives the total torsional resistance:

$$T = q \oint r\, dt \tag{3-44}$$

From Figure 3.6a it can be seen that $r\,dt$ in the integral is equal to twice the area of the shaded triangle formed by r and dt. Summing these areas around the whole cross section results in

$$\oint r\, dt = 2A_0 \tag{3-45}$$

where A_0 is the cross-sectional area bounded by the center line of the shear flow. This parameter A_0 is a measure of the lever arm of the circulating shear flow and will be called the lever arm area. Substituting $2A_0$ from Eq. (3-45) into Eq. (3-44) gives

$$q = \frac{T}{2A_0} \tag{3-46}$$

Equation (3-46) was first derived by Bredt (1896).

3.1.4.2   *Torsion Equations*

A shear element isolated from the wall of a tube of bulky cross section (Figure 3.6b) may be subjected to a warping action in addition to the pure shear action discussed previously. This warping action will be taken into account in Chapter 8, Section 8.2. If the warping action is neglected, then this shear element becomes identical to the shear element in Figure 3.2 that is subjected to pure shear only. As a result, the three equilibrium equations [Eqs. (3-4) to (3-6)] derived for the element shear in Figure 3.2 become valid. Substituting q from Eq. (3-46) into Eqs. (3-4), (3-5), and (3-6), we obtain the three equilibrium equations for torsion:

$$T = \frac{\overline{N}_l}{p_0}(2A_0)\tan\alpha \tag{3-47}$$

$$T = n_t(2A_0)\cot\alpha \tag{3-48}$$

$$T = (\sigma_d h)(2A_0)\sin\alpha\cos\alpha \tag{3-49}$$

Notice in Eq. (3-47) that $\overline{N}_l = n_l p_0$. This is because n_l, which is the longitudinal force per unit length, must be multiplied by the whole perimeter of the shear flow, p_0, to arrive at the total longitudinal force due to torsion, $\overline{N}_l$.

Assuming the yielding of the longitudinal and transverse steel, then $\overline{N}_l = \overline{N}_{ly}$ and $n_t = n_{ty}$. Dividing Eqs. (3-47) by (3-48) to eliminate T we have

$$\tan\alpha = \sqrt{\frac{n_{ty}}{\left(\overline{N}_{ly}/p_0\right)}} \tag{3-50}$$

Multiplying Eqs. (3-47) and (3-48) to eliminate α gives

$$T_y = 2A_0\sqrt{\left(\overline{N}_{ly}/p_0\right)n_{ty}} \tag{3-51}$$

3.1.5   SUMMARY OF BASIC EQUILIBRIUM EQUATIONS

A summary of the basic equilibrium equations for bending, element shear, beam shear, and torsion is given in Table 3.1. The table includes 18 equations: 3 for bending and 5 each for the three cases of element shear, beam shear, and torsion. Comparison of the three sets of five equations for element shear, beam shear, and torsion shows that they are basically the same. The five equations for beam shear are simply the five equations for element shear multiplied by a length of d_v. The five equations for torsion are simply those for element shear multiplied by the area $2A_0$. Hence it is only necessary to understand the geometric and algebraic relationships of the set of five equations for element shear.

Within each set of five equations for element shear, beam shear, and torsion the three in the left column are derived directly from the three equilibrium conditions. The geometric relationships of these three equations are described by the second type of expression for the Mohr circle in Figure 3.3c.

The two equations of each set in the right column are derived from the basic equations in the left column, assuming that the steel in both the longitudinal and transverse directions has yielded. In the typical case of element shear the angle α at

TABLE 3.1 Summary of Basic Equilibrium Equations

	Basic Equations	At yielding

Bending

$M = A_s f_s (jd)$

$\alpha = 90°$
$M_y = A_s f_y (jd)$

Element shear (q = shear flow)

$q = n_l \tan \alpha$
$q = n_t \cot \alpha$
$q = (\sigma_d h)\sin \alpha \cos \alpha$

$\tan \alpha = \sqrt{n_{ty}/n_{ly}}$
$q_y = \sqrt{n_{ly} n_{ty}}$

Beam shear ($V = q d_v$ and $\overline{N}_l = n_l d_v$)

$V = \overline{N}_l \tan \alpha$
$V = n_t d_v \cot \alpha$
$V = (\sigma_d h) d_v \sin \alpha \cos \alpha$

$\tan \alpha = \sqrt{n_{ty}/(\overline{N}_{ly}/d_v)}$
$V_y = d_v \sqrt{(\overline{N}_{ly}/d_v) n_{ty}}$

Torsion ($T = q(2 A_0)$ and $\overline{N}_l = n_l p_0$)

$T = (\overline{N}_l/p_0)(2 A_0)\tan \alpha$
$T = n_t (2 A_0)\cot \alpha$
$T = (\sigma_d h)(2 A_0)\sin \alpha \cos \alpha$

$\tan \alpha = \sqrt{n_{ty}/(\overline{N}_{ly}/p_0)}$
$T_y = 2 A_0 \sqrt{(\overline{N}_{ly}/p_0) n_{ty}}$

yield is obtained by eliminating q from the first two basic equations, whereas the shear flow at yield q_y is obtained by eliminating α.

The equations shown in Table 3.1 for pure bending, pure shear, and pure torsion are all derived in a consistent and logical manner. Such clarity in concept makes the interaction of these actions relatively simple. These interaction relationships will be enunciated in Section 3.2.

3.2   Interaction Relationships

3.2.1   SHEAR–BENDING INTERACTION

A model of beam subjected to shear and bending is shown in Figure 3.7a. The moment M creates a tensile force M/d_v in the bottom stringer and an equal compressive force in the top stringer. The shear force V, however, is acting on a shear element as shown in Figure 3.7b. It induces a total tensile force of $V \cot \alpha$ in the longitudinal steel (Figure 3.7c). Due to symmetry, the top and bottom stringers should each resist one half of the tensile force, $V \cot \alpha / 2$, Figure 3.7a. In the transverse direction, the shear force V will produce a transverse force per unit length n_t in the transverse steel (Figure 3.7d).

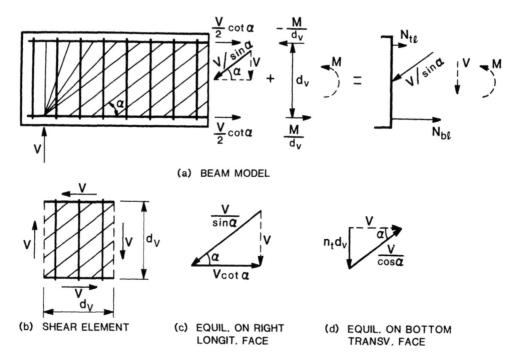

(a) BEAM MODEL

(b) SHEAR ELEMENT   (c) EQUIL. ON RIGHT   (d) EQUIL. ON BOTTOM
                                 LONGIT. FACE   TRANSV. FACE

Figure 3.7   Equilibrium for shear –bending interaction.

Failure of a beam under shear and bending may occur in two modes:

First Failure Mode

The first failure mode occurs due to yielding of the bottom stringer and the transverse steel.

Looking at Figure 3.7a, and d, the force in the bottom stringer N_{bl} and in the force per unit length in the transverse steel n_t are

$$N_{bl} = \frac{M}{d_v} + \frac{V}{2}\cot\alpha \tag{3-52}$$

$$n_t = \frac{V}{d_v}\tan\alpha \tag{3-53}$$

Substituting α from Eq. (3-53) into Eq. (3-52) to eliminate α, we have

$$\frac{M}{N_{bl}d_v} + \frac{V^2}{d_v^2(2N_{bl}/d_v)n_t} = 1 \tag{3-54}$$

Equation (3-54) expresses the interaction relationship between M and V. Assume that yielding occurs in the bottom stringer and in the transverse steel. Then $N_{bl} = N_{bly}$ and $n_t = n_{ty}$. Also define the pure bending strength M_0 and the pure shear strength

V_0 according to Table 3.1 as follows:

$$M_0 = N_{bly} d_v \tag{3-55}$$

$$V_0 = d_v \sqrt{\left(\frac{2 N_{tly}}{d_v}\right) n_{ty}} \tag{3-56}$$

where

$$N_{tly} = \text{force in the top stringer at yielding}$$

$$N_{bly} = \text{force in the bottom stringer at yielding}$$

Notice that the pure shear strength V_0 is defined based on the top stringer force at yield N_{tly}, rather than the bottom stringer force at yield N_{bly}. This definition gives the lowest positive value for V_0, assuming the top stringer force at yield is less than the bottom stringer force at yield. The total longitudinal force due to shear, $\overline{N}_l$, is then equal to $2 N_{tly}$. To replace N_{bly} by N_{tly} in Eq. (3-54) we introduce the yield force ratio R:

$$R = \frac{N_{tly}}{N_{bly}} \tag{3-57}$$

Substituting the definitions of Eqs. (3-55), (3-56), and (3-57) into Eq. (3-54), the interaction equation for M and V is now expressed in a nondimensionalized form:

$$\frac{M}{M_0} + \left(\frac{V}{V_0}\right)^2 R = 1 \tag{3-58}$$

Equation (3-58) is plotted in Figure 3.8 for $R = 0.25, 0.5,$ and 1.

Second Failure Mode

The second failure mode occurs due to yielding of the top stringer and the transverse steel.

Looking at Figure 3.7a and d, the force in the top stringer N_{tl} and in the force per unit length in the transverse steel n_t are

$$N_{tl} = -\frac{M}{d_v} + \frac{V}{2} \cot \alpha \tag{3-59}$$

$$n_t = \frac{V}{d_v} \tan \alpha \tag{3-60}$$

Substituting α from Eq. (3-60) into Eq. (3-59) to eliminate α, we have

$$-\frac{M}{N_{tl} d_v} + \frac{V^2}{d_v^2 (2 N_{tl}/d_v) n_t} = 1 \tag{3-61}$$

Assuming that yielding occurs in the top stringer and in the transverse steel, then $N_{tl} = N_{tly}$ and $n_t = n_{ty}$. Substituting the definitions of M_0, V_0, and R from Eqs.

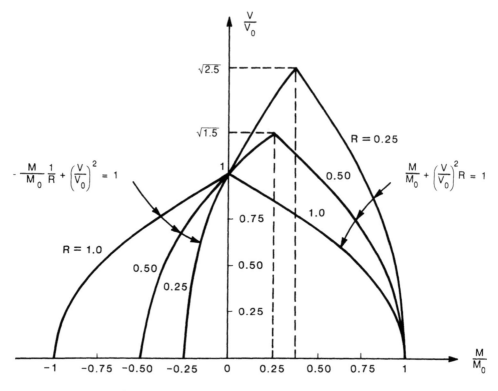

Figure 3.8 Shear–bending interaction curves.

(3-55), (3-56), and (3-57) into Eq. (3-61), the nondimensionalized interaction equation for M and V becomes

$$-\left(\frac{M}{M_0}\right)\frac{1}{R} + \left(\frac{V}{V_0}\right)^2 = 1 \tag{3-62}$$

Equation (3-62) is also plotted in Figure 3.8 for $R = 0.25$, 0.5, and 1. A series of complete interaction curves for shear and bending is now clearly illustrated in Figure 3.8.

Take, for example, the case of $R = 0.5$, where the top stringer has one half the yield capacity of the bottom stringer. The moment will then vary from $M/M_0 = -0.5$ to 1 depending on the magnitude of V/V_0. When $M/M_0 = -0.5$, this negative moment introduces a tensile force in the top stringer equal to the yield force. Consequently, the shear strength V, which is based on the top stringer strength, becomes zero. When the moment is increased (in an algebraic sense toward the right), the tensile force in the top stringer due to bending decreases. The remaining tensile capacity in the top stringer is now available to resist shear, resulting in an increase of the shear strength. When the moment is increased to zero (i.e., $M/M_0 = 0$), the full capacity of the top stringer is available to resist shear, and the shear strength V becomes the pure shear strength V_0 (i.e., $V/V_0 = 1$).

When the moment becomes positive, it induces a compressive stress in the top stringer that can be used to reduce the tensile stress due to shear. As a result, the shear strength continues to increase until the second mode of failure (yielding of top stringer) is changed into the first mode of failure (yielding of bottom stringer). When $M/M_0 = 0.25$, the condition is reached where the bottom stringer and the top stringer yield simultaneously. This peak point provides the highest possible shear strength. Beyond this point, the shear strength decreases with increasing moment, because failure is now caused by the tensile yielding of the bottom stringer. In the bottom stringer the tensile stresses due to shear and bending are additive. When the moment reaches the pure bending strength ($M/M_0 = 1$), the bottom stringer will yield under the moment itself, leaving no capacity for shear. The shear strength then becomes zero ($V/V_0 = 0$).

The point of maximum shear, which corresponds to the simultaneous yielding of top and bottom stringers, is the intersection point of the two curves for the first and second modes of failure. The locations of these peak points can be obtained by solving the two interaction equations, Eqs. (3-58) and (3-62). Multiplying Eq. (3-62) by R and adding it to Eq. (3-58), we derive V/V_0 by eliminating M/M_0:

$$\frac{V}{V_0} = \sqrt{\frac{1+R}{2R}} \tag{3-63}$$

Multiplying Eq. (3-62) by R and subtracting it from Eq. (3-58), we derive M/M_0 by eliminating N/N_0:

$$\frac{M}{M_0} = \frac{1-R}{2} \tag{3-64}$$

For the case of $R = 0.5$, Eqs. (3-64) and (3-63) illustrate that the peak point is located at $M/M_0 = 0.25$ and $V/V_0 = \sqrt{1.5}$. These values are indicated in Figure 3.8.

3.2.2   TORSION–BENDING INTERACTION

A model of beam subjected to torsion and bending is shown in Figure 3.9. The moment M creates a tensile force M/d_v in the bottom stringer and an equal compressive force in the top stringer. The torsional moment, however, induces a total tensile force of $(Tp_0/2A_0)\cot\alpha$ in the longitudinal steel. Due to symmetry, the top and bottom stringers should each resist one half of the total tensile force, $(Tp_0/4A_0)\cot\alpha$. In the transverse direction, the torsional moment will also produce a transverse force per unit length, $(T/2A_0)\tan\alpha$, in the hoop steel.

Failure of a beam under torsion and bending also occurs in two modes:

First Failure Mode

The first failure mode occurs due to yielding of the bottom stringer and the transverse steel.

Reviewing Figure 3.9 reveals that the force in the bottom stringer N_{bl} and the force per unit length in the transverse steel n_t are

$$N_{bl} = \frac{M}{d_v} + \frac{Tp_0}{4A_0}\cot\alpha \tag{3-65}$$

$$n_t = \frac{T}{2A_0}\tan\alpha \tag{3-66}$$

Substituting α from Eq. (3-66) into Eq. (3-65) to eliminate α results in

$$\frac{M}{N_{bl}d_v} + \frac{T^2}{4A_0^2(2N_{bl}/p_0)n_t} = 1 \tag{3-67}$$

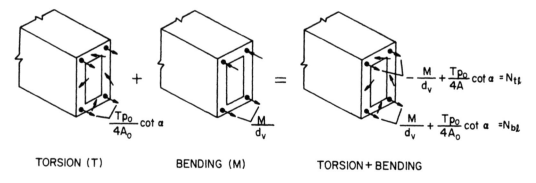

Figure 3.9 Superposition of stringer forces due to torsion and bending.

Equation (3-67) expresses the interaction relationship between M and T. Assume that yielding occurs in the bottom stringer and in the transverse steel, then $N_{bl} = N_{bly}$ and $n_t = n_{ty}$. Also define the pure torsional strength T_0 according to Table 3.1 as

$$T_0 = 2A_0\sqrt{\left(\frac{2N_{tly}}{p_0}\right)n_{ty}} \qquad (3\text{-}68)$$

Notice that the pure torsional strength T_0 is defined based on the top stringer force at yield N_{tly}, rather than the bottom stringer force at yield N_{bly}. This definition gives the lowest positive value for T_0, assuming the top stringer force at yield is less than the bottom stringer force at yield. The total longitudinal force due to torsion $\overline{N}_l$ is then equal to $2N_{tly}$.

Substituting the definitions of M_0, R, and T_0 from Eqs. (3-55), (3-57), and (3-68) into Eq. (3-67), the interaction equation for M and T is expressed in a nondimensionalized form:

$$\frac{M}{M_0} + \left(\frac{T}{T_0}\right)^2 R = 1 \qquad (3\text{-}69)$$

Second Failure Mode

The second failure mode occurs due to yielding of the top stringer and the transverse steel.

Looking at Figure 3.9, the force in the top stringer N_{tl} and the force per unit length in transverse steel n_t are

$$N_{tl} = -\frac{M}{d_v} + \frac{Tp_0}{4A_0}\cot\alpha \qquad (3\text{-}70)$$

$$n_t d_v = \frac{T}{2A_0}\tan\alpha \qquad (3\text{-}71)$$

Substituting α from Eq. (3-71) into Eq. (3-70) to eliminate α, we have

$$-\frac{M}{N_{tl}d_v} + \frac{T^2}{4A_0^2(2N_{tl}/p_0)n_t} = 1 \qquad (3\text{-}72)$$

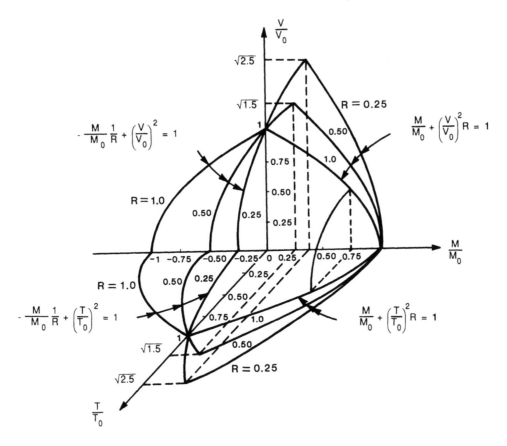

Figure 3.10 Torsion–bending interaction curves (also includes V-M curves from Figure 3.8).

Assuming that yielding occurs in the top stringer and in the transverse steel, then $N_{tl} = N_{tly}$ and $n_t = n_{ty}$. Substituting the definitions of M_0, R, and T_0 from Eqs. (3-55), (3-57), and (3-68) into Eq. (3-72), the nondimensionalized interaction equation for M and T becomes

$$-\left(\frac{M}{M_0}\right)\frac{1}{R} + \left(\frac{T}{T_0}\right)^2 = 1 \tag{3-73}$$

When the nondimensionalized interaction equation for the first mode of failure in torsion and moment [Eq. (3-69)] is compared to that in shear and moment [Eq. (3-58)], it can be seen that they are identical if the nondimensionalized ratio T/T_0 is replaced by V/V_0. Similar observation can also be seen when the equations for the second mode of failure [Eqs. (3-73) and (3-62)] are compared. Therefore, the interaction curves for torsion and moment can be illustrated by Figure 3.8 if the axis V/V_0 is replaced by an axis T/T_0. Both the torsion–bending interaction curves and the shear–bending interaction curves are plotted in Figure 3.10 using the three axes V/V_0, T/T_0, and M/M_0. The shear–bending curves are shown in the vertical plane and the torsion–bending curves in the horizontal plane.

In Section 3.2.1 we discussed the peak point of maximum shear. Similar logic can be used to find the peak point of maximum torsional moment, which corresponds also to the simultaneous yielding of top and bottom stringers. This intersection point of two curves for the first and second modes of failure is located at $T/T_0 = \sqrt{(1 + R)/2R}$ and $M/M_0 = (1 - R)/2$. For the case of $R = 0.5$, this peak point is located at $T/T_0 = \sqrt{1.5}$ and $M/M_0 = 0.25$, which are indicated in Figure 3.10.

Now that the two sets of shear–bending and torsion–bending interaction curves are obtained, we can derive the interaction relationships of shear, torsion, and bending (Elfgren, 1972). These relationships can be represented by a series of interaction surfaces bridging the two sets of interaction curves in Figure 3.10.

3.2.3   SHEAR–TORSION–BENDING INTERACTION

The interaction relationship for shear, torsion, and bending will be derived for a box section as shown in Figure 3.11. When both shear and torsion are present, the shear flows q on the four walls of the box section will be different. The shear flow q_V on each of the two webs due to the applied shear force V is $V/2d_v$ (Figure 3.11a) and the shear flow q_T on each wall due to the applied torsional moment T is $T/2A_0$ (Figure 3.11b). The shear flows due to shear and torsion are additive on the left wall and are subtractive on the right wall. On the top and bottom walls, only shear flow due to torsion exists. Utilizing the subscripts l, r, t, and b to indicate left, right, top, and bottom walls, respectively, the shear flows in the four walls are

$$q_l = \frac{V}{2d_v} + \frac{T}{2A_0} \tag{3-74}$$

$$q_r = -\frac{V}{2d_v} + \frac{T}{2A_0} \tag{3-75}$$

$$q_t = q_b = \frac{T}{2A_0} \tag{3-76}$$

The shear flows in the four walls are also shown in Figure 3.12. It is assumed in Figure 3.12 that the center line of the shear flow coincides with both the center line of the hoop bars and the center line connecting the centroids of the longitudinal bars. The inclination angle α of the concrete struts should also be different in the four walls. According to Eq. (3-5) the value of α is

$$\cot \alpha = \frac{q}{n_t} \tag{3-77}$$

Substituting the four qs from Eqs. (3-74) to (3-76) into Eq. (3-77) we have

$$\cot \alpha_l = \frac{1}{n_t} \left(\frac{V}{2d_v} + \frac{T}{2A_0} \right) \tag{3-78}$$

$$\cot \alpha_r = \frac{1}{n_t} \left(-\frac{V}{2d_v} + \frac{T}{2A_0} \right) \tag{3-79}$$

$$\cot \alpha_t = \cot \alpha_b = \frac{1}{n_t} \left(\frac{T}{2A_0} \right) \tag{3-80}$$

These four angles are also shown in Figure 3.12.

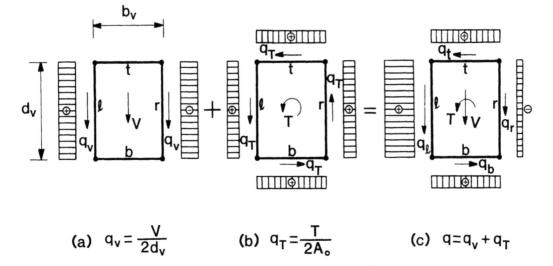

(a) $q_v = \dfrac{V}{2d_v}$ (b) $q_T = \dfrac{T}{2A_o}$ (c) $q = q_v + q_T$

Figure 3.11   Shear flow due to shear and torsion in a box section.

A rectangular box section subjected to shear, torsion, and bending may fail in three modes:

First Failure Mode

The first failure mode is caused by yielding in the bottom stringer and in the transverse reinforcement on the side where shear flows due to shear and torsion are additive (i.e., left wall).

Equating the external moment to the internal moment about the top wall (Figure 3.12) gives

$$M = N_{bl}d_v - q_l d_v (\cot \alpha_l) \frac{d_v}{2} - q_r d_v (\cot \alpha_r) \frac{d_v}{2} - q_b b_v (\cot \alpha_b) d_v \quad (3\text{-}81)$$

Substituting q_l, q_r, and q_b from Eqs. (3-74) through (3-76) and $\cot \alpha_l$, $\cot \alpha_r$, and $\cot \alpha_b$ from Eqs. (3-78) through (3-80) into Eq. (3-81) gives

$$M = N_{bl}d_v - \frac{d_v^2}{2n_t}\left(\frac{V}{2d_v} + \frac{T}{2A_0}\right)^2 - \frac{d_v^2}{2n_t}\left(-\frac{V}{2d_v} + \frac{T}{2A_0}\right)^2 - \frac{b_v d_v}{n_t}\left(\frac{T}{2A_0}\right)^2 \quad (3\text{-}82)$$

When the square terms are multiplied in Eq. (3-82) we notice that the two mixed terms of VT can be canceled. The two V^2 terms and the three T^2 terms can be grouped. Making these simplifications and dividing the whole equation by $N_{bl}d_v$ result in

$$\frac{M}{N_{bl}d_v} + \left(\frac{V}{2d_v}\right)^2 \frac{d_v}{N_{bl}} \frac{1}{n_t} + \left(\frac{T}{2A_0}\right)^2 \frac{(d_v + b_v)}{N_{bl}} \frac{1}{n_t} = 1 \quad (3\text{-}83)$$

Assume yielding in the bottom stringer and in the transverse steel. Then $N_{bl} = N_{bly}$

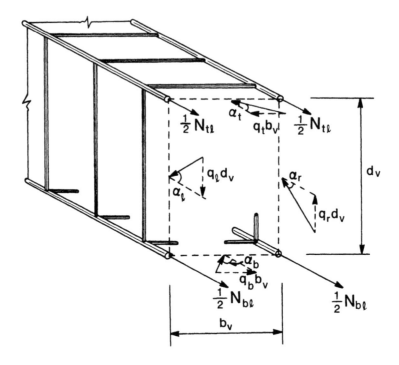

Figure 3.12   Forces in a box section subjected to shear, torsion, and bending.

and $n_t = n_{ty}$. Also recall the definitions

$$M_0 = N_{bly}d_v$$

$$V_0 = 2d_v\sqrt{\left(\frac{N_{tly}}{d_v}\right)n_{ty}} \qquad \text{for two webs of a box}$$

$$T_0 = 2A_0\sqrt{\left(\frac{2N_{tly}}{p_0}\right)n_{ty}} \qquad \text{noting } d_v + b_v = p_0/2$$

$$R = \frac{N_{tly}}{N_{bly}}$$

Substituting these definitions of M_0, V_0, T_0, and R into Eq. (3-83) we obtain a nondimensionalized interaction relationship for M, V, and T in the first mode:

$$\frac{M}{M_0} + \left(\frac{V}{V_0}\right)^2 R + \left(\frac{T}{T_0}\right)^2 R = 1 \qquad (3\text{-}84)$$

Equation (3-84) shows that the interaction of V and T is circular for a constant M. A typical V-T interaction curve for $M/M_0 = 0.75$ and $R = 1$ is shown in Figure 3.10. When M is varied, Eq. (3-84) represents an interaction surface. This surface intersects the vertical V-M plane to form an interaction curve expressed by Eq. (3-58) and intersects the horizontal T-M plane to form an interaction curve expressed by Eq. (3-69).

Second Failure Mode

The second failure mode is caused by yielding in the top stringer and in the transverse reinforcement on the side where shear flows due to shear and torsion are additive (i.e., left wall).

Equating the external moment to the internal moment about the bottom wall (Figure 3.12) gives

$$M = -N_{tl}d_v + q_l d_v (\cot \alpha_l) \frac{d_v}{2} + q_r d_v (\cot \alpha_r) \frac{d_v}{2} + q_t b_v (\cot \alpha_t) d_v \quad (3\text{-}85)$$

Substituting q_l, q_r, and q_t from Eqs. (3-74) through (3-76) and $\cot \alpha_l$, $\cot \alpha_r$, and $\cot \alpha_t$ from Eqs. (3-78) through (3-80) into Eq. (3-85) and simplifying give

$$-\frac{M}{N_{tl}d_v} + \left(\frac{V}{2d_v}\right)^2 \frac{d_v}{N_{tl}} \frac{1}{n_t} + \left(\frac{T}{2A_0}\right)^2 \frac{(d_v + b_v)}{N_{tl}} \frac{1}{n_t} = 1 \quad (3\text{-}86)$$

Assume yielding in the top stringer and in the transverse steel. Then $N_{tl} = N_{tly}$ and $n_t = n_{ty}$. Substituting the definitions of M_0, V_0, T_0, and R into Eq. (3-86) we obtain a nondimensionalized interaction relationship for M, V, and T in the second mode:

$$-\left(\frac{M}{M_0}\right) \frac{1}{R} + \left(\frac{V}{V_0}\right)^2 + \left(\frac{T}{T_0}\right)^2 = 1 \quad (3\text{-}87)$$

Equation (3-87) shows that the interaction of V and T is also circular for a constant M in the second mode of failure (Figure 3.10). This equation represents an interaction surface that intersects the vertical V-M plane to form an interaction curve expressed by Eq. (3-62) and intersects the horizontal T-M plane to form an interaction curve expressed by Eq. (3-73).

The intersection of the two failure surfaces for the first and second modes of failure will form a peak interaction curve between V and T. The expression of this peak curve can be obtained by solving Eqs. (3-87) and (3-84) to eliminate the M term:

$$\left(\frac{V}{V_0}\right)^2 + \left(\frac{T}{T_0}\right)^2 = \frac{(1 + R)}{2R} \quad (3\text{-}88)$$

The location of this curve can be obtained by solving Eqs. (3-87) and (3-84) to eliminate the V and T terms:

$$\frac{M}{M_0} = \frac{1 - R}{2} \quad (3\text{-}89)$$

Equation (3-89) is, of course, the same as Eq. (3-64). The plane formed by the peak curve at $M/M_0 = (1 - R)/2$ will be designated as the peak plane. Equation (3-88) for $R = 0.25$, 0.5, and 1 on the peak planes is plotted as a series of dotted curves in Figure 3.13.

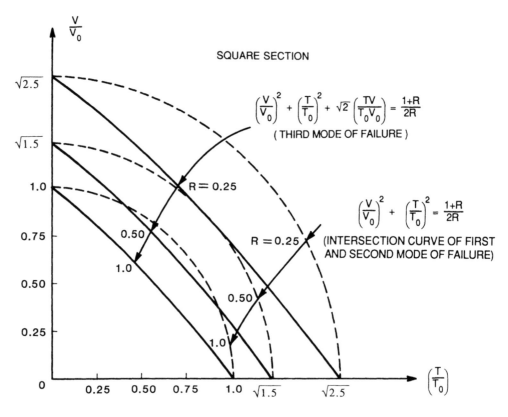

Figure 3.13 Interaction of shear and torsion on peak plane.

Third Failure Mode

The third failure mode is caused by yielding in the top bar (not top stringer), in the bottom bar, and in the transverse reinforcement, all on the side where shear flows due to shear and torsion are additive (i.e., left wall).

Taking the moment about the right side wall (Figure 3.12), where shear flows due to shear and torsion are subtractive, will furnish the equilibrium equation

$$0 = \frac{1}{2}(N_{bl} + N_{tl})b_v - q_l d_v (\cot \alpha_l) b_v - q_t b_v (\cot \alpha_t)\frac{b_v}{2} - q_b b_v (\cot \alpha_b)\frac{b_v}{2} \quad (3\text{-}90)$$

Substituting q_l, q_t, and q_b from Eqs. (3-74) and (3-76) and $\cot \alpha_l$, $\cot \alpha_t$, and $\cot \alpha_b$ from Eq. (3-78) and (3-80) into Eq. (3-90) and simplifying give

$$\frac{N_{bl} + N_{tl}}{2} = \frac{d_v}{n_t}\left(\frac{V}{2d_v} + \frac{T}{2A_0}\right)^2 + \frac{b_v}{n_t}\left(\frac{T}{2A_0}\right)^2 \quad (3\text{-}91)$$

When the square terms are multiplied in Eq. (3-91), we notice the appearance of a mixed term for VT. Grouping the two T^2 terms and dividing the whole equation by N_{tl} results in

$$\frac{N_{bl} + N_{tl}}{2N_{tl}} = \left(\frac{V}{2d_v}\right)^2\frac{d_v}{N_{tl}}\frac{1}{n_t} + \left(\frac{T}{2A_0}\right)^2\frac{(d_v + b_v)}{N_{tl}}\frac{1}{n_t} + 2\left(\frac{V}{2d_v}\right)\left(\frac{T}{2A_0}\right)\frac{d_v}{N_{tl}}\frac{1}{n_t}$$

$$(3\text{-}92)$$

Assume the yielding of the top bar, the bottom bar, and the transverse steel in the left wall. Then $N_{tl} = N_{tly}$, $N_{bl} = N_{bly}$, and $n_t = n_{ty}$. Substituting the definitions of V_0 and T_0 into Eq. (3-92) and noticing that

$$\frac{N_{bly} + N_{tly}}{2 N_{tly}} = \frac{(1 + R)}{2R} \quad \text{and} \quad d_v + b_v = \frac{p_0}{2}$$

we obtain a nondimensionalized interaction relationship for M, V, and T in the third mode:

$$\left(\frac{V}{V_0}\right)^2 + \left(\frac{T}{T_0}\right)^2 + \left(\frac{VT}{V_0 T_0}\right) 2\sqrt{\frac{2d_v}{p_0}} = \frac{(1 + R)}{2R} \tag{3-93}$$

It should be pointed out that the third mode of failure is independent of the bending moment M. The mixed VT term is a function of the shape of the cross section. For a square section, $2\sqrt{2d_v/p_0} = \sqrt{2}$ and Eq. (3-93) becomes

$$\left(\frac{V}{V_0}\right)^2 + \left(\frac{T}{T_0}\right)^2 + \sqrt{2}\left(\frac{VT}{V_0 T_0}\right) = \frac{(1 + R)}{2R} \tag{3-94}$$

Equation (3-94) is a function of R and represents a series of cylindrical interaction surfaces perpendicular to the V-T plane. The intersection of each cylindrical surface with its peak plane will produce a curve, which is plotted as a solid curve in Figure 3.13. This solid curve, representing the third interaction surface, is much lower than the corresponding dotted peak interaction curve formed by the intersection of the first and second interaction surfaces. It can be concluded, therefore, that the third mode of failure will always govern in the vicinity of the peak planes.

Figure 3.14 illustrates the three interaction surfaces in a perspective manner for the case of $R = 1/3$. The shaded area in the vicinity of the peak plane is where the third interaction surface replaces the first two interaction surfaces and, therefore, Eq. (3-93) governs. Figure 3.14 also shows that the longitudinal steel in one wall will yield when the member fails on one of the three interaction surfaces. On the interaction curves of any two interaction surfaces, however, longitudinal steel in two adjacent walls will yield. At the two peak points where all three surfaces intersect, all the longitudinal steel will yield.

It should be pointed out that the relationship between shear and torsion is not linear. This is because the plasticity truss model allows the α angle to vary in each of the four walls of a member, such that both the longitudinal and transverse steel will yield and the capacity of the member is maximized. As a result, the plasticity truss model always provides an upper bound solution.

If the compatibility condition of a R/C member is considered, some of the steel may not yield and the predicted capacity may not be reached. Such situations will be discussed in Chapters 7 and 8. However, tests have shown that the interaction relationships predicted by the plasticity truss model could be conservative even when the steel does not yield in one direction. Perhaps this theoretical unconservatism is counteracted by the strain hardening of the yielded rebars and the tensile strength of concrete, both of which are neglected in the analysis.

It should also be pointed out that the interaction surfaces based on the plasticity truss model have been shown by tests to be unconservative near the region of pure torsion. The theoretical torque may overestimate the test results by 20%. This

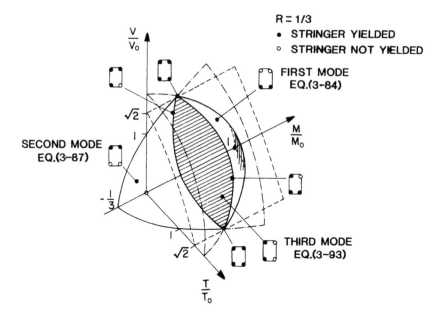

Figure 3.14 Interaction surface for shear, torsion, and bending.

unconservatism is caused by the overestimation of the lever arm area A_0 if it is calculated according to the center line of the hoop bars. Detailed discussion of this problem will be given in Chapter 8.

3.2.4   AXIAL-TENSION–SHEAR–BENDING INTERACTION

The effect of axial tension on the yield strength of a member easily can be included in the interaction relationship. Only a brief discussion is required in this section. The effect of axial tension will be illustrated by its interaction with shear and bending, and by the modification of the derivation in Section 3.2.1. In the truss model, it is assumed that the axial tension is resisted only by the longitudinal steel bars. Consequently, the axial tension does not destroy the internal equilibrium of the beam truss action under shear and bending. The only addition to the equilibrium condition is the tensile forces in the bottom and top stringers. In other words, Eqs. (3-52) and (3-59) for shear and bending interaction should include a simple new term due to the axial tension force P:

$$N_{bl} = \frac{P}{2} + \frac{M}{d_v} + \frac{V}{2}\cot\alpha \qquad (3\text{-}95)$$

$$N_{tl} = \frac{P}{2} - \frac{M}{d_v} + \frac{V}{2}\cot\alpha \qquad (3\text{-}96)$$

The other equilibrium equations remain valid for the forces in the transverse steel bars and in the concrete struts.

Using the same rational approach as in Section 3.2.1, except including the new term for P, the P-M-V interaction relationships for the two modes of failure can be

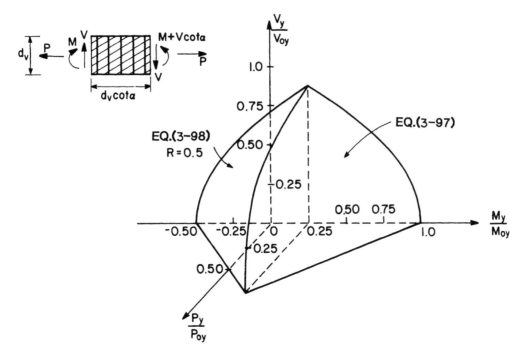

Figure 3.15 Nondimensional interaction surface for axial load, shear, and bending.

derived. For the first mode of failure we have

$$\frac{P}{P_0} + \frac{M}{M_0} + \left(\frac{V}{V_0}\right)^2 R = 1 \qquad (3\text{-}97)$$

The second mode of failure will give

$$\frac{P}{P_0}\left(\frac{1}{R}\right) - \frac{M}{M_0}\left(\frac{1}{R}\right) + \left(\frac{V}{V_0}\right)^2 = 1 \qquad (3\text{-}98)$$

Equations (3-97) and (3-98) are shown in Figure 3.15 as two interaction surfaces for $R = 0.5$. It can be seen that P and M have a straight line relationship, but the relationship between P and V is nonlinear.

3.3   Analysis of Beams

3.3.1   BEAMS SUBJECTED TO MIDSPAN CONCENTRATED LOAD

A truss model of a beam subjected to a concentrated load at midspan is shown in Figure 3.16a. The truss model is made up of two parallel top and bottom stringers, a series of transverse steel bars spaced uniformly at a spacing of s, and a series of diagonal concrete struts. The beam can be divided into two types of regions (main and local) depending on the α angle of the diagonal concrete struts. The main region is

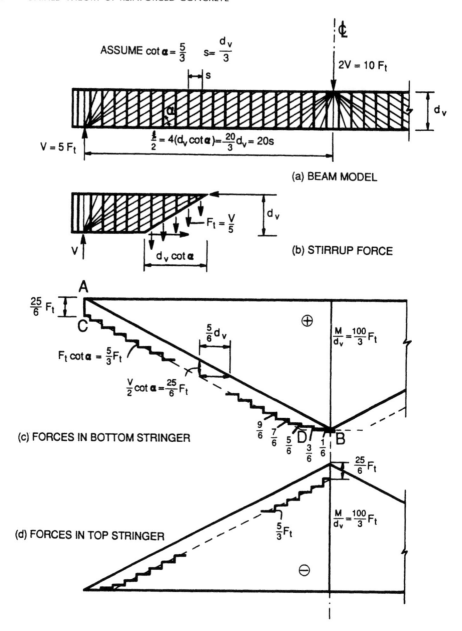

Figure 3.16 Truss model for a beam with midspan concentrated load.

the one where the α angle is a constant, so that the series of diagonal struts are parallel to each other. Regularity of the truss in this region makes simple analysis possible based on the sectional actions of M, V, T, and P. Analysis of elements from such regions has been made in Sections 3.1 and 3.2. The local region is the one where the α angle varies. Such regions lie in the vicinity of the concentrated loads, including the end reaction region and the midspan region under the load.

In the end reaction region, the local effect is shown in Figure 3.16a as the "fanning" of the compression struts from the application point of the concentrated load. In this case the angle varies from 90° to α and each concrete strut has a narrow triangular shape. Because the angle of the principal compressive stress is assumed to coincide with the angle of a concrete strut, the compressive forces will radiate upward from the load application point to form truss action with the forces in the top stringer and in the transverse steel bars. Similarly, Figure 3.16a also shows the "fanning" of the compression struts under the concentrated load at midspan. Because the load application point is on the top surface, the compressive forces radiate downward and form truss action with the forces in the bottom stringer and in the transverse steel bars.

Now let us analyze how the forces vary in the transverse and longitudinal reinforcement. To simplify the analysis we select the following three geometric relationships: $\cot \alpha = 5/3$, $s = d_v/3$, and $l/2 = 4(d_v \cot \alpha) = (20/3)d_v = 20s$. The α angle is the smallest allowed by the CEB code and should be the most economical. The spacing s and the half-span $l/2$ are typical in practice.

Transverse Stirrup Force F_t

In the truss model theory for beams, we have assumed that the shear force in a cross section is uniformly distributed over the depth of the beam, i.e., shear flow is constant in the transverse direction. In a beam subjected to concentrated load at midspan, the shear force should also be constant along the length of the beam within one half span, i.e., shear flow should also be constant in the longitudinal direction. Because any element in the web of the beam is subjected to an identical pure shear condition, the stirrup force should be constant along its own length (transverse direction), as well as throughout the length of the beam (longitudinal direction).

By definition, the force in one stirrup, F_t, should be $n_t s$. Recalling $n_t = (V/d_v)\tan \alpha$ from the beam shear equation in Table 3.1 we have

$$F_t = V\left(\frac{s}{d_v}\right)\tan \alpha \tag{3-99}$$

Adopting the assumed geometric relationship of $\tan \alpha = 3/5$ and $s = d_v/3$, a very simple equation for F_t results:

$$F_t = \frac{V}{5} \tag{3-100}$$

This simple value of F_t can also be obtained by taking vertical equilibrium of the free body shown in Figure 3.16b. Because F_t is a constant throughout the beam, we will use it as a reference force to measure other forces in the beam.

Forces in the Bottom Stringer

Forces in the bottom stringer are contributed by the bending moment and the shear force according to Eq. (3-52): $N_{bl} = M/d_v + (V/2)\cot \alpha$. Because the bending moment varies linearly along the length of a beam subjected to a midspan load, the force caused by bending should have a triangular shape in one half span as shown by the solid line AB in Figure 3.16c. Adopting the length of one half span as $(20/3)d_v$, the maximum stringer force at midspan is $V(l/2)/d_v = 5F_t(20/3)d_v/d_v = (100/3)F_t$.

Because the shear force V is a constant along the beam, the stringer force due to shear should also be a constant and is equal to $(V/2)\cot \alpha = (5F_t/2)(5/3) = (25/6)F_t$. The sum of the two stringer forces due to bending and shear is then represented by the dotted line CD, which is displaced vertically from the solid line by a distance of $(25/6)F_t$. In actuality, of course, the stirrups are not uniformly smeared, but are concentrated at discrete points with spacing s. Therefore, the stringer force contributed by shear should change at each stirrup and should have a stepped shape as indicated. Each step of change should introduce a stringer force of $F_t \cot \alpha$. For the main region of the beam, $\cot \alpha = 5/3$ and each step is $(5/3)F_t$. When the midspan is approached, however, $\cot \alpha$ gradually decreases, and the last five steps in the local region decrease in the following sequence: $(9/6)F_t$, $(7/6)F_t$, $(5/6)F_t$, $(3/6)F_t$, and $(1/6)F_t$. This stepped curve near the midspan can be approximated conservatively by a horizontal dotted line DB, which is commonly used in design.

Shift Rule

As previously stated, the dotted line CD in Figure 3.16c for bottom stringer force is displaced downward by a distance $(25/6)F_t$ from the solid line AB. It is also possible to view the dotted line CD as having displaced horizontally toward the support by a distance $(5/6)d_v$ from the solid line AB. This distance can be determined from the geometry of similar triangles. Because $(25/6)F_t$ is $1/8$ of $(100/3)F_t$ at midspan, then $1/8$ of one half span, $(20/3)d_v$, is $(5/6)d_v$. This shifting of the diagram has been recognized in many codes throughout the years and is known as the *shift rule*. In determining the cutoff points for longitudinal bars, such codes require that the design moment diagram be shifted a distance d (effective depth) toward the support. This detailing measure is an indirect recognition of the additional longitudinal tensile force demanded by the applied shear force, which has not been taken into account in the design of bottom longitudinal steel. Such a provision requiring a shift of a distance d is obviously conservative, because it is greater than the theoretically required shifting of $(5/6)d_v$ that has been calculated from the smallest generally accepted α angle.

Forces in the Top Stringer

Using the same logic, the forces in top stringer are plotted in Figure 3.16d. It can be seen that the compressive forces in the top stringer due to bending are reduced by the tensile forces due to shear.

3.3.2   BEAMS SUBJECTED TO UNIFORMLY DISTRIBUTED LOAD

The same model beam analyzed in Section 3.3.1 will now be subjected to a uniformly distributed load (Figure 3.17a). According to the plasticity truss model the following two assumptions are made:

1 The forces in all the truss members are transferred only at the levels of the top and bottom stringers. There will be no transfer of forces between the stirrups and the diagonal concrete struts in the main body (or web) of the beam between the top and bottom stringers. Hence, the tensile force in a stirrup and the compressive force in a concrete strut are uniform throughout their lengths.

2 The stirrups and the concrete struts should have infinite plasticity to deform freely. This second assumption is necessary, because the first assumption implies that no bond or compatibility exists between the diagonal concrete struts and the stirrups in the main body. The first assumption is applicable only if the second assumption is valid.

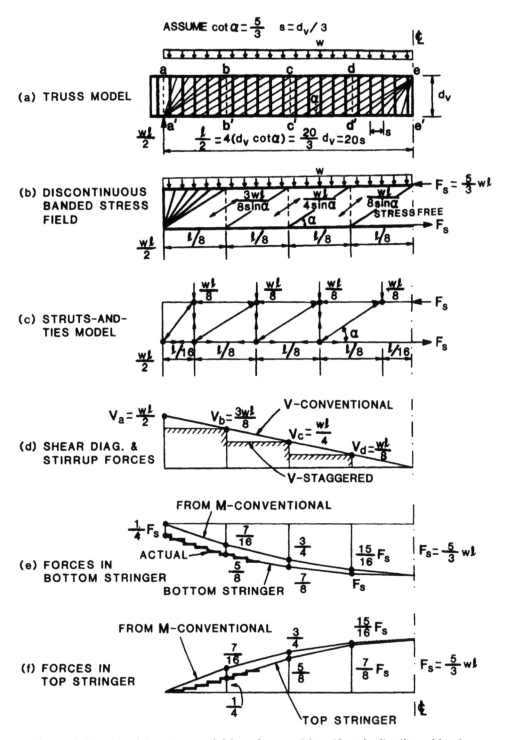

Figure 3.17 Plasticity truss model for a beam with uniformly distributed load.

Stress Flow

Based on these two assumptions, the stresses in the beam would flow from the midspan to the support in a manner shown by the discontinuous banded stress field in

Figure 3.17b and the struts-and-ties model in Figure 3.17c. Each strut or tie in the latter figure is located in the center line of each band of stress field in the former figure. The force in each strut or tie represents the resultant of the stresses in the band. Analysis of the stress flow in the stirrups (to be discussed next) will result in a staggered shear diagram as shown in Figure 3.17d.

Stirrup Forces and Staggered Shear Diagram

Following the stress flow from midspan to the support, we start with the applied load w from points d to e (Figure 3.17a and b). The total load within this distance is $wl/8$. It is resisted by the truss actions of the concrete struts and the top stringer. The total force in the concrete struts from d to e should be $wl/8 \sin \alpha$. This strut force is transmitted diagonally from the top stringer to the bottom stringer within line $c'd'$ while crossing the section d-d' en route. At the bottom stringer level between c' and d', the diagonal force in the concrete struts is resisted by a stringer force, as well as a stirrup force of $wl/8$. This stirrup force is equal to the shear force at section d-d', which is designated as V_d. Because the stirrup force is distributed uniformly between points c' and d', the required shear capacity should also be constant from c' to d'. This constitutes the first step of the staggered shear diagram plotted in Figure 3.17d.

In the second step, the uniform stirrup force at the bottom stringer from c' to d' is transmitted vertically to the top stringer within c and d. This stirrup force at the top stringer level remains at $wl/8$. At this level between c and d, however, the beam will receive an additional load of $wl/8$ from the externally applied load. The combined vertical force of $wl/4$ between c and d should then be resisted by the truss actions of the concrete struts and the top stringer. The force in the concrete struts from c to d should be $wl/4 \sin \alpha$. This strut force is transmitted diagonally to the bottom stringer within b' and c', while crossing section c-c' en route. At the bottom stringer level between b' and c', the diagonal force in the concrete struts is resisted by a stringer force as well as a stirrup force of $wl/4$. This stirrup force is equal to the shear force at section c-c' and is designated as V_c. Because this stirrup force is uniform between b' and c', the shear diagram should also be constant. This constitutes the second step of the staggered shear diagram in Figure 3.17d.

Using the same logic, the stirrup force in the region from a' to b' should be $3wl/8$, which is equal to the shear force at section b-b'. This third step completes the entire staggered shear diagram. It should be noted that the plasticity truss model predicts zero forces in the stirrup near the midspan from d to e. In practice, however, a minimum amount of stirrup should always be provided.

Figure 3.17d gives also the conventional shear diagram of triangular shape for a uniformly distributed loading. Design based on the staggered shear diagram requires considerably less stirrups than that using the conventional shear diagram.

Forces in Top Stringers

The stress flow from midspan to support also induces forces in the bottom and top stringers. These forces are plotted in Figure 3.17e and f, respectively. Because the stirrup force is no longer constant along the length of this beam, the stringer force at midspan, F_s, will be used as a reference force in studying the variation of forces in the stringers. Noting that the bending moment at midspan is $wl^2/8$ and that l/d_v has been selected as $40/3$, the stringer force at midspan is

$$F_s = \frac{wl^2}{8d_v} = \frac{5}{3}wl \qquad (3\text{-}101)$$

Because the vertical force from the external load between d and e is $wl/8$, the total force increment in the top stringer from d to e is $(wl/8)\cot\alpha = (5/24)wl = (1/8)F_s$. Therefore, the top stringer force at point d is $F_s - (1/8)F_s = (7/8)F_s$, as shown in Figure 3.17e. Further down the route of stress flow, the total vertical load on the top stringer from c to d is $wl/4$ ($wl/8$ from external load plus $wl/8$ from the stirrups), and the total force increment in the top stringer from c to d is $(wl/4)\cot\alpha = (5/12)wl = (1/4)F_s$. Then the top stringer force at c is $(7/8)F_s - (1/4)F_s = (5/8)F_s$. Similar logic would give the top stringer force at b to be $(5/8)F_s - (3/8)F_s = (1/4)F_s$. These values of top stringer forces at points e, d, c, and b are connected by straight lines in Figure 3.17f. In actuality, of course, the stringer forces should change at the discrete points of the stirrups. In the region b to c, for example, each force increment is $(3wl/8)(1/5)\cot\alpha = (1/8)wl = (3/40)F_s$.

In the local "fanning" region between a and b, however, each discrete vertical force at the stirrup locations is $(wl/2)(1/5) = wl/10$ and the increments of the top stringer force are $(wl/10)\cot\alpha$, depending on the variable angle α. The cotangent of α changes from b to a in the following sequence: $(9/10)(5/3)$, $(7/10)(5/3)$, $(5/10)(5/3)$, $(3/10)(5/3)$, and $(1/10)(5/3)$. Therefore, the force increments change from b to a as follows: $(9/100)F_s$, $(7/100)F_s$, $(5/100)F_s$, $(3/100)F_s$, and $(1/100)F_s$.

The force diagram for the top stringer derived from the plasticity truss model is compared to the force diagram calculated from the conventional moment diagram in Figure 3.17f. It can be seen that the former is smaller than the latter due to the shear effect.

Forces in the Bottom Stringer

Now looking at the bottom stringer in Figure 3.17a, the vertical force from d' to e' is zero. Therefore, no increment of force will occur in the bottom stringer between these two points and the stringer force remains a constant of F_s from d' to e' (Figure 3.17e). In the next region from c' to d', however, the vertical force is $wl/8$. The total force increment in the bottom stringer is $(wl/8)\cot\alpha = (5/24)wl = (1/8)F_s$. Therefore, the bottom stringer force at c' is $F_s - (1/8)F_s = (7/8)F_s$. Similar reasoning gives the bottom stringer forces at points b' and a' to be $(5/8)F_s$ and $(1/4)F_s$, respectively.

Comparing these values from the plasticity truss model with those obtained from the conventional moment diagram, Figure 3.17e shows that the shear effect increases the bottom stringer force. Therefore, the "shift rule" for detailing of bottom reinforcement remains valid.

Comments

In conclusion, the plasticity truss model provides a very clear concept of stress flow in the beam and the resulting forces in the stirrups, stringers, and concrete struts. However, this model satisfies only the equilibrium condition, not the strain compatibility condition. Because the second assumption of infinite plastic deformation cannot be guaranteed for concrete, a design based on the staggered shear diagram may not be conservative. In Chapter 6, Mohr compatibility truss model based on the elastic behavior of concrete and steel will be presented. As far as the material property is concerned, the plasticity truss model provides an upper bound solution, whereas the compatibility truss model yields the lower bound solution. For a nonlinear material of limited plasticity, such as concrete, the solution should lie between these two solutions.

3.4 Applications in CEB-FIP Model Code

3.4.1 BEAM SHEAR WITH INCLINED WEB REINFORCEMENT

The CEB-FIP Code (1978) includes provisions for the design of beams with inclined web reinforcement. We will first analyze such beams using the plasticity truss model.

In Section 3.1.3 we studied a beam subjected to a concentrated load at midspan (Figure 3.5a). The beam was reinforced with vertical transverse steel bars. In this section we shall generalize the problem by assuming that the beam is reinforced with steel bars inclined at an angle β to the longitudinal axis as shown in Figure 3.18a. The beam in Figure 3.5 is, therefore, a special case of the beam in Figure 3.18 when $\beta = 90°$.

A model of the isolated beam element is shown in Figure 3.18b. The top and bottom stringers are separated from the main body of the beam element so that the stringers are resisting the bending moment and the main body is carrying the shear force. The previous two assumptions for the model remain valid. First, the shear flow q in the main body is distributed uniformly over the depth, so that $V = qd_v$. Second, the shear flow q in the main body is also distributed uniformly along the length of the main body so that the transverse stresses (f_t) and the stresses in the diagonal concrete struts (σ_d) vary uniformly along their lengths.

Equilibrium of forces on the vertical face of the beam element is shown by the force polygon on the left face (Figure 3.18b). The shear force V pointing upward is resolved into a longitudinal steel force $\overline{N}_l$, a diagonal concrete force $(\sigma_d h)d_v \cos \alpha$, and an inclined steel force $n_t d_v \cot \beta$. From this force polygon the shear force V can be related to the longitudinal steel force $\overline{N}_l$ by

$$V = \frac{\overline{N}_l}{(\cot \alpha - \cot \beta)} \tag{3-102}$$

Similarly, equilibrium of forces on the horizontal face of the beam element is shown by the force triangle on the top face (Figure 3.18b). The shear force V pointing leftward is resolved into an inclined steel force $n_t d_v$ and a diagonal concrete force $(\sigma_d h)d_v \sin \alpha$. From this force triangle the shear force V can be related to the inclined steel force $n_t d_v$ by geometry as follows:

$$V = z(\cot \alpha + \cot \beta) \tag{3-103}$$

Because

$$z = n_t d_v \sin \beta \tag{3-104}$$

inserting z from Eq. (3-104) into Eq. (3-103) gives

$$V = n_t d_v \sin \beta(\cot \alpha + \cot \beta) \tag{3-105}$$

The shear force V can be related to the diagonal concrete stress σ_d using either the force polygon on the vertical face or the force triangle on the horizontal face. From geometry of the force triangles on the horizontal face, which is simpler, we obtain

$$z = (\sigma_d h)d_v \sin^2 \alpha \tag{3-106}$$

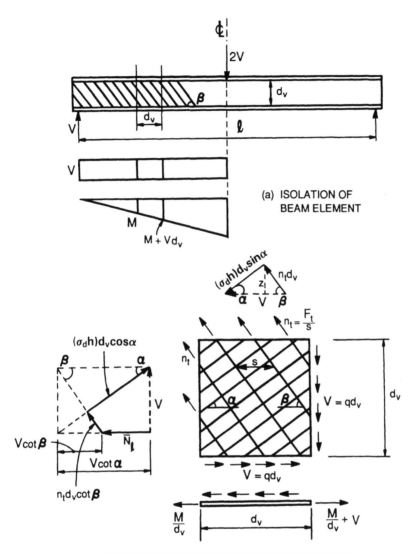

(a) ISOLATION OF BEAM ELEMENT

(b) EQUILIBRIUM OF FORCES IN BEAM ELEMENT

Figure 3.18   Beam shear with inclined reinforcement.

Inserting z from Eq. (3-106) into Eq. (3-103) gives

$$V = (\sigma_d h)d_v \sin^2 \alpha (\cot \alpha + \cot \beta) \qquad (3\text{-}107)$$

This set of three equations, Eqs. (3-102), (3-105), and Eq. (3-107), is identical to the set of three equations, Eqs. (3-38) to (3-40), for vertical stirrups if $\beta = 90°$.

Assume the yielding of the longitudinal and inclined steel bars, $\bar{N}_l = \bar{N}_{ly}$, $n_t = n_{ty}$, and $V = V_y$. Equating Eqs. (3-102) to (3-105) and eliminating V, we have

$$\cot \alpha = \sqrt{\frac{\bar{N}_{ly}}{n_{ty}d_v \sin \beta} + \cot^2 \beta} \qquad (3\text{-}108)$$

To solve V from Eqs. (3-102) and (3-105) we eliminate α in the following manner. First, α is expressed explicitly from Eq. (3-105):

$$\cot \alpha = \frac{V_y}{n_t d_v \sin \beta} - \cot \beta \qquad (3\text{-}109)$$

Substituting $\cot \alpha$ from Eq. (3-109) into Eq. (3-102) results in a quadratic equation:

$$\frac{V_y^2}{n_{ty} d_v \sin \beta} - 2V_y \cot \beta - \bar{N}_{ly} = 0 \qquad (3\text{-}110)$$

Solving the quadratic equation gives

$$V_y = \sqrt{\bar{N}_{ly}(n_{ty} d_v \sin \beta) + (n_{ty} d_v \cos \beta)^2} + (n_{ty} d_v \cos \beta) \qquad (3\text{-}111)$$

When $\beta = 90°$, $\sin \beta = 1$, and $\cos \beta = \cot \beta = 0$, Eqs. (3-108) and (3-111) become

$$\cot \alpha = \sqrt{\frac{\bar{N}_{ly}}{n_{ty} d_v}} \qquad (3\text{-}112)$$

$$V_y = d_v \sqrt{\left(\bar{N}_{ly}/d_v\right) n_{ty}} \qquad (3\text{-}113)$$

Equations (3-112) and (3-113) are, of course, identical to Eqs. (3-41) and (3-42) for vertical stirrups.

3.4.2   BASIC PRINCIPLES OF CEB SHEAR AND TORSION PROVISIONS

The 1978 CEB-FIP Model Code (CEB Code for short) includes two design methods. The first is the "Standard Method," which is based on the traditional 45° truss model, modified by the addition of an empirical term attributed to the "contribution of concrete." This method follows basically the same philosophy used by the ACI Code. It is intended for the rapid design of small or medium-size structural members, especially those used in buildings. The second, so-called accurate method, however, is based on the variable-angle plasticity truss model presented in this chapter. It is also modified by the addition of a term for contribution of concrete. This second method is intended for the design of large girders, particularly those employed in bridges and guideways. In this section, we shall introduce only the "accurate method."

Three strength criteria must be considered in the design of reinforced concrete beams. First, the web reinforcement must have enough strength to resist the applied shear force. Second, the longitudinal force caused by the applied shear must be determined. This longitudinal force should be added to the longitudinal forces caused by the bending moment and the axial load in the calculation of the longitudinal steel area. Third, the diagonal concrete struts should have sufficient strength to resist the compressive stress caused by the applied shear, until the yielding of the web reinforcement. This purpose is achieved by specifying a maximum shear strength in

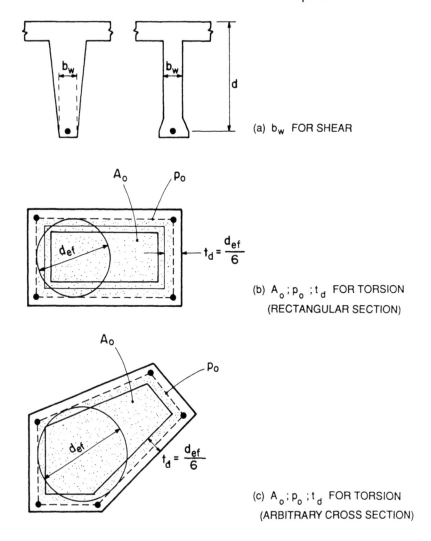

(a) b_w FOR SHEAR

(b) $A_o ; P_o ; t_d$ FOR TORSION
(RECTANGULAR SECTION)

(c) $A_o ; P_o ; t_d$ FOR TORSION
(ARBITRARY CROSS SECTION)

Figure 3.19   Definitions of symbols in CEB Code.

connection with an effective compressive strength of concrete. From a design point of view, the specified maximum shear strength is utilized to determine the minimum cross section of a member.

Four simplifications are also made in the design provisions:

1 At the failure of the diagonal concrete struts, the stress σ_d is specified as $\zeta f_c'$, with the softening coefficient ζ taken as a constant 0.6 for shear and 0.5 for torsion.

2 The lever arm of the bending couple d_v is taken as $0.9d$.

3 The width of the web, b_w, to resist shear is taken as the minimum width as shown in Figure 3.19a.

4 The parameters A_0, p_0, and t_d for torsion are calculated from the center line of shear flow defined by the polygon connecting the centroids of the longitudinal corner bars (Figure 3.19b and c; see later definitions). Based on these simplifica-

tions, the three design equations for shear and the three for torsion are grouped as follows.

3.4.2.1 Beam Shear

Vertical Stirrups ($\beta = 90°$)

From the three equilibrium equations [Eqs. (3-39), (3-38), and (3-40)], we have

Transverse steel

$$V_n = \frac{A_v f_{vy}}{s}(0.9d)\cot \alpha \qquad (3.114)$$

Longitudinal steel

$$V_n = \overline{N}_{IV} \tan \alpha \qquad (3-115)$$

Diagonal concrete

$$V_{n,\,\text{max}} = 0.6 f_c' b_w d \sin \alpha \cos \alpha \qquad (3-116)$$

Inclined Bars ($\beta \neq 90°$)

From Eqs. (3-105), (3-102), and (3-107),

$$V_n = \frac{A_v f_{vy}}{s}(0.9d)\sin \beta (\cot \alpha + \cot \beta) \qquad (3-117)$$

$$V_n = \frac{\overline{N}_{IV}}{(\cot \alpha - \cot \beta)} \qquad (3-118)$$

$$V_{n,\,\text{max}} = 0.6 f_c' b_w d \sin^2 \alpha (\cot \alpha + \cot \beta) \qquad (3-119)$$

where

V_n = nominal shear capacity at failure

$V_{n,\,\text{max}}$ = maximum nominal shear capacity at failure

A_v = cross-sectional area of transverse bars; two legs when a stirrup is used

$\overline{N}_{IV}$ = total longitudinal steel force due to shear

f_{vy} = yield strength of stirrups

b_w = width of web, taken at the narrowest level as defined in Figure 3.19a

α = angle of inclination of the diagonal concrete struts with respect to the longitudinal reinforcement, limited to a range of $30.9° < \alpha < 59.1°$

In deriving Eq. (3-119) from Eq. (3-107), d_v was taken simply as d, because the effective stress of $0.6 f_c'$ was only an approximate value.

Comparing Eq. (3-119) to Eq.(3-116), we see that they are identical when $\cot \beta = 0$, i.e., $\beta = 90°$. If β is reduced, then $V_{n,\,\text{max}}$ is increased. This means that the inclined bars are more advantageous than the vertical bars, and the advantage increases with decreasing β angle. To limit such advantage to a maximum of 50%, CEB Code specifies

$$V_{n,\,\text{max}} \not> 0.9 f_c' b_w d \sin \alpha \cos \alpha \qquad (3.120)$$

3.4.2.2 *Torsion*

Applying the new symbol definitions and simplifications to the three equilibrium equations for torsion [Eqs. (3-48), (3-47), and (3-49)], we have

Hoop steel

$$T_n = \frac{A_t f_{ty}}{s}(2A_0)\cot\alpha$$

(3-121)

Longitudinal steel

$$T_n = \frac{\overline{N}_{lT}}{p_0}(2A_0)\tan\alpha$$

(3-122)

Diagonal concrete

$$T_{n,\,\max} = 0.5f_c' t_d(2A_0)\sin\alpha\cos\alpha$$

(3-123)

where

T_n = nominal torsional capacity at failure

$T_{n,\,\max}$ = maximum nominal torsional capacity at failure

A_t = cross-sectional area of one hoop bar (one leg when a stirrup is used)

f_{ty} = yield strength of hoop steel

$\overline{N}_{lT}$ = total longitudinal steel force due to torsion

A_0 = area within the center line of shear flow; center line of shear flow is defined by the polygon connecting the centroids of the corner longitudinal bars (Figure 3.19b and c)

p_0 = perimeter of the preceding polygon, Figure 3.19b and c

t_d = thickness of shear flow zone, defined as one sixth of the diameter of the largest circle that can be contained within the polygon (Figure 3.19b and c)

The definitions of A_0, p_0, and t_d as shown in Figure 3.19b and c were found, unfortunately, to be grossly in error in many situations. The errors exhibit themselves in two ways: First, the area A_0 is frequently underestimated, because the dotted polygon defined by the centroids of the longitudinal corner bars (Figure 3.19b and c) could not represent the real center line of shear flow with reasonable accuracy. The CEB definition of A_0 was originally determined in an empirical manner by testing large box sections intended for the design of bridges. For such large box sections the concrete cover and the size of the steel bars are relatively small in comparison to other overall dimensions. For smaller size sections normally used in buildings, however, the concrete cover and the size of the steel bars become quite significant with respect to the overall dimensions. Therefore, the area within the dotted polygon becomes considerably smaller than the area within the real center line of the shear flow.

Second, the CEB definition of t_d frequently underestimates the real thickness of the shear flow zone. In many cases, the shear flow zone determined by the CEB thickness t_d lies completely outside the dotted polygon. One reason for this error could be attributed to the fact that a thorough understanding of the softening phenomenon in the shear flow zone had not been developed when the definition was originally proposed.

In view of the preceding two errors, the maximum torque calculated by Eq. (3-123) could be grossly underestimated. In some cases, the calculated torque of a reinforced concrete beam could be smaller than the cracking torque of a plain concrete beam of the same size. These situations are, of course, unacceptable.

The source of the difficulties lies in the fact that the CEB definitions of t_d, A_0, and p_0 do not consider the strain compatibility condition. Rational derivations of t_d, A_0, and p_0 will be provided in Chapter 8, taking into account the strain compatibility condition and the softened constitutive law of concrete. Simplified design equations are also given in Section 8.5.4 as follows:

$$t_d = \frac{4T_n}{A_c f_c'}$$

(3-124)

$$A_0 = A_c - \frac{t_d}{2} p_c$$

(3-125)

$$p_0 = p_c - 4t_d$$

(3-126)

where

A_c = area within the outer boundary of a cross section

p_c = perimeter of the outer boundary of a cross section

3.4.3 MODIFICATIONS BY CONCRETE CONTRIBUTION

Tests have shown that the basic equilibrium equations based on the plasticity truss model [Eqs. (3-117) to (3-119) for shear and Eqs. (3-121) to (3-123) for torsion] are sufficiently accurate for design when the percentage of web reinforcement is high. However, for small percentage of web reinforcement, the shear and torsional capacities exceed those predicted by the theory. This excess capacity is attributed to the contribution of concrete. This "concrete contribution" includes not only the tensile strength of concrete and the aggregate interlock along the cracks, but also other factors, such as the dowel resistance of reinforcing bars. Based on this philosophy, the nominal strengths V_n and T_n are each made up of two terms:

$$V_n = V_c + V_s$$

(3-127)

$$T_n = T_c + T_s$$

(3-128)

where

V_c, T_c = empirical shear and torsional resistance, respectively, contributed by concrete

V_s, V_s = shear and torsional resistances, respectively, attributed to transverse steel in the web and calculated by the plasticity truss model theory

3.4.3.1   *Beam Shear*

Transverse Steel

Adding the empirical term of concrete contribution V_{cv} to Eq. (3-117) gives

$$V_n = V_{cv} + \frac{A_v f_{vy}}{s}(0.9d)\sin\beta(\cot\alpha + \cot\beta) \qquad (3\text{-}129)$$

When

$$V_n \le V_c \qquad V_{cv} = V_c = 2.5\tau_R b_w d \qquad (3\text{-}130)$$

$$V_n \ge 3V_c \qquad V_{cv} = 0 \qquad (3\text{-}131)$$

$$V_c < V_n < 3V_c \quad \text{use linear interpolation for } V_{cv}$$

where

$$\tau_R = \text{allowable shear stress, defined as } f_t'/4, \qquad (3\text{-}132)$$

$$f_t' = 0.214(f_c')^{2/3} \qquad (3\text{-}133)$$

where f_c' is the cylinder compressive strength of concrete and f_t' is the tensile strength of concrete. Both f_c' and f_t' have a SI unit of megapascal.

Equation (3-129) is nondimensionalized and plotted in Figure 3.20 for the case of $\beta = 90°$ and $\cot\alpha = 5/3$. The linear interpolation of V_{cv} from Eqs. (3-130) to (3-131) is clearly illustrated. The accurate method represented by Eq. (3-129) is also compared to the standard method, which assumes an α angle of 45° and a constant V_c value.

Longitudinal Steel

From Eq. (3-110) for $\beta \ne 90°$,

$$\bar{N}_{ly} = \frac{V_y^2}{n_{ty}d_v \sin\beta} - 2V_y \cot\beta \qquad (3\text{-}134)$$

Assume that the bottom longitudinal steel receives one half of the total longitudinal force. Then taking $d_v = d$ and $V_y = V_n$, and noticing $n_{ty} = A_v f_{vy}/s$ result in

$$N_{bly} = \frac{V_n^2 s}{2 A_v f_{vy} d \sin\beta} - V_n \cot\beta \qquad (3\text{-}135)$$

For $\beta = 90°$,

$$N_{bly} = \frac{V_n^2 s}{2 A_v f_{vy} d} \qquad (3\text{-}136)$$

Equations (3-135) or (3-136) are used in the CEB Code for the design of the bottom longitudinal steel in lieu of Eqs. (3-118) or (3-115). This means that the

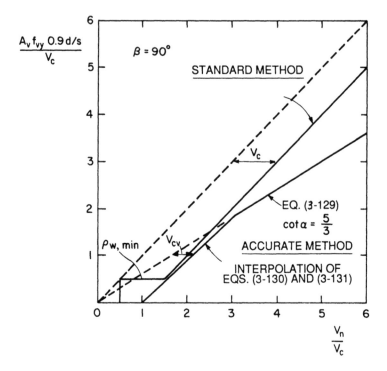

Figure 3.20 Comparison of standard method and accurate method in 1978 CEB-FIP Model Code.

longitudinal steel has been increased to compensate for the reduction of transverse steel due to concrete contribution. In other words, the transverse and longitudinal steels alone are capable of resisting the applied shear according to the plasticity truss model, without resorting to the contribution of concrete. Because the transverse steel is more expensive than the longitudinal steel, the CEB Code design method is aimed at being economical while maintaining the principle of the plasticity truss model. Such design, of course, could easily result in an α angle less than the lower limit of 30.9°. This limit for α must be checked.

The bottom longitudinal steel force N_{bly} is used to calculate the area of the bottom steel bars due to shear. This area should be added to those required by the bending moment and the axial load. It is interesting to point out that the effect of shear on the top compression stringer is not recognized by the CEB Code. In other words, the top longitudinal steel force N_{tly} due to shear is neglected in the design of longitudinal steel in the flexural compression zone.

3.4.3.2 Torsion

Transverse Steel

By adding the empirical term of concrete contribution T_{cv} to Eq. (3-121) we have

$$T_n = T_{cv} + \frac{A_t f_{ty}}{s}(2A_0)\cot\alpha \qquad (3\text{-}137)$$

When

$$T_n \leq T_c \qquad T_{cv} = T_c = 2.5\tau_R t_d (2A_0) \qquad (3\text{-}138)$$

$$T_n \geq 3T_c \qquad T_{cv} = 0 \qquad (3\text{-}139)$$

$$T_c < T_n < 3T_c \quad \text{use linear interpolation for } T_{cv}$$

τ_R has been given in Eqs. (3-132) and (3-133).

Longitudinal Steel

In the CEB Code the design of longitudinal steel area to resist torsion is based on the longitudinal steel force obtained from Eq. (3-122)

$$\overline{N}_{lT} = \frac{T_n p_0}{2A_0} \cot \alpha \qquad (3\text{-}140)$$

Equation (3-140) has been derived strictly from the equilibrium of plasticity truss model. On the one hand, this design method does not take advantage of the concrete contribution to reduce the amount of longitudinal steel. On the other hand, it does not increase the amount of longitudinal steel to compensate for the reduction of web reinforcement due to concrete contribution, as is done in the design of longitudinal steel for shear. The difference in design philosophy for shear and torsion perhaps reflects the relative importance of shear over torsion.

Equation (3-140) provides the total longitudinal steel force due to torsion alone. In combined torsion and bending, the part of the longitudinal steel force located in the flexural tension zone is additive to the flexural tensile steel force. The other portion of the longitudinal steel force located in the flexural compression zone, however, should be subtractive to the flexural compression steel force.

3.4.4   COMBINED SHEAR AND TORSION

Design of Reinforcement

A logical way to design a beam subjected to shear and torsion is to follow the shear flows as shown in Figure 3.11. First, the shear flows in each wall of a cross section are determined individually for shear and for torsion. Second, the shear flows in each wall are combined. Finally, each wall is designed according to its own combined shear flow. This method would be quite appropriate for the design of the transverse and longitudinal steel in large cross sections normally used in bridges. It is overly tedious, however, for smaller cross sections.

For members under combined shear and torsion, the CEB Code requires that the transverse and longitudinal steel due to torsion should be added to the transverse and longitudinal steel due to shear. This provision is, of course, very conservative, because on one side of the cross section the shear flows due to shear and torsion are subtractive (Figure 3.11). However, its simplicity is very appealing in the design of smaller cross sections.

For smaller cross sections, a less conservative, but still simple, method has been proposed by Collins and Mitchell (1980) for the design of longitudinal steel. In this method, an equivalent longitudinal tensile force $\overline{N}_{l_{eq}}$ is calculated:

$$\overline{N}_{l_{eq}} = \cot\alpha\sqrt{V_n^2 + \left(\frac{T_n p_0}{2A_0}\right)^2} \qquad (3\text{-}141)$$

In the flexural tension zone, the longitudinal steel area can be determined by

$$A_{bl} = \frac{1}{f_{ly}}\left(\frac{M_n}{d_v} + \frac{\overline{N}_{l_{eq}}}{2}\right) \qquad (3\text{-}142)$$

and in the flexural compression zone the longitudinal steel area is

$$A_{tl} = \frac{1}{f_{ly}}\left(-\frac{M_n}{d_v} + \frac{\overline{N}_{l_{eq}}}{2}\right) \qquad (3\text{-}143)$$

In the region of high moment when $M_n/d_v > \overline{N}_{l_{eq}}/2$, no longitudinal steel is required in the flexural compression zone.

Check Cross Section

For a cross section subjected to shear and torsion, the CEB Code simply specifies a linear interaction between shear and torsion, resulting in

$$\frac{V_n}{V_{n,\max}} + \frac{T_n}{T_{n,\max}} \leq 1 \qquad (3\text{-}144)$$

The maximum shear and torsional capacities $V_{n,\max}$ and $T_{n,\max}$ are calculated from Eqs. (3-119) and (3-123), respectively.

The linear interaction between shear and torsion appears to be reasonable. First, Eq. (3-144) is based on the compression failure of the diagonal concrete struts. Because concrete does not exhibit much plastic deformation, it would be prudent not to rely on any gain derived from plastic stress redistribution. Second, Figure 3.13 shows that at the peak planes, which is the optimum for design, the interaction curve between shear and torsion is close to a straight line for the third mode of failure.

3.4.5   PRACTICAL CONSIDERATIONS

The following specifications are given in the CEB Code to assist in the practical design of reinforced concrete structures:

(1) Maximum Spacing of Web Reinforcement

To ensure the development of shear strength designed according to the preceding method, the spacing of the web reinforcement must be limited. The CEB provisions for beam shear are

$$s_{\max} = 0.5d \leq 30\text{ cm (12 in.)} \quad \text{for } V_n \leq \tfrac{2}{3}V_{n,\max} \qquad (3\text{-}145)$$

$$s_{\max} = 0.3d \leq 20\text{ cm (8 in.)} \quad \text{for } V_n > \tfrac{2}{3}V_{n,\max} \qquad (3\text{-}146)$$

Physically, the transition from Eqs. (3-146) to (3-145) means that the effective strength of concrete σ_d in Eq. (3-119) has been reduced from $0.6f'_c$ to $0.4f'_c$ when the spacing of web reinforcement is increased from $0.3d$ to $0.5d$.

For torsion, the maximum spacing of the hoop reinforcement is

$$s_{max} = \frac{p_0}{8} \quad \text{or} \quad 30 \text{ cm (12 in.)} \tag{3-147}$$

It is interesting to point out that the maximum spacings for both shear and torsion are not a function of the angle α.

(2) Maximum Spacing of Longitudinal Bars

There is no specific requirement to limit the spacing of longitudinal steel in order to ensure the development of shear or torsional strengths. The maximum spacing of 30 cm (12 in.) is usually adopted to prevent the development of excessive crack widths.

(3) Minimum Web Reinforcement

The web reinforcement should satisfy a specified minimum reinforcement ratio $\rho_{w, min}$ as given in Table 3.2. The ratio ρ_w is calculated by

$$\rho_w = \frac{A_v}{b_w s \sin \beta} \tag{3-148}$$

(4) Shear and Torsion Near Supports

The CEB Code states that the shear and torsional stresses within the distance d from the face of a direct support need not be checked, but the web reinforcement calculated at the distance d should be continued up to the support.

(5) "Staggering Concept" for Shear Design

As discussed in Section 3.3.2, a shear design of web reinforcement based on the "staggering concept" may be unconservative. A conservative model of beams subjected to uniform loading—the Mohr compatibility truss model—is given in Chapter 6, Section 6.2.

3.5   Comments on Equilibrium (Plasticity) Truss Model

The equilibrium (plasticity) truss model presented in this chapter summarizes the major advances in reinforced concrete theory achieved during the decade of the 1970s. A milestone in this development was the adoption of this model by the 1978

TABLE 3.2 Minimum Shear Reinforcement Ratio $\rho_{w, min}$

| | f_y, MPa (ksi) | | |
| | S200 | S400 | S500 |
f'_c, MPa (psi)	(31.4)	(58.0)	(72.5)
C12 to C20 (1740–2900)	0.0016	0.0009	0.0007
C25 to C35 (3625–5075)	0.0024	0.0013	0.0011
C40 to C50 (5800–7250)	0.0030	0.0016	0.0013

CEB Code. However, serious weaknesses exist in these provisions due to its historical limitations. The difficulties stemmed primarily from a lack of understanding of the strain compatibility condition for shear and torsion, as well as the biaxial behavior of reinforced concrete membrane elements. These problems had been extensively studied in the 1980s and the new information is presented in Chapters 6 to 8.

The pros and cons of the equilibrium (plasticity) truss model will now be summarized:

Advantages

1 The equilibrium truss model theory satisfies completely the equilibrium condition. It provides three equilibrium equations that are conceptionally identical in element shear, beam shear, and torsion.

2 From a design point of view, the three equilibrium equations can be used directly to design the three components of the truss model, namely, the transverse steel, the longitudinal steel, and the diagonal concrete struts.

3 The three equilibrium equations satisfy the Mohr circle.

4 The model provides a very clear concept of the interaction of bending, shear, torsion, and axial load.

5 The model could explain the "shift rule" (Section 3.3.1) in a logical manner.

Deficiencies

1 The equilibrium truss model does not take into account the strain compatibility condition. As a result, it cannot predict the shear or torsional deformations of a member.

2 The model cannot predict the strains in the steel or concrete. Consequently, the yielding of steel or the crushing of concrete cannot be rationally determined, and the modes of failure cannot be discerned.

3 The model is unable to produce a method to correctly determine the thickness of the shear flow zone t_d in a torsional member. As such, the lever arm area A_0 and the perimeter of the shear flow p_0 cannot be logically defined.

4 The staggering concept of shear design may not be conservative.

4
Stresses in Membrane Elements

4.1 Stress Transformation

4.1.1 PRINCIPLE OF TRANSFORMATION

In Chapter 3 an equilibrium truss model was presented for elements that are subjected to pure shear and in which the tensile strength of concrete in the direction of cracking is assumed to be zero. In this chapter, however, the elements will be subjected to in-plane normal stresses in addition to in-plane shear stresses, and the concrete will also exhibit a small average tensile strength even after cracking. Such elements are called membrane elements and the in-plane stresses are known as the membrane stresses.

The stresses in a membrane element are best analyzed by the *principle of stress transformation*. Figure 4.1a shows an element in the stationary *l-t* coordinate system, defined by the directions of the longitudinal and transverse steel. The three stress components on the element, σ_l, σ_t, and τ_{lt} (or τ_{tl}) are shown in their positive directions.

For the shear stress τ_{lt}, the first subscript *l* represents the face on which the stress acts, and each face is defined by its outward normal. The second subscript *t* represents the direction of the stress itself. The upward shear stress on the right face is positive because this stress is acting on the positive *l* face and pointing in the positive *t* direction. The downward shear stress on the left face is also positive because this stress is acting on the negative *l* face and pointing in the negative *t* direction. In the case of the shear stress τ_{tl}, the rightward stress on the top *t* face is positive, because it is acting on the positive *t* face and pointing in the positive *l* direction. This approach to define the sign of a stress will be called the basic sign convention (Figure 4.1a).

For the two normal stresses σ_l and σ_t, the single subscript represents both the direction of face and stress. As such, the tensile stress will always be positive, because the direction of the stress always coincides with the outward normal. Compressive stresses will always be negative, because their direction is always opposite to the outward normal.

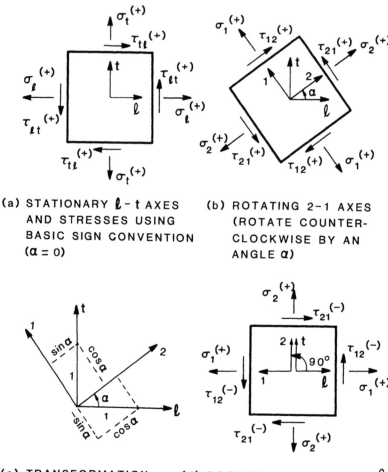

(a) STATIONARY ℓ-t AXES AND STRESSES USING BASIC SIGN CONVENTION ($\alpha = 0$)

(b) ROTATING 2-1 AXES (ROTATE COUNTER-CLOCKWISE BY AN ANGLE α)

(c) TRANSFORMATION GEOMETRY

(d) ROTATE 2-1 AXES BY $90°$ (STRESSES ARE WITH RESPECT TO 2-1 AXES)

Figure 4.1   Transformation of stresses.

To find the three stress components in various directions, we introduce a rotating 2-1 coordinate system as shown in Figure 4.1b. The 2-1 axes have been rotated counterclockwise by an angle of α with respect to the stationary l-t axes. The three stress components in this rotating coordinate system are σ_2, σ_1, and τ_{21} (or τ_{12}). They are all shown in their positive directions. The relationship between the rotating stress components σ_2, σ_1, and τ_{21} and the stationary stress components σ_l, σ_t, and τ_{lt} is the stress transformation. This relationship is, of course, a function of the angle α.

The relationship between the rotating 2-1 axes and the stationary l-t axes is shown by the transformation geometry in Figure 4.1c. A positive unit length on the l axis will have projections of $\cos \alpha$ and $-\sin \alpha$ on the 2 and 1 axis, respectively. A positive unit length on the t axis, however, should give projections of $\sin \alpha$ and $\cos \alpha$. Hence, the rotation matrix $[R]$ is

$$[R] = \begin{bmatrix} \cos \alpha & \sin \alpha \\ -\sin \alpha & \cos \alpha \end{bmatrix} \tag{4-1}$$

The relationship between the stresses in the 2-1 coordinate $[\sigma_{21}]$ and the stresses in the *l-t* coordinate $[\sigma_{lt}]$ is

$$[\sigma_{21}] = [R][\sigma_{lt}][R]^T \tag{4-2}$$

or

$$\begin{bmatrix} \sigma_2 & \tau_{21} \\ \tau_{12} & \sigma_1 \end{bmatrix} = \begin{bmatrix} \cos\alpha & \sin\alpha \\ -\sin\alpha & \cos\alpha \end{bmatrix} \begin{bmatrix} \sigma_l & \tau_{lt} \\ \tau_{tl} & \sigma_t \end{bmatrix} \begin{bmatrix} \cos\alpha & -\sin\alpha \\ \sin\alpha & \cos\alpha \end{bmatrix} \tag{4-3}$$

Performing the matrix multiplications and noticing that $\tau_{tl} = \tau_{lt}$ and $\tau_{12} = \tau_{21}$ result in the following three equations:

$$\sigma_2 = \sigma_l \cos^2\alpha + \sigma_t \sin^2\alpha + \tau_{lt}2\sin\alpha\cos\alpha \tag{4-4}$$

$$\sigma_1 = \sigma_l \sin^2\alpha + \sigma_t \cos^2\alpha - \tau_{lt}2\sin\alpha\cos\alpha \tag{4-5}$$

$$\tau_{21} = (-\sigma_l + \sigma_t)\sin\alpha\cos\alpha + \tau_{lt}(\cos^2\alpha - \sin^2\alpha) \tag{4-6}$$

Equations (4-4) to (4-6) can be expressed in matrix form by one equation:

$$\begin{bmatrix} \sigma_2 \\ \sigma_1 \\ \tau_{21} \end{bmatrix} = \begin{bmatrix} \cos^2\alpha & \sin^2\alpha & 2\sin\alpha\cos\alpha \\ \sin^2\alpha & \cos^2\alpha & -2\sin\alpha\cos\alpha \\ -\sin\alpha\cos\alpha & \sin\alpha\cos\alpha & (\cos^2\alpha - \sin^2\alpha) \end{bmatrix} \begin{bmatrix} \sigma_l \\ \sigma_t \\ \tau_{lt} \end{bmatrix} \tag{4-7}$$

This 3×3 matrix in Eq. (4-7) is the *transformation matrix* for transforming the stresses in the stationary *l-t* coordinate to the stresses in the rotating 2-1 coordinate.

4.1.2   MOHR CIRCLE

The trigonometric functions of α in Eqs. (4-4) to (4-6) can be written in the double angle 2α by

$$\cos^2\alpha = \tfrac{1}{2}(1 + \cos 2\alpha) \tag{4-8}$$

$$\sin^2\alpha = \tfrac{1}{2}(1 - \cos 2\alpha) \tag{4-9}$$

$$\sin\alpha\cos\alpha = \tfrac{1}{2}\sin 2\alpha \tag{4-10}$$

$$\cos^2\alpha - \sin^2\alpha = \cos 2\alpha \tag{4-11}$$

Substituting Eqs. (4-8) to (4-11) into Eq. (4-4) to (4-6) gives the three transformation equations in terms of the double angle 2α as follows:

$$\sigma_2 = \frac{\sigma_l + \sigma_t}{2} + \frac{\sigma_l - \sigma_t}{2}\cos 2\alpha + \tau_{lt}\sin 2\alpha \tag{4-12}$$

$$\sigma_1 = \frac{\sigma_l + \sigma_t}{2} - \frac{\sigma_l - \sigma_t}{2}\cos 2\alpha - \tau_{lt}\sin 2\alpha \tag{4-13}$$

$$\tau_{21} = -\frac{\sigma_l - \sigma_t}{2}\sin 2\alpha + \tau_{lt}\cos 2\alpha \tag{4-14}$$

Equations (4-12) to (4-14) were recognized by Otto Mohr to be algebraically analogous to a set of equations describing a circle in the σ-τ coordinate as shown in Figure 4.2a. This circle has its center on the σ axis at some distance from the origin. The ordinates and the abscissas of a set of two diametrically opposite points $B\ (\sigma_2, \tau_{21})$ and $D\ (\sigma_1, -\tau_{21})$ on the circle can be expressed in terms of a set of two diametrically opposite reference points $A\ (\sigma_l, \tau_{lt})$ and $C\ (\sigma_t, -\tau_{lt})$. These two sets of points are located on the circle at an angle ϕ apart. The geometric relationships of the two sets of points are:

$$\sigma_2 = \frac{\sigma_l + \sigma_t}{2} + \frac{\sigma_l - \sigma_t}{2}\cos\phi - \tau_{lt}\sin\phi \tag{4-15}$$

$$\sigma_1 = \frac{\sigma_l + \sigma_t}{2} - \frac{\sigma_l - \sigma_t}{2}\cos\phi + \tau_{lt}\sin\phi \tag{4-16}$$

$$\tau_{21} = \frac{\sigma_l - \sigma_t}{2}\sin\phi + \tau_{lt}\cos\phi \tag{4-17}$$

The circular property of Eqs. (4-15) to (4-17) can be proven as follows. Equations (4-15) and (4-16) can first be written in one equation, where $\sigma_{2,1}$ means either σ_2 or σ_1:

$$\left(\sigma_{2,1} - \frac{\sigma_l + \sigma_t}{2}\right) = \pm\left(\frac{\sigma_l - \sigma_t}{2}\cos\phi - \tau_{lt}\sin\phi\right) \tag{4-18}$$

Squaring Eq. (4-18) gives

$$\left(\sigma_{2,1} - \frac{\sigma_l + \sigma_t}{2}\right)^2 = \left(\frac{\sigma_l - \sigma_t}{2}\right)^2\cos^2\phi + \tau_{lt}^2\sin^2\phi - (\sigma_l - \sigma_t)\tau_{lt}\sin\phi\cos\phi$$

$$\tag{4-19}$$

Also squaring Eq. (4-17) gives

$$\tau_{21}^2 = \left(\frac{\sigma_l - \sigma_t}{2}\right)^2\sin^2\phi + \tau_{lt}^2\cos^2\phi + (\sigma_l - \sigma_t)\tau_{lt}\sin\phi\cos\phi \tag{4-20}$$

Adding Eqs. (4-19) and (4-20) results in

$$\left(\sigma_{2,1} - \frac{\sigma_l + \sigma_t}{2}\right)^2 + \tau_{21}^2 = \left(\frac{\sigma_l - \sigma_t}{2}\right)^2 + \tau_{lt}^2 = r^2 \tag{4-21}$$

Equation (4-21) is the equation of a circle with a radius of r and centered on the σ axis at a distance of $(\sigma_l + \sigma_t)/2$ from the origin.

Comparing the transformation relationship of Eqs. (4-12) to (4-14) to the geometric relationship of Eqs. (4-15) to (4-17) reveals that the equations are similar but not identical. Two differences can be noted:

1 All the angles in the transformation equations are twice those in the geometric relationship.

2 All the signs of the angles are opposite in the two sets of equations.

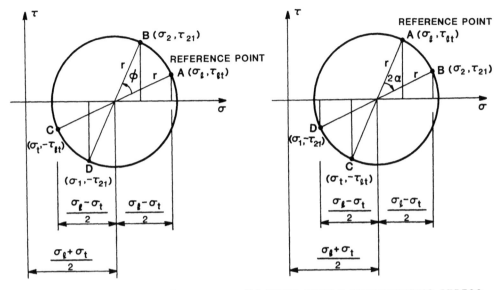

(a) GEOMETRIC RELATIONSHIP OF
MOHR CIRCLE

(b) MOHR CIRCLE REPRESENTING STRESS
TRANSFORMATION RELATIONSHIP

Figure 4.2 Mohr Circles.

In order to bring these two sets of equations into direct correspondence, we can eliminate these two differences by taking

$$\phi = -2\alpha \qquad (4\text{-}22)$$

The resulting Mohr circle is shown in Figure 4.2b.

In Mohr's graphical representation of stress transformation, the angles in the Mohr circle are always twice as great and measured in the opposite direction compared to the actual stress field in an element. One must always remember that although the algebraic analogy is precisely true, the geometric analogy is not, due to the double-angle relationship.

In the Mohr circle (Figure 4.2b), the point C presents considerable confusion with regard to the negative sign of the shear stress. This confusion can now be explained. With respect to the stationary l-t axes in the actual stress field, the top t face is 90° from the reference l face. τ_{tl} on the top t face must be positive as shown in Figure 4.1a. However, the same shear stress must be negative with respect to the rotating 2-1 axes, when the 2-1 axes have rotated 90° as shown in Figure 4.1d. This is because the rightward shear stress τ_{21} is acting on a positive 2 face and pointing in the negative 1 direction. Translating this 90° rotation (α) to the graphical expression of the Mohr circle, the rotating point B has now rotated 180° ($\phi = -2\alpha$) from the reference point A to point C. Point C, therefore, should have the coordinate of $(\sigma_2, -\tau_{21})$, rather than the coordinate of $(\sigma_1, -\tau_{lt})$ indicated. Because imagining a rotating 2-1 coordinate on top of a stationary l-t coordinate is inconvenient in the Mohr circle when 2α exceeds 180°, we return point C to the l-t coordinate as indicated in the figure.

In summary, a mixed notation has been used for point C in the Mohr circle. The subscripts of the stresses come from the stationary l-t axes, whereas the signs of the stresses are taken with respect to the rotating 2-1 axes. This inconsistency can be remedied by reversing the sign convention of τ_{tl} as defined in Figure 4.1a, i.e., by taking

$$\tau_{tl} = -\tau_{lt} \tag{4-23}$$

Utilizing Eq. (4-23) permits us not to consider the rotating 2-1 coordinate and to consider the stresses only in terms of the stationary *l-t* coordinate. The change of α angle is simply a change of faces within the stationary *l-t* coordinate, rather than a change of a rotating coordinate. The definition for a positive or negative shear stress has also been changed. Shear stress causing the counterclockwise rotation of the element is considered positive, whereas shear stress causing a clockwise rotation of the element is negative.

It should always be remembered that Eq. (4-23) is used only for convenience in connection with the Mohr circle, and should not be applied to the transformation relationship. The change of sign for τ_{tl} is cosmetic rather than substantive. This cosmetic remedy has been applied in a new sign convention developed specifically for reinforced concrete. The new sign convention will be presented in the next section.

4.1.3   SIGN CONVENTION FOR SHEAR IN REINFORCED CONCRETE

The left support of a typical reinforced concrete beam is shown in Figure 4.3a. Two concrete elements, A and B, are indicated and are shown separately in Figure 4.3b and c. Element A is taken in the *l-t* coordinate and element B in the principal 2-1 coordinate. The stresses on these two elements have been indicated in a qualitative sense with regard to sign. We will now study how to draw a Mohr circle to represent the stresses on these two elements.

We will first draw a Mohr circle based on the modified basic sign convention shown in Figure 4.4a. This sign convention is the same as the basic sign convention in Figure 4.1a, except it is modified by $\tau_{tl} = -\tau_{lt}$, the remedy given in Eq. (4-23). Notice that the direction of the shear stress τ_{tl} on the top face is defined as negative rather than positive. When this sign convention is applied to element A, the signs of the stresses are indicated in Figure 4.4b.

The stresses in Figure 4.4b are represented graphically by the Mohr circle in Figure 4.4c. The reference point A represents the stresses on the l face, $-\sigma_l$ and $-\tau_{lt}$. The point C gives the stresses $-\sigma_t$ and $+\tau_{lt}$ on the t face at $2\alpha = 180°$ away. The point B, representing the principal compressive stress condition $(-\sigma_2, 0)$ on element B, is located at an angle 2α from the reference point A. Point D, which is 180° from point B, gives the principal tensile stress condition $(+\sigma_1, 0)$.

Application of the modified basic sign convention (Figure 4.4a) to the reinforced concrete beam elements still reveals two disadvantages. First, the shear stress τ_{lt} on the reference l face is negative. Second, the angle 2α in the Mohr circle rotates in a clockwise direction, which is opposite to the counterclockwise direction of the α angle in the actual stress field.

To overcome these two undesirable characteristics, we reverse the sign convention of the shear stresses in Figure 4.4a to obtain a new sign convention as shown in Figure 4.5a. This new sign convention is suitable for reinforced concrete and will be called the RC sign convention. This sign convention addresses both the undesirable characteristics. First, the sign of the shear stress τ_{lt} becomes positive on the reference l face as shown in Figure 4.5b. Second, the angle 2α in the Mohr circle also rotates counterclockwise, Figure 4.5c, as in the stress field shown in Figure 4.3a.

The definition for a positive shear stress will, of course, also be changed. In the RC

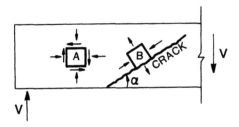

(a) ELEMENTS IN A TYPICAL BEAM

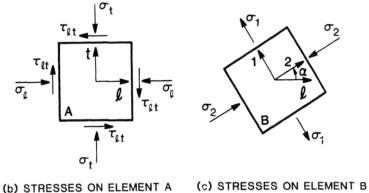

(b) STRESSES ON ELEMENT A
IN ℓ–t COORDINATE

(c) STRESSES ON ELEMENT B
IN PRINCIPAL 2–1
COORDINATE

Figure 4.3   Stresses on elements in typical reinforced concrete beams.

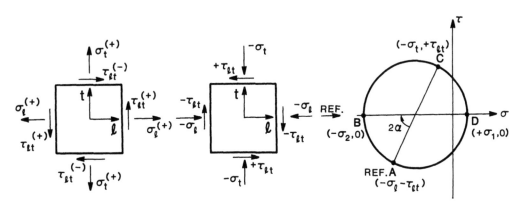

(a) MODIFIED BASIC SIGN
CONVENTION ($\tau_{t\ell} = -\tau_{\ell t}$)

(b) SIGN OF STRESSES
IN BEAM ELEMENT

(c) MOHR CIRCLE USING
MODIFIED BASIC SIGN
CONVENTION

Figure 4.4   Mohr stress circle for typical beam element using modified basic sign convention.

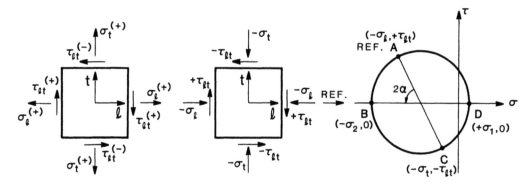

(a) RC SIGN CONVENTION   (b) SIGN OF STRESSES   (c) MOHR CIRCLE USING
                              IN BEAM ELEMENT   RC SIGN CONVENTION

Figure 4.5 Sign convention for reinforced concrete.

sign convention (Figure 4.5a) shear stress causing clockwise rotation of the element is positive, whereas shear stress causing a counterclockwise rotation is negative.

By reversing the sign of the shear stresses τ_{lt} and τ_{21}, Eqs. (4-4) to (4-6) become

$$\sigma_2 = \sigma_l \cos^2 \alpha + \sigma_t \sin^2 \alpha - \tau_{lt} 2 \sin \alpha \cos \alpha \tag{4-24}$$

$$\sigma_1 = \sigma_l \sin^2 \alpha + \sigma_t \cos^2 \alpha + \tau_{lt} 2 \sin \alpha \cos \alpha \tag{4-25}$$

$$\tau_{21} = (\sigma_l - \sigma_t)\sin \alpha \cos \alpha + \tau_{lt}(\cos^2 \alpha - \sin^2 \alpha) \tag{4-26}$$

Equations (4-24) to (4-26) can be expressed in matrix form by one equation:

$$\begin{bmatrix} \sigma_2 \\ \sigma_1 \\ \tau_{21} \end{bmatrix} = \begin{bmatrix} \cos^2 \alpha & \sin^2 \alpha & -2 \sin \alpha \cos \alpha \\ \sin^2 \alpha & \cos^2 \alpha & 2 \sin \alpha \cos \alpha \\ \sin \alpha \cos \alpha & -\sin \alpha \cos \alpha & (\cos^2 \alpha - \sin^2 \alpha) \end{bmatrix} \begin{bmatrix} \sigma_l \\ \sigma_t \\ \tau_{lt} \end{bmatrix} \tag{4-27}$$

This 3×3 matrix in Eq. (4-27) is the transformation matrix based on the RC sign convention.

4.1.4 PRINCIPAL STRESSES

Principal stresses are defined as the normal stresses on the face oriented in such a way that the shear stress vanishes. To find the principal stresses we will first express the double-angle transformation equations [Eqs. (4-12) to (4-14)] using the RC sign convention. Reverse all the signs of the shear stresses τ_{lt} and τ_{21}, and Eqs. (4-12) to (4-14) become

$$\sigma_2 = \frac{\sigma_l + \sigma_t}{2} + \frac{\sigma_l - \sigma_t}{2} \cos 2\alpha - \tau_{lt} \sin 2\alpha \tag{4-28}$$

$$\sigma_1 = \frac{\sigma_l + \sigma_t}{2} - \frac{\sigma_l - \sigma_t}{2} \cos 2\alpha + \tau_{lt} \sin 2\alpha \tag{4-29}$$

$$\tau_{21} = \frac{\sigma_l - \sigma_t}{2} \sin 2\alpha + \tau_{lt} \cos 2\alpha \tag{4-30}$$

The stresses σ_2 and σ_1 in Eqs. (4-28) and (4-29) become the principal stresses when the shear stress τ_{21} in Eq. (4-30) vanishes. Setting τ_{21} in Eq. (4-30) equals to zero gives the angle α that defines the two principal directions:

$$\cot 2\alpha = \frac{\sigma_l - \sigma_t}{-2\tau_{lt}} \tag{4-31}$$

From geometric relationship shown in Figure 4.6 we find $\sin 2\alpha$ and $\cos 2\alpha$ as follows:

$$\sin 2\alpha = \frac{\tau_{lt}}{\sqrt{((\sigma_l - \sigma_t)/2)^2 + \tau_{lt}^2}} \tag{4-32}$$

$$\cos 2\alpha = \frac{-((\sigma_l - \sigma_t)/2)}{\sqrt{((\sigma_l - \sigma_t)/2)^2 + \tau_{lt}^2}} \tag{4-33}$$

Substituting $\sin 2\alpha$ and $\cos 2\alpha$ from Eqs. (4-32) and (4-33) into Eqs. (4-28) and (4-29) results in

$$\sigma_2 = \frac{\sigma_l + \sigma_t}{2} - \sqrt{\left(\frac{\sigma_l - \sigma_t}{2}\right)^2 + \tau_{lt}^2} \tag{4-34}$$

$$\sigma_1 = \frac{\sigma_l + \sigma_t}{2} + \sqrt{\left(\frac{\sigma_l - \sigma_t}{2}\right)^2 + \tau_{lt}^2} \tag{4-35}$$

The expressions of these two principal stresses in Eqs. (4-34) and (4-35) can actually be obtained directly from Eq. (4-21) by setting τ_{21} equal to zero and taking the square root:

$$\sigma_{2,1} - \frac{\sigma_l + \sigma_t}{2} = \mp\sqrt{\left(\frac{\sigma_l - \sigma_t}{2}\right)^2 + \tau_{lt}^2} = \mp r \tag{4-36}$$

Equation (4-36) is the same as Eqs. (4-34) and (4-35). However, Eq. (4-36) shows clearly that the square root term is the radius of the Mohr circle.

Equations (4-34) and (4-35) can be expressed graphically by the Mohr circle in Figure 4.7c. For this example we choose an element that is subjected to biaxial tension and shear (Figure 4.7a). The principal tensile stress σ_1 is much larger in magnitude than the principal compressive stress σ_2 (Figure 4.7b). σ_1 and σ_2 are represented by the two points of the circle lying on the abscissa. σ_1 is the sum of the average stress $(\sigma_l + \sigma_t)/2$ and the radius r; σ_2 is the difference of these two terms.

4.2   Stress Transformation in Terms of Principal Stresses

4.2.1   TRANSFORMATION EQUATIONS

In Section 4.1.3 we derived Eq. (4-27), which provides the transformation matrix based on the RC sign convention. This matrix transforms a set of stresses in the *l-t*

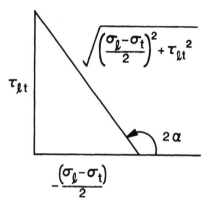

Figure 4.6 Geometric relationships for trigonometric function.

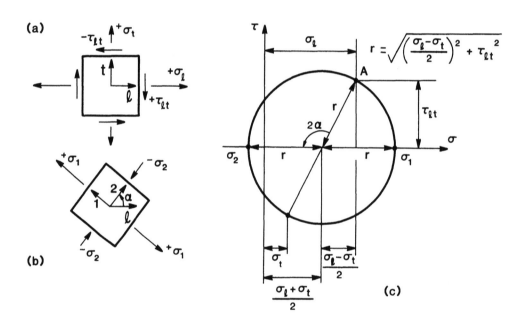

Figure 4.7 Graphical expression of principal stresses.

coordinate (σ_l, σ_t, and τ_{lt}) into a set of stresses in the 2-1 coordinate (σ_2, σ_1, and τ_{21}) (Figure 4.8a). Now, let us reverse the process and find the transformation matrix that transforms the stresses in the 2-1 coordinate (σ_2, σ_1, and τ_{21}) into the stresses in the *l-t* coordinate (σ_l, σ_t, and τ_{lt}) (Figure 4.8b).

To do so it is only necessary to change all the angles α in the matrix of Eq. (4-27) into $-\alpha$. It can be seen that all the trigonometric functions remain the same, except that $\sin(-\alpha) = -\sin\alpha$. By changing the sign of the four terms with $\sin\alpha$, we have

$$\begin{bmatrix} \sigma_l \\ \sigma_t \\ \tau_{lt} \end{bmatrix} = \begin{bmatrix} \cos^2\alpha & \sin^2\alpha & 2\sin\alpha\cos\alpha \\ \sin^2\alpha & \cos^2\alpha & -2\sin\alpha\cos\alpha \\ -\sin\alpha\cos\alpha & \sin\alpha\cos\alpha & (\cos^2\alpha - \sin^2\alpha) \end{bmatrix} \begin{bmatrix} \sigma_2 \\ \sigma_1 \\ \tau_{21} \end{bmatrix} \quad (4\text{-}37)$$

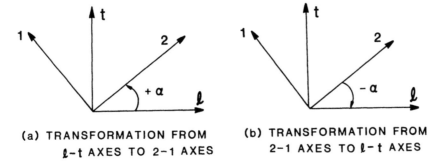

(a) TRANSFORMATION FROM
l-t AXES TO 2-1 AXES

(b) TRANSFORMATION FROM
2-1 AXES TO l-t AXES

Figure 4.8 Reversed transformation.

If we define the 2-1 axes as the principal axes, then τ_{21} must vanish. Inserting $\tau_{21} = 0$ into Eq. (4-37) results in three equations:

$$\sigma_l = \sigma_2 \cos^2 \alpha + \sigma_1 \sin^2 \alpha \tag{4-38}$$

$$\sigma_t = \sigma_2 \sin^2 \alpha + \sigma_1 \cos^2 \alpha \tag{4-39}$$

$$\tau_{lt} = (-\sigma_2 + \sigma_1)\sin \alpha \cos \alpha \tag{4-40}$$

It can be seen that the transformation equations are considerably simplified when the three stresses in the l-t coordinate are expressed in terms of the principal stresses σ_2 and σ_1.

Equations (4-38) to (4-40) can also be looked at from a more general point of view. Because they represent the equilibrium condition, they are also the equilibrium equations. These three equations contain six variables, namely, σ_l, σ_t, τ_{lt}, σ_2, σ_1, and α. When any three of the six variables are given, the other three can be solved by these three equilibrium equations.

It is interesting to note that the four normal stresses σ_l, σ_t, σ_2, and σ_1 have a simple relationship. Adding Eqs. (4-38) and (4-39) gives

$$\sigma_l + \sigma_t = \sigma_2 + \sigma_1 \tag{4-41}$$

That means any normal stress can be determined when the other three are given. For example,

$$\sigma_2 = \sigma_l + \sigma_t - \sigma_1 \tag{4-42}$$

4.2.2   FIRST TYPE OF EXPRESSION

In the case of a concrete element, σ_1 and σ_2 are the principal tensile stress and principal compressive stress, respectively, of concrete, whereas σ_l, σ_t, and τ_{lt} are the three stresses on the concrete element in the l-t coordinate. In concrete, σ_1 is smaller by an order of magnitude when compared to the other four stresses. Naturally, our interest will be focused on the relationship among the four stresses σ_l, σ_t, σ_2, and τ_{lt}, while considering σ_1 as a small given stress. With this aim in mind, the three transformation equations [Eqs. (4-38), (4-39), and (4-40)] can be changed into three explicit relationships among the four stresses σ_l, σ_t, σ_2, and τ_{lt}.

Inserting $\sigma_1 \sin^2 \alpha = \sigma_1 - \sigma_i \cos^2 \alpha$ into Eq. (4-38) gives

$$(-\sigma_l + \sigma_1) = (-\sigma_2 + \sigma_1)\cos^2 \alpha \tag{4-43}$$

Similarly, inserting $\sigma_1 \cos^2 \alpha = \sigma_1 - \sigma_1 \sin^2 \alpha$ into Eq. (4-39) gives

$$(-\sigma_t + \sigma_1) = (-\sigma_2 + \sigma_1)\sin^2 \alpha \tag{4-44}$$

Equation (4-40) remains the same:

$$\tau_{lt} = (-\sigma_2 + \sigma_1)\sin \alpha \cos \alpha \tag{4-45}$$

Equations (4-43) to (4-45) are the first type of expression for the equilibrium condition. In this type of expression, the three stresses σ_l, σ_t, and τ_{lt} are each related individually to the principal compressive stress σ_2. These three equations are convenient for the *analysis* of membrane elements as will be seen later.

Equations (4-43) to (4-45) represent the first type of geometric relationship in the Mohr circle as shown in Figure 4.9a and b. The geometric relationships of the half Mohr circle defined by ABD are illustrated in Figure 4.9b. If BD in Figure 4.9b is taken as unity, then $ED = \cos^2 \alpha$, $BE = \sin^2 \alpha$, and $AE = \sin \alpha \cos \alpha$. These three trigonometric values are actually the ratios of the three stresses $(-\sigma_l + \sigma_1)$, $(-\sigma_t + \sigma_1)$, and τ_{lt}, respectively, divided by the sum of the principal stresses $(-\sigma_2 + \sigma_1)$.

When σ_1 is taken as zero, then Eqs. (4-43) to (4-45) degenerate into Eqs. (3-15) to (3-17), which have been derived on the basis of a simple truss model with zero tensile strength of concrete. It should be recalled that all the equations in Chapter 3 are derived directly from the physical truss model, while assuming all the stresses to have absolute values. That is to say, both the tensile stress in the steel bars and the compressive stress in the concrete struts, σ_d, are taken as positive. In this chapter, however, all the equations are derived from stress transformation, adopting a definite sign convention for the stresses. In the RC sign convention, tensile stress is taken as positive and the compressive stress in the concrete struts, σ_2, is negative. Hence, $-\sigma_2$ in Eqs. (4-43) to (4-45) will always be positive and is the same as σ_d in Eqs. (3-15) to (3-17).

The difference in sign between the stresses σ_l in Eqs. (4-43) [or σ_t in Eq. (4-44)] and the stresses $\rho_l f_l$ in Eqs. (3-15) [or $\rho_t f_t$ in Eq. (3-16)] is more subtle. The smeared tensile stresses of the steel, $\rho_l f_l$ and $\rho_t f_t$, are derived from a pure shear element. The stresses σ_l and σ_t, on the other hand, are derived from a more general membrane element. For the special case of a pure shear element, σ_l and σ_t represent the compressive stresses on the concrete and, therefore, are equal and opposite to the tensile stress due to steel, $\rho_l f_l$ and $\rho_t f_t$. More explanation will be given in Section 4.3 when the reinforced concrete membrane element is introduced.

Knowing $\sigma_l = -\rho_l f_l$, $\sigma_t = -\rho_t f_t$, and $-\sigma_2 = \sigma_d$ for the case of pure shear element, we observed that Figure 4.9a is the same as Figure 3.3a, when $\sigma_1 = 0$. The geometric relationships in figure 4.9b are also the same as those in Figure 3.3b.

Dividing Eq. (4-44) by Eq. (4-43) we have

$$\tan^2 \alpha = \frac{-\sigma_t + \sigma_1}{-\sigma_l + \sigma_1} \tag{4-46}$$

Multiplying Eqs. (4-43) and (4-44) gives

$$(-\sigma_l + \sigma_1)(-\sigma_t + \sigma_1) = (-\sigma_2 + \sigma_1)^2 \sin^2 \alpha \cos^2 \alpha \tag{4-47}$$

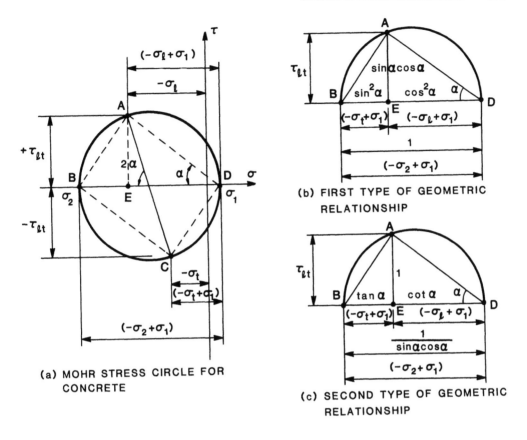

(a) MOHR STRESS CIRCLE FOR CONCRETE

(b) FIRST TYPE OF GEOMETRIC RELATIONSHIP

(c) SECOND TYPE OF GEOMETRIC RELATIONSHIP

Figure 4.9   Geometric relationships in Mohr stress circle.

Squaring Eq. (4-45) gives

$$\tau_{lt}^2 = (-\sigma_2 + \sigma_1)^2 \sin^2 \alpha \cos^2 \alpha \qquad (4\text{-}48)$$

Equating Eqs. (4-47) and (4-48) results in

$$\tau_{lt} = \pm\sqrt{(-\sigma_l + \sigma_1)(-\sigma_t + \sigma_1)} \qquad (4\text{-}49)$$

Equations (4-49) and (4-46) degenerate into Eqs. (3-21) and (3-22) when $\sigma_1 = 0$ and when the equations are restricted to the case of a pure shear element.

4.2.3   SECOND TYPE OF EXPRESSION

The three equilibrium equations [Eqs. (4-43) to (4-45)] can also be expressed in another form. The three stresses, σ_l, σ_t, and σ_2, can each be related individually to the shear stress τ_{lt}. Substituting $(-\sigma_2 + \sigma_1)$ from Eq. (4-45) into Eqs. (4-43) and (4-44) gives

$$(-\sigma_l + \sigma_1) = \tau_{lt} \cot \alpha \qquad (4\text{-}50)$$

$$(-\sigma_t + \sigma_1) = \tau_{lt} \tan \alpha \qquad (4\text{-}51)$$

Equation (4-45) itself can be written as

$$(-\sigma_2 + \sigma_1) = \tau_{lt} \frac{1}{\sin\alpha\cos\alpha} \qquad (4\text{-}52)$$

Equations (4-50) to (4-52) are the second type of expression for the equilibrium condition. They are convenient for the *design* of membrane elements as will be seen later.

Equations (4-50) to (4-52) represent the second type of geometric relationship in the Mohr circle as shown in Figure 4.9c. When AE in Figure 4.9c is taken as unity, then $ED = \cot\alpha$, $BE = \tan\alpha$, and $BD = 1/\sin\alpha\cos\alpha$. These three trigonometric values are actually the ratios of the three stresses $(-\sigma_l + \sigma_1)$, $(-\sigma_t + \sigma_1)$, and $(-\sigma_2 + \sigma_1)$, respectively, divided by the shear stress τ_{lt}.

When $\sigma_1 = 0$ and for the case of pure shear elements, Figure 4.9a and c become the same as Figure 3.3a and c, respectively.

Dividing Eq. (4-51) by Eq. (4-50) gives Eq. (4-46), and multiplying Eq. (4-51) by Eq. (4-50) produces Eq. (4-49).

4.3   Equilibrium of Reinforced Concrete Membrane Elements

4.3.1   TRANSFORMATION TYPE OF EQUILIBRIUM EQUATIONS

We will now study a concrete element reinforced orthogonally with longitudinal and transverse steel bars as shown in Figure 4.10a. The three stress components σ_l, σ_t, and τ_{lt} are now the applied stresses on the reinforced concrete element viewed as a whole. Their directions are based on the RC sign convention. The stresses on the concrete strut itself are denoted as σ_{lc}, σ_{tc}, and τ_{ltc}, and are shown in Figure 4.10b. The longitudinal and transverse steel provide the smeared stresses of $\rho_l f_l$ and $\rho_t f_t$ as shown in Figure 4.10c.

It is important to recognize the difference between the two sets of stresses: σ_l, σ_t, and τ_{lt} for the reinforced concrete element and σ_{lc}, σ_{tc}, and τ_{ltc} for the concrete struts. The principal stresses for σ_l, σ_t, and τ_{lt} will be defined as σ_2 and σ_1 based on the 2-1 axes shown in Figure 4.10d. The angle between the 2 axis and the reference l axis will be denoted as α_2. The subscript 2 is added to α to emphasize that it is measured from the 2 axis.

The principal stresses for σ_{lc}, σ_{tc}, and τ_{ltc} will be defined as σ_d and σ_r based on the d-r axes shown in Figure 4.10e. The angle between the d axis and the reference l axis will be denoted as α. To be exact, we should have added a subscript d to indicate that α is measured from the d axis. However, this d subscript is omitted for simplicity.

Both sets of stresses (σ_l, σ_t, τ_{lt} and σ_{lc}, σ_{tc}, τ_{ltc}) satisfy the transformation equations, Eqs. (4-38) to (4-40). However, the angle α_2 depends only on the relative ratios of the applied stresses σ_l, σ_t, and τ_{lt}. When these stresses increase proportionally, the α_2 angle does not change. For this reason, it is referred to as the fixed angle. In contrast, the angle α depends on the relative amounts of reinforcement in the

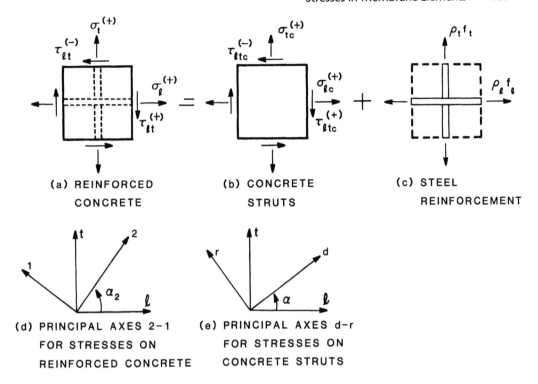

(a) REINFORCED
CONCRETE

(b) CONCRETE
STRUTS

(c) STEEL
REINFORCEMENT

(d) PRINCIPAL AXES 2–1
FOR STRESSES ON
REINFORCED CONCRETE

(e) PRINCIPAL AXES d–r
FOR STRESSES ON
CONCRETE STRUTS

Figure 4.10 Stress condition in reinforced concrete.

longitudinal and transverse directions. When unequal amounts of reinforcement are provided, the angle α will rotate with increasing proportional loading. Consequently, it is referred to as the rotating angle.

In summing the concrete stresses and the steel stresses in the l and t directions, we make a fundamental assumption. It is assumed that the steel reinforcement can take only axial stresses. Any possible dowel action is neglected. Hence, the superposition principle for concrete and steel becomes valid and we can write

$$\sigma_l = \sigma_{lc} + \rho_l f_l \qquad (4\text{-}53)$$

$$\sigma_t = \sigma_{tc} + \rho_t f_t \qquad (4\text{-}54)$$

$$\tau_{lt} = \tau_{ltc} \qquad (4\text{-}55)$$

Because the stresses on the concrete (σ_{lc}, σ_{tc}, and τ_{ltc}) must satisfy the principle of transformation, they can be expressed in terms of the principal stresses σ_d and σ_r in the d-r coordinate according to Eqs. (4-38) to (4-40):

$$\sigma_{lc} = \sigma_d \cos^2 \alpha + \sigma_r \sin^2 \alpha \qquad (4\text{-}56)$$

$$\sigma_{tc} = \sigma_d \sin^2 \alpha + \sigma_r \cos^2 \alpha \qquad (4\text{-}57)$$

$$\tau_{ltc} = (-\sigma_d + \sigma_r)\sin \alpha \cos \alpha \qquad (4\text{-}58)$$

Substituting Eqs. (4-56) to (4-58) into Eq. (4-53) to (4-55) results in the general equilibrium equations for reinforced concrete:

$$\sigma_l = \sigma_d \cos^2 \alpha + \sigma_r \sin^2 \alpha + \rho_l f_l \qquad (4\text{-}59)$$

$$\sigma_t = \sigma_d \sin^2 \alpha + \sigma_r \cos^2 \alpha + \rho_t f_t \qquad (4\text{-}60)$$

$$\tau_{lt} = (-\sigma_d + \sigma_r)\sin \alpha \cos \alpha \qquad (4\text{-}61)$$

It is important to appreciate the state of stress in a reinforced concrete element as represented by these three equilibrium equations. The understanding should focus on three aspects. First, we can look at the reinforced concrete element as a whole. The state of stress due to the three applied external stresses σ_l, σ_t, and τ_{lt} is shown by the Mohr circle in Figure 4.11a. It can be seen that point A represents the reference l face with stresses of σ_l and τ_{lt}; the principal stress σ_2 is located at an angle of $2\alpha_2$ away. Second, the axial smeared stresses of the steel reinforcement are shown in Figure 4.11c. The longitudinal and transverse stresses are $\rho_l f_l$ and $\rho_t f_t$, respectively. There is no Mohr circle for steel stresses, because the steel bars are assumed to be incapable of resisting shear (or dowel) stresses. Finally, the state of stress in the concrete struts is shown in Figure 4.11b. The stresses in the l-t coordinate are determined by $\sigma_{lc} = \sigma_l - \rho_l f_l$ and $\sigma_{tc} = \sigma_t - \rho_t f_t$. The stresses on the reference l face, represented by point A, are $(\sigma_l - \rho_l f_l)$ and τ_{lt}. The principal stress σ_d is located at an angle of 2α away from point A.

Adding Eqs. (4-59) and (4-60) gives

$$\sigma_d + \sigma_r = \sigma_l + \sigma_t - (\rho_l f_l + \rho_t f_t) \qquad (4\text{-}62)$$

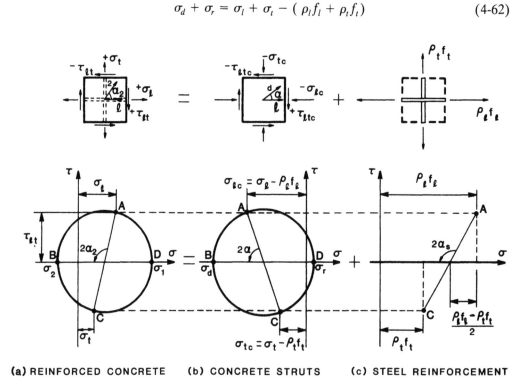

(a) REINFORCED CONCRETE   (b) CONCRETE STRUTS   (c) STEEL REINFORCEMENT

Figure 4.11 Mohr stress circles for reinforced concrete.

Because the stress in the cracking direction σ_r is small and can be considered a given value, the compressive stress in the concrete strut, σ_d, can be calculated from the externally applied stresses σ_l and σ_t and the steel stresses $\rho_l f_l$ and $\rho_t f_t$ according to Eq. (4-62).

4.3.2   FIRST TYPE OF EQUILIBRIUM EQUATIONS

In the three equilibrium equations [Eqs. (4-59) to (4-61)], the tensile stress of concrete, σ_r, is smaller by an order of magnitude when compared to the other internal stresses σ_d, $\rho_l f_l$, and $\rho_t f_t$. Therefore, our focus will be on the relationships among the stresses $(\sigma_l - \rho_l f_l)$, $(\sigma_t - \rho_t f_t)$, τ_{lt}, and σ_d, while considering σ_r as a small value of secondary importance. With this in mind, we can relate the three stresses $(\sigma_l - \rho_l f_l)$, $(\sigma_t - \rho_t f_t)$, and τ_{lt} to the compressive stress in the concrete struts, σ_d.

Inserting $\sigma_r \sin^2 \alpha = \sigma_r - \sigma_r \cos^2 \alpha$ into Eq. (4-59) gives

$$-\sigma_l + \rho_l f_l + \sigma_r = (-\sigma_d + \sigma_r)\cos^2 \alpha \tag{4-63}$$

Similarly, inserting $\sigma_r \cos^2 \alpha = \sigma_r - \sigma_r \sin^2 \alpha$ into Eq. (4-60) gives

$$-\sigma_t + \rho_t f_t + \sigma_r = (-\sigma_d + \sigma_r)\sin^2 \alpha \tag{4-64}$$

Equation (4-61) remains the same:

$$\tau_{lt} = (-\sigma_d + \sigma_r)\sin \alpha \cos \alpha \tag{4-65}$$

Equations (4-63) to (4-65) are the first type of expression for the equilibrium condition. These three equations are convenient for the analysis of reinforced concrete membrane elements.

Equations (4-63) to (4-65) represent the first type of geometric relationship in the Mohr circle as shown in Figure 4.12a. The geometric relationships of the half Mohr circle defined by ABD are illustrated in Figure 4.12b. If BD in Figure 4.12b is taken as unity, then $ED = \cos^2 \alpha$, $BE = \sin^2 \alpha$, and $AE = \sin \alpha \cos \alpha$. These three trigonometric values are actually the ratios of the three stresses $(-\sigma_l + \rho_l f_l + \sigma_r)$, $(-\sigma_t + \rho_t f_t + \sigma_r)$, and τ_{lt}, respectively, divided by the sum of the principal stresses $(-\sigma_d + \sigma_r)$.

Dividing Eq. (4-64) by Eq. (4-63) we have

$$\tan^2 \alpha = \frac{-\sigma_t + \rho_t f_t + \sigma_r}{-\sigma_l + \rho_l f_l + \sigma_r} \tag{4-66}$$

Multiplying Eqs. (4-63) and (4-64) gives

$$(-\sigma_l + \rho_l f_l + \sigma_r)(-\sigma_t + \rho_t f_t + \sigma_r) = (-\sigma_d + \sigma_r)^2 \sin^2 \alpha \cos^2 \alpha \tag{4-67}$$

Squaring Eq. (4-65) gives

$$\tau_{lt}^2 = (-\sigma_d + \sigma_r)^2 \sin^2 \alpha \cos^2 \alpha \tag{4-68}$$

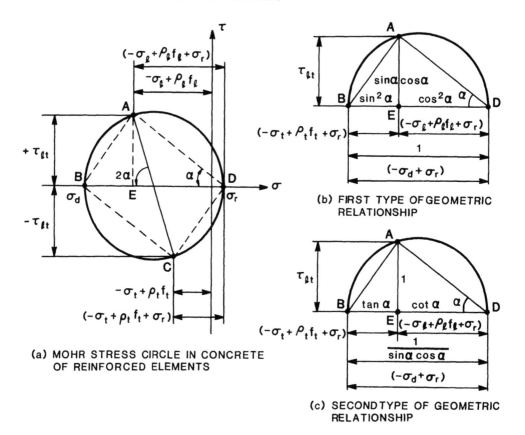

(a) MOHR STRESS CIRCLE IN CONCRETE
OF REINFORCED ELEMENTS

(b) FIRST TYPE OF GEOMETRIC
RELATIONSHIP

(c) SECOND TYPE OF GEOMETRIC
RELATIONSHIP

Figure 4.12 Geometric relationships in Mohr stress circle for concrete in reinforced elements.

Equating Eqs. (4-67) and (4-68) results in

$$\tau_{lt} = \pm\sqrt{(-\sigma_l + \rho_l f_l + \sigma_r)(-\sigma_t + \rho_t f_t + \sigma_r)} \qquad (4\text{-}69)$$

For a simple truss model, $\sigma_r = 0$. Equations (4-63) to (4-65) are simplified as

$$-\sigma_l + \rho_l f_l = (-\sigma_d)\cos^2\alpha \qquad (4\text{-}70)$$

$$-\sigma_t + \rho_t f_t = (-\sigma_d)\sin^2\alpha \qquad (4\text{-}71)$$

$$\tau_{lt} = (-\sigma_d)\sin\alpha\cos\alpha \qquad (4\text{-}72)$$

In the case of a reinforced concrete element subjected to pure shear, $\sigma_l = \sigma_t = 0$. Then

$$\rho_l f_l = (-\sigma_d)\cos^2\alpha \qquad (4\text{-}73)$$

$$\rho_t f_t = (-\sigma_d)\sin^2\alpha \qquad (4\text{-}74)$$

$$\tau_{lt} = (-\sigma_d)\sin\alpha\cos\alpha \qquad (4\text{-}75)$$

Equations (4-73) to (4-75) are identical to Eqs. (3-15) to (3-17). The difference in the sign of σ_d is due to the sign convention defined. In Eqs. (4-73) to (4-75) the RC sign convention requires the compressive stress in the concrete struts to be negative. In Eqs. (3-15) to (3-17), however, all stresses have been taken as absolute values without sign.

The state of stresses in a reinforced concrete element subjected to pure shear ($\sigma_l = \sigma_t = 0$) and assuming zero tensile stress ($\sigma_r = 0$) is shown by the Mohr circles in Figure 4.13. Looking at the element as a whole, only the shear stress τ_{lt} exists on the reference l face represented by point A. The principal compressive stress σ_2 is oriented at $\alpha_2 = 45°$ (Figure 4.13a). The compressive stresses on the concrete in the l-t coordinate are equal and opposite to the smeared stresses of the steel (Figure 4.13b). That is to say, $\sigma_{lc} = -\rho_l f_l$ and $\sigma_{tc} = -\rho_t f_t$. When $\rho_l f_l \neq \rho_t f_t$, the angle α is not 45°. If $\rho_l f_l = \rho_t f_t$, then $\alpha = 45°$.

4.3.3   SECOND TYPE OF EQUILIBRIUM EQUATIONS

The three equilibrium equations [Eqs. (4-63) to (4-65)] can also be expressed in another form. Substituting $(-\sigma_d + \sigma_r)$ from Eq. (4-65) into Eqs. (4-63) and (4-64) gives

$$-\sigma_l + \rho_l f_l + \sigma_r = \tau_{lt} \cot \alpha \qquad (4\text{-}76)$$

$$-\sigma_t + \rho_t f_t + \sigma_r = \tau_{lt} \tan \alpha \qquad (4\text{-}77)$$

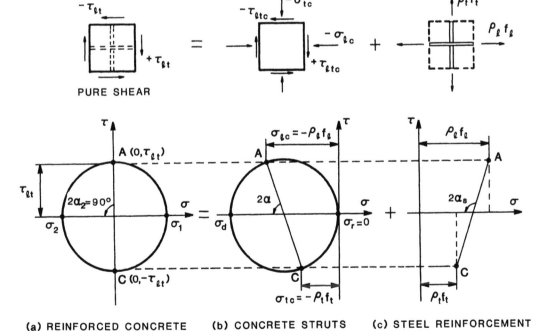

(a) REINFORCED CONCRETE   (b) CONCRETE STRUTS   (c) STEEL REINFORCEMENT

(ASSUME $\sigma_r = 0$)

Figure 4.13 Reinforcement concrete element subjected to pure shear and using simple truss model.

Equation (4-65) itself can be written as

$$(-\sigma_d + \sigma_r) = \tau_{lt}\frac{1}{\sin\alpha\cos\alpha} \tag{4-78}$$

Equations (4-76) to (4-78) are the second type of expression for equilibrium condition. They are convenient for the design of reinforced concrete membrane elements.

Equations (4-76) to (4-78) represent the second type of geometric relationship in the Mohr circle as shown in Figure 4.12c. When AE in Figure 4.12c is taken as unity, then $ED = \cot\alpha$, $BE = \tan\alpha$, and $BD = 1/\sin\alpha\cos\alpha$. These three trigonometric values are actually the ratios of the three stresses $(-\sigma_l + \rho_l f_l + \sigma_r)$, $(-\sigma_t + \rho_t f_t + \sigma_r)$, and $(-\sigma_d + \sigma_r)$, respectively, divided by the shear stress τ_{lt}.

Dividing Eq. (4-77) by Eq. (4-76) gives Eq. (4-66), and multiplying Eq. (4-76) by Eq. (4-77) produces Eq. (4-69).

In design, the small tensile stress of concrete is often neglected, i.e., $\sigma_r = 0$. Then Eqs. (4-76) to (4-78) become

$$\rho_l f_l = \sigma_l + \tau_{lt}\cot\alpha \tag{4-79}$$

$$\rho_t f_t = \sigma_t + \tau_{lt}\tan\alpha \tag{4-80}$$

$$(-\sigma_d) = \tau_{lt}\frac{1}{\sin\alpha\cos\alpha} \tag{4-81}$$

Furthermore, for the case of pure shear, $\sigma_l = \sigma_t = 0$. Then

$$\rho_l f_l = \tau_{lt}\cot\alpha \tag{4-82}$$

$$\rho_t f_t = \tau_{lt}\tan\alpha \tag{4-83}$$

$$(-\sigma_d) = \tau_{lt}\frac{1}{\sin\alpha\cos\alpha} \tag{4-84}$$

Equations (4-82) to (4-84) are identical to Eqs. (3-12) to (3-14). When the tensile stress of concrete is neglected ($\sigma_r = 0$) and for the case of pure shear elements ($\sigma_l = \sigma_t = 0$), Figure 4.12a and c become the same as Figure 3.3a and c, respectively.

4.3.4   EQUILIBRIUM EQUATIONS IN TERMS OF DOUBLE ANGLE

The equilibrium equations can also be expressed in terms of the relationships between the angles α_2, α, and an imaginary angle α_s to be defined. Subtracting Eq. (4-77) from Eq. (4-76) gives

$$-(\sigma_l - \sigma_t) = \tau_{lt}(\cot\alpha - \tan\alpha) - (\rho_l f_l - \rho_t f_t) \tag{4-85}$$

Dividing Eq. (4-85) by $2\tau_{lt}$ gives

$$\frac{(\sigma_l - \sigma_t)}{-2\tau_{lt}} = \frac{1}{2}(\cot\alpha - \tan\alpha) + \frac{(\rho_l f_l - \rho_t f_t)}{-2\tau_{lt}} \tag{4-86}$$

Notice in Eq. (4-86) that the first term on the right-hand side is

$$\tfrac{1}{2}(\cot\alpha - \tan\alpha) = \cot 2\alpha \qquad (4\text{-}87)$$

and the term on the left side of Eq. (4-86) can be written according to Eq. (4-31) as

$$\frac{(\sigma_l - \sigma_t)}{-2\tau_{lt}} = \cot 2\alpha_2 \qquad (4\text{-}88)$$

The angle in Eq. (4-88) is α_2, because the externally applied stresses σ_l, σ_t, and τ_{lt} are acting on the reinforced concrete element as a whole.

Let us now define an imaginary angle α_s for the second term on the right-hand side of Eq. (4-86), such that

$$\frac{(\rho_l f_l - \rho_t f_t)}{-2\tau_{lt}} = \cot 2\alpha_s \qquad (4\text{-}89)$$

Substituting Eqs. (4-87) to (4-89) into Eq. (4-86) we arrived at a very simple equation, relating the three angles α_2, α, and α_s:

$$\cot 2\alpha_2 = \cot 2\alpha + \cot 2\alpha_s \qquad (4\text{-}90)$$

Equation (4-90) is very useful in understanding the relationship among the three diagrams in Figure 4.11. The cotangent of the angle $2\alpha_2$ for the reinforced concrete element is simply the sum of the cotangent of 2α for the concrete struts and the cotangent of $2\alpha_s$ for the steel reinforcement.

It should again be emphasized that the angle α_s is imaginary, because the steel reinforcement does not resist a shear stress τ_{lt}. The vertical axis τ in Figure 4.11c does not have a real physical meaning and no Mohr circle can be prepared for the steel reinforcement. However, the invention of the α_s angle provides a very convenient way to understand and to check the three diagrams in Figure 4.11a, b, and c.

4.4   Plasticity Truss Model for Membrane Elements

In the plasticity truss model, the design of reinforced concrete membrane elements is based on the yielding of both the longitudinal and transverse steels. Assuming $f_l = f_{ly}$ and $f_t = f_{ty}$, then σ_l, σ_t, and τ_{lt} will be denoted as σ_{ly}, σ_{ty}, and τ_{lty}. Taking the case where the tensile stress of concrete is neglected ($\sigma_r = 0$), the three equilibrium equations [Eqs. (4-79) to (4-81)] become

$$\rho_l f_{ly} = \sigma_{ly} + \tau_{lty}\cot\alpha \qquad (4\text{-}91)$$

$$\rho_t f_{ty} = \sigma_{ty} + \tau_{lty}\tan\alpha \qquad (4\text{-}92)$$

$$(-\sigma_d) = \tau_{lty}\frac{1}{\sin\alpha\cos\alpha} \qquad (4\text{-}93)$$

Adding Eqs. (4-91) and (4-92) and recalling $\cot \alpha + \tan \alpha = 1/\sin \alpha \cos \alpha$ gives

$$\rho_l f_{ly} + \rho_t f_{ty} = \sigma_{ly} + \sigma_{ty} + \tau_{lty}\frac{1}{\sin \alpha \cos \alpha} \tag{4-94}$$

The total steel requirement in two directions, $\rho_l f_{ly} + \rho_t f_{ty}$, is plotted against the angle α in Figure 4.14 according to Eq. (4-94). It can be seen that a minimum total steel can be achieved by designing α equal to 45°. This angle of 45° also provides the best crack control as will be studied in Chapter 5, Section 5.3.

Designing a membrane element based on $\alpha = 45°$, Eqs. (4-91) and (4-92) become

$$\rho_l f_{ly} = \sigma_{ly} + \tau_{lty} \tag{4-95}$$

$$\rho_t f_{ty} = \sigma_{ty} + \tau_{lty} \tag{4-96}$$

If the element is subjected to a compressive transverse stress σ_{ty}, the magnitude of which is greater than the shear stress τ_{lty}, Eq. (4-96) gives $\rho_t f_{ty} < 0$. Such design is, of course, not valid. The best we can do is to set $\rho_t f_{ty} = 0$. Inserting this condition into Eq. (4-92) we obtain an angle α that is greater than 45°:

$$\tan \alpha = \frac{-\sigma_{ty}}{\tau_{lty}} \tag{4-97}$$

Substituting this α angle into Eq. (4-91) we obtain the required longitudinal steel:

$$\rho_l f_{ly} = \sigma_{ly} + \frac{\tau_{lty}^2}{-\sigma_{ty}} \tag{4-98}$$

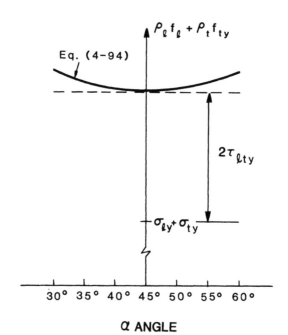

Figure 4.14   Total reinforcement in element as a function of angle α.

If $\rho_l f_{ly}$ is still negative, then no reinforcement is required in both the longitudinal and transverse directions.

If the element is subjected to a compressive longitudinal stress σ_{ly}, the magnitude of which is greater than the shear stress τ_{lty}, Eq. (4-95) gives $\rho_l f_{ly} < 0$. Because this design is also not valid, we set $\rho_l f_{ly} = 0$. Inserting this condition into Eq. (4-91) we obtain an angle α that is less than 45°:

$$\cot \alpha = \frac{-\sigma_{ly}}{\tau_{lty}} \qquad (4\text{-}99)$$

Substituting this α angle into Eq. (4-92) we obtain the required transverse steel:

$$\rho_t f_{ty} = \sigma_{ty} + \frac{\tau_{lty}^2}{-\sigma_{ly}} \qquad (4\text{-}100)$$

If $\rho_t f_{ty}$ is still negative, then no reinforcement is required in both the longitudinal and transverse directions. In practice, of course, a minimum amount of steel should always be provided in both directions.

EXAMPLE PROBLEM 4.1

A membrane element is subjected to a set of three membrane stresses as shown in Figure 4.15a, i.e., $\sigma_l = 310$ psi (tension), $\sigma_t = -310$ psi (compression), and $\tau_{lt} = 537$ psi. Design the reinforcement in the longitudinal and transverse directions according to the plasticity truss model, i.e., assuming the yielding of steel in the two directions, $f_{ly} = f_{ty} = 60,000$ psi. Calculate the stresses in the steel bars, in the concrete struts, and in the membrane element, and express them in terms of Mohr circles. Assume equal percentages of steel in both directions, i.e., $\rho_l = \rho_t$, and neglect the tensile strength of concrete $\sigma_r = 0$.

SOLUTION

The fixed angle α_2 for the reinforced concrete element is calculated directly from the applied membrane stresses. From Eq. (4-88),

$$\cot 2\alpha_2 = \frac{(\sigma_l - \sigma_t)}{-2\tau_{lt}} = \frac{310 - (-310)}{-2(537)} = -0.5773$$

$$2\alpha_2 = 120° \quad \text{or} \quad \alpha_2 = 60° \quad \text{or} \quad \tan \alpha_2 = 0.5773$$

The Mohr circle for the applied stresses σ_l, σ_t, and τ_{lt} is shown in Figure 4.15b. The angle $2\alpha_2 = 120°$ is indicated.

The reinforcement can be designed according to the basic equations of the plasticity truss model [Eqs. (4-91) and (4-92)]. Because $\rho_l f_{ly} = \rho_t f_{ty}$, the rotating angle α can be found by equating these two equations:

$$\sigma_{ly} + \tau_{lty} \cot \alpha = \sigma_{ty} + \tau_{lty} \tan \alpha$$

$$\tau_{lty}(\tan \alpha - \cot \alpha) = \sigma_{ly} - \sigma_{ty}$$

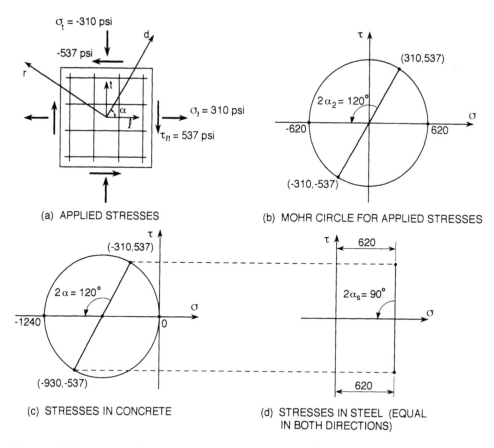

Figure 4.15 Stresses (psi) in a membrane element (Example 4.1) with equal steel stresses in two directions.

Noticing that $(\tan \alpha - \cot \alpha)$ is equal to $-\cot 2\alpha$, the angle α is expressed as

$$\cot 2\alpha = \frac{\sigma_{ly} - \sigma_{ty}}{-2\tau_{lty}} = \frac{310 - (-310)}{-2(537)} = -0.5773$$

$$2\alpha = 120° \quad \text{or} \quad \alpha = 60° \quad \text{or} \quad \tan \alpha = 0.5773$$

Substituting α back into Eq. (4-91) or Eq. (4-92) we have

$$\rho_l f_{ly} = \sigma_{ly} + \tau_{lty} \cot \alpha = 310 + 537(0.5773) = 620 \text{ psi}$$

$$\rho_t f_{ty} = 620 \text{ psi}$$

$$\rho_l = \rho_t = \rho = \frac{620}{60000} = 0.01033$$

If the element has a thickness h of 12 in., the steel area per unit length, A_l/s and A_t/s, is

$$\frac{A_l}{s} = \frac{A_t}{s} = \rho h = 0.01033(12) = 0.1240 \text{ in.}^2/\text{in.}$$

We could use two layers of No. 7 bars at 9-in. spacing in both directions. $A_l/s = A_t/s$ = $2(0.60)/9 = 0.133$ in.^{2}/in. > 0.124 in.^{2}/in.   **O.K.**

The smeared steel stresses $\rho_l f_{ly}$ and $\rho_t f_{ty}$ are indicated in Figure 4.15d. The stresses in the concrete struts are shown by the Mohr circle in Figure 4.15c using the additional values

$$\sigma_{ly} - \rho_l f_{ly} = 310 - 620 = -310 \text{ psi}$$

$$\sigma_{ty} - \rho_t f_{ty} = -310 - 620 = -930 \text{ psi}$$

$$\sigma_d = \frac{-\tau_{lty}}{\sin \alpha \cos \alpha} = \frac{-537}{(0.866)(0.500)} = -1240 \text{ psi}$$

In conclusion, Figure 4.15a, b, and c clearly illustrate that the rotating angle α is equal to the fixed angle α_2 when the smeared steel stresses in both directions, $\rho_l f_{ly}$ and $\rho_t f_{ty}$, are equal. If all the steel bars are assumed to yield as in the plasticity truss model, then $\alpha = \alpha_2$ can occur only when the membrane element is reinforced with equal percentages of steel in both directions, as shown in this example. This interesting case is a direct consequence of Eq. (4-90) for the double-angle relationship among the steel bars, the concrete struts, and the reinforced concrete element as a whole. When $\cot 2\alpha_s$ for the steel is zero, Eq. (4-90) requires that $\cot 2\alpha$ for the concrete is equal to $\cot 2\alpha_2$ for the whole element.

5

Strains in Membrane Elements

5.1   Strain Transformation

5.1.1   PRINCIPLE OF TRANSFORMATION

In Chapter 4 we studied the principle of transformation for stresses. This same principle will now be applied to strains. In order to avoid as much repetition as possible, the reader is advised to first study Chapter 4.

A strain is defined as a displacement per unit length. The definitions of strains ε_l, ε_t, and γ_{lt} (or γ_{tl}) in the l-t coordinate are illustrated in Figure 5.1a and b. They are indicated as positive using the basic sign convention. The strain ε_l indicated in Figure 5.1a is positive, because both the displacement and the unit length are in the positive l direction. For the same reason, the strain ε_t is positive in the t direction.

For the shear strain γ_{lt} in Figure 5.1b the first subscript l indicates the original unit length in the l direction and the second subscript t indicates the displacement in the t direction. Because both the unit length and the displacement are in their positive directions, the shear strain γ_{lt} indicated is positive. Similarly, γ_{tl} indicated is also positive.

To find the three strain components in various directions, we introduce a rotating d-r coordinate system as shown in Figure 5.1c. The d-r axes have been rotated counterclockwise by an angle of α with respect to the stationary l-t coordinate system. The three strain components in this rotating coordinate system are ε_d, ε_r, and γ_{dr} (or γ_{rd}). The relationship between the rotating strain components ε_d, ε_r, and γ_{dr} and the stationary strain components ε_l, ε_t, and γ_{lt} is the strain transformation.

The relationship between the rotating d-r axes and the stationary l-t axes is shown by the transformation geometry in Figure 5.1c. A positive unit length on the l axis will have projections of $\cos \alpha$ and $-\sin \alpha$ on the d and r axis, respectively. A positive unit length on the t axis, however, should give projections of $\sin \alpha$ and $\cos \alpha$. Hence, the rotation matrix $[R]$ is

$$[R] = \begin{bmatrix} \cos \alpha & \sin \alpha \\ -\sin \alpha & \cos \alpha \end{bmatrix} \tag{5-1}$$

149

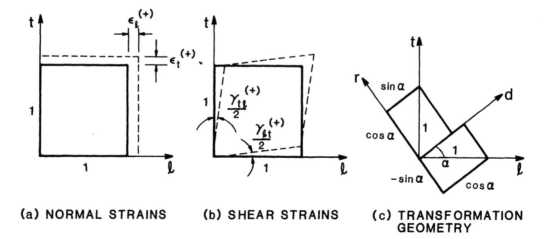

(a) NORMAL STRAINS   **(b) SHEAR STRAINS**   **(c) TRANSFORMATION GEOMETRY**

Figure 5.1   Definitions of strains and transformation geometry.

The relationship between the strains in the d-r coordinate $[\varepsilon_{dr}]$ and the strain in the l-t coordinate $[\varepsilon_{lt}]$ is

$$[\varepsilon_{dr}] = [R][\varepsilon_{lt}][R]^T \tag{5-2}$$

or

$$\begin{bmatrix} \varepsilon_d & \gamma_{dr}/2 \\ \gamma_{rd}/2 & \varepsilon_r \end{bmatrix} = \begin{bmatrix} \cos\alpha & \sin\alpha \\ -\sin\alpha & \cos\alpha \end{bmatrix} \begin{bmatrix} \varepsilon_l & \gamma_{lt}/2 \\ \gamma_{tl}/2 & \varepsilon_t \end{bmatrix} \begin{bmatrix} \cos\alpha & -\sin\alpha \\ \sin\alpha & \cos\alpha \end{bmatrix} \tag{5-3}$$

Performing the matrix multiplications and noticing that $\gamma_{tl} = \gamma_{lt}$ and $\gamma_{rd} = \gamma_{dr}$ result in the following three equations:

$$\varepsilon_d = \varepsilon_l \cos^2\alpha + \varepsilon_t \sin^2\alpha + \frac{\gamma_{lt}}{2}(2\sin\alpha\cos\alpha) \tag{5-4}$$

$$\varepsilon_r = \varepsilon_l \sin^2\alpha + \varepsilon_t \cos^2\alpha - \frac{\gamma_{lt}}{2}(2\sin\alpha\cos\alpha) \tag{5-5}$$

$$\frac{\gamma_{dr}}{2} = (-\varepsilon_l + \varepsilon_t)\sin\alpha\cos\alpha + \frac{\gamma_{lt}}{2}(\cos^2\alpha - \sin^2\alpha) \tag{5-6}$$

Eqs. (5-4) to (5-6) can be expressed in the matrix form by one equation:

$$\begin{bmatrix} \varepsilon_d \\ \varepsilon_r \\ \gamma_{dr}/2 \end{bmatrix} = \begin{bmatrix} \cos^2\alpha & \sin^2\alpha & 2\sin\alpha\cos\alpha \\ \sin^2\alpha & \cos^2\alpha & -2\sin\alpha\cos\alpha \\ -\sin\alpha\cos\alpha & \sin\alpha\cos\alpha & (\cos^2\alpha - \sin^2\alpha) \end{bmatrix} \begin{bmatrix} \varepsilon_l \\ \varepsilon_t \\ \gamma_{lt}/2 \end{bmatrix} \tag{5-7}$$

This 3×3 matrix in Eq. (5-7) is the transformation matrix for transforming the strains in the stationary l-t coordinate to the strains in the rotating d-r coordinate.

Equations (5-4) to (5-6) can be illustrated strictly by geometry in Figure 5.2. Figure 5.2a gives the geometric relationships between the three strain components ε_d, ε_r, and $\gamma_{dr}/2$ in the d-r coordinate and the two normal strains ε_l and ε_t in the l-t

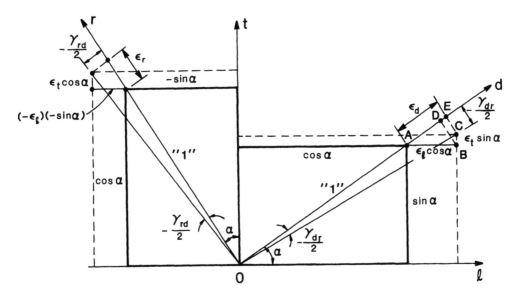

$$\epsilon_d = (\cos\alpha)(\epsilon_\ell)(\cos\alpha) + (\sin\alpha)(\epsilon_t)(\sin\alpha) = \epsilon_\ell \cos^2\alpha + \epsilon_t \sin^2\alpha$$

$$\epsilon_r = (-\sin\alpha)(-\epsilon_\ell)(\sin\alpha) + (\cos\alpha)(\epsilon_t)(\cos\alpha) = \epsilon_\ell \sin^2\alpha + \epsilon_t \cos^2\alpha$$

$$\frac{\gamma_{dr}}{2} = (\cos\alpha)(\epsilon_\ell)(-\sin\alpha) + (\sin\alpha)(\epsilon_t)(\cos\alpha) = (-\epsilon_\ell + \epsilon_t)\sin\alpha\cos\alpha$$

(a)

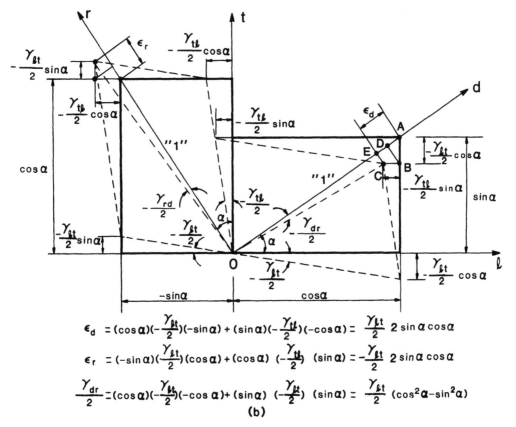

$$\epsilon_d = (\cos\alpha)(-\frac{\gamma_{\ell t}}{2})(-\sin\alpha) + (\sin\alpha)(-\frac{\gamma_{t\ell}}{2})(-\cos\alpha) = \frac{\gamma_{\ell t}}{2} 2\sin\alpha\cos\alpha$$

$$\epsilon_r = (-\sin\alpha)(\frac{\gamma_{\ell t}}{2})(\cos\alpha) + (\cos\alpha)(-\frac{\gamma_{t\ell}}{2})(\sin\alpha) = -\frac{\gamma_{\ell t}}{2} 2\sin\alpha\cos\alpha$$

$$\frac{\gamma_{dr}}{2} = (\cos\alpha)(-\frac{\gamma_{\ell t}}{2})(-\cos\alpha) + (\sin\alpha)(-\frac{\gamma_{\ell t}}{2})(\sin\alpha) = \frac{\gamma_{\ell t}}{2}(\cos^2\alpha - \sin^2\alpha)$$

(b)

Figure 5.2 (a) Transformation relationship (for ε_l and ε_t). (b) Transformation relationship (for γ_{lt}).

coordinate. Take ε_d for example: It is contributed by both ε_l and ε_t. Taking a unit diagonal length OA on the d axis, the projections of this unit length on the l and t axes are $\cos\alpha$ and $\sin\alpha$. The displacements along the l and t axes are then $AB = \varepsilon_l\cos\alpha$ and $BC = \varepsilon_t\sin\alpha$. The projections of these two displacements on the d axis are $AD = \varepsilon_l\cos^2\alpha$ and $DE = \varepsilon_t\sin^2\alpha$. The sum of these two displacement projections, AE, is the strain ε_d due to ε_l and ε_t, because it is measured from an original length OA of unity.

The geometric relationships between the three strain components ε_d, ε_r, and $\gamma_{dr}/2$ in the d-r coordinate and the shear strain $\gamma_{lt}/2$ in the l-t coordinate is shown in Figure 5.2b. Again, take ε_d for example. The projections of the diagonal unit length OA on the l and t axes are $\cos\alpha$ and $\sin\alpha$. For a shear strain $-\gamma_{lt}/2$, indicated by the dotted lines, the angular displacements AB and BC along the l and t axes are then $(-\gamma_{lt}/2)\cos\alpha$ and $(-\gamma_{lt}/2)\sin\alpha$. The projections of these two displacements AD and DE on the d axis are $(-\gamma_{lt}/2)\cos\alpha(-\sin\alpha)$ and $(-\gamma_{lt}/2)\sin\alpha(-\cos\alpha)$. The sum of these two displacement projections, $AE = (\gamma_{lt}/2)2\sin\alpha\cos\alpha$, is the strain ε_d due to the shear strain γ_{lt}, because it has an original length OA of unity.

Summing the strain ε_d due to ε_l and ε_t (Figure 5.2a) and that due to γ_{lt} (Figure 5.2b), the total ε_d is expressed by Eq. (5-4). The expressions for ε_t and $\gamma_{lt}/2$ in Eqs. (5-5) and (5-6) can similarly be demonstrated by direct geometric relationships.

5.1.2   RC SIGN CONVENTION

Equations (5-4) to (5-6) are derived based on the basic sign convention using a rotational d-r coordinate and a stationary l-t coordinate. Similar to the stress transformation discussed in Section 4.1.3, this basic sign convention has the same two weaknesses when applied to the strain transformation:

1 The shear strain γ_{lt} is always negative in the truss model for reinforced concrete. As shown in Figure 5.2b, γ_{lt} is always negative when concrete struts are in compression, i.e., shortened.

2 The clockwise direction of rotation in the Mohr strain circle is opposite to the counterclockwise direction of rotation in the actual stress field.

To overcome these two weaknesses, we reverse the sign for γ_{lt} and γ_{tl} in Eqs. (5-4) to (5-6), which gives

$$\varepsilon_d = \varepsilon_l\cos^2\alpha + \varepsilon_t\sin^2\alpha - \frac{\gamma_{lt}}{2}(2\sin\alpha\cos\alpha) \tag{5-8}$$

$$\varepsilon_r = \varepsilon_l\sin^2\alpha + \varepsilon_t\cos^2\alpha + \frac{\gamma_{lt}}{2}(2\sin\alpha\cos\alpha) \tag{5-9}$$

$$\frac{\gamma_{dr}}{2} = (\varepsilon_l - \varepsilon_t)\sin\alpha\cos\alpha + \frac{\gamma_{lt}}{2}(\cos^2\alpha - \sin^2\alpha) \tag{5-10}$$

In matrix form we have

$$\begin{bmatrix} \varepsilon_d \\ \varepsilon_r \\ \gamma_{dr}/2 \end{bmatrix} = \begin{bmatrix} \cos^2\alpha & \sin^2\alpha & -2\sin\alpha\cos\alpha \\ \sin^2\alpha & \cos^2\alpha & 2\sin\alpha\cos\alpha \\ \sin\alpha\cos\alpha & -\sin\alpha\cos\alpha & (\cos^2\alpha - \sin^2\alpha) \end{bmatrix} \begin{bmatrix} \varepsilon_l \\ \varepsilon_t \\ \gamma_{lt}/2 \end{bmatrix} \tag{5-11}$$

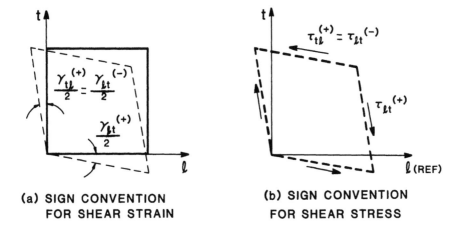

(a) SIGN CONVENTION FOR SHEAR STRAIN

(b) SIGN CONVENTION FOR SHEAR STRESS

Figure 5.3 RC sign convention for shear strains and shear stresses in the Mohr circle.

This 3×3 matrix in Eq. (5-11) is the transformation matrix based on the RC sign convention. It transforms the strains in the l-t coordinate into the strains in the d-r coordinate.

The RC sign convention for shear strain is illustrated in Figure 5.3a. This sign convention for shear strain γ_{lt} is consistent with the sign convention for shear stress τ_{lt}, as indicated in Figure 5.3b.

5.1.3   PRINCIPAL STRAINS

Recalling the double-angle trigonometric relationship of Eqs. (4-8) to (4-11),

$$\cos^2 \alpha = \tfrac{1}{2}(1 + \cos 2\alpha) \tag{5-12}$$

$$\sin^2 \alpha = \tfrac{1}{2}(1 - \cos 2\alpha) \tag{5-13}$$

$$\sin \alpha \cos \alpha = \tfrac{1}{2}\sin 2\alpha \tag{5-14}$$

$$\cos^2 \alpha - \sin^2 \alpha = \cos 2\alpha \tag{5-15}$$

Substituting these equations into Eqs. (5-8) to (5-10) results in

$$\varepsilon_d = \frac{\varepsilon_l + \varepsilon_t}{2} + \frac{\varepsilon_l - \varepsilon_t}{2}\cos 2\alpha - \frac{\gamma_{lt}}{2}\sin 2\alpha \tag{5-16}$$

$$\varepsilon_r = \frac{\varepsilon_l + \varepsilon_t}{2} - \frac{\varepsilon_l - \varepsilon_t}{2}\cos 2\alpha + \frac{\gamma_{lt}}{2}\sin 2\alpha \tag{5-17}$$

$$\frac{\gamma_{dr}}{2} = \frac{\varepsilon_l - \varepsilon_t}{2}\sin 2\alpha + \frac{\gamma_{lt}}{2}\cos 2\alpha \tag{5-18}$$

Equations (5-16) to (5-18) represent the Mohr strain circle in the ε-$(\gamma/2)$ coordinate as shown in Figure 5.4. The circle has its center on the ε axis at a distance

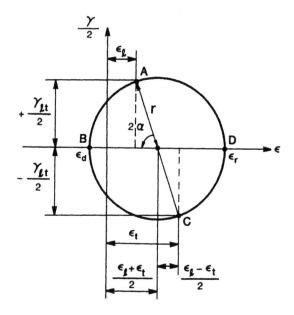

Figure 5.4   Mohr circle for strains.

$(\varepsilon_l + \varepsilon_t)/2$ from the origin. The radius of the circle is

$$r = \sqrt{\left(\frac{\varepsilon_l - \varepsilon_t}{2}\right)^2 + \left(\frac{\gamma_{lt}}{2}\right)^2} \tag{5-19}$$

The principal strain ε_d is oriented at an angle α from the reference l direction. This minimum principal strain is represented by point B in the Mohr circle, which is located at an angle of 2α from the reference point A. The angle 2α can be found by setting $\gamma_{dr} = 0$ in Eq. (5-18):

$$\cot 2\alpha = \frac{\varepsilon_l - \varepsilon_t}{-\gamma_{lt}} \tag{5-20}$$

Substituting this 2α into Eqs. (5-16) and (5-17) results in the principal strains

$$\varepsilon_d = \frac{\varepsilon_l + \varepsilon_t}{2} - \sqrt{\left(\frac{\varepsilon_l - \varepsilon_t}{2}\right)^2 + \left(\frac{\gamma_{lt}}{2}\right)^2} \tag{5-21}$$

$$\varepsilon_r = \frac{\varepsilon_l + \varepsilon_t}{2} + \sqrt{\left(\frac{\varepsilon_l - \varepsilon_t}{2}\right)^2 + \left(\frac{\gamma_{lt}}{2}\right)^2} \tag{5-22}$$

The relationships in Eqs. (5-19) to (5-22) can all be observed graphically in terms of the Mohr circle in Figure 5.4.

Point C on the Mohr circle, which is located at 180° from the reference point A, represents the normal and shear strains in the t direction. The shear strain $\gamma_{tl}/2$ in the t direction is negative in the Mohr circle. Utilizing the same remedy discussed in association with Eq. (4-23), we define $\gamma_{tl}/2 = -\gamma_{lt}/2$. This change of sign is indicated in Figures 5.3 and 5.4.

5.2 Strain Transformation in Terms of Principal Strains

5.2.1 TRANSFORMATION EQUATIONS

In Section 5.1.2 we have derived Eq. (5-11), which provides the transformation matrix for strains based on the RC sign convention. This matrix transforms a set of strains in the *l-t* coordinate (ε_l, ε_t, and γ_{lt}) into a set of strains in the *d-r* coordinate (ε_d, ε_r, and γ_{dr}). Now, let us reverse the process and find the transformation matrix that transforms the strains in the *d-r* coordinate (ε_d, ε_r, and γ_{dr}) into the strains in the *l-t* coordinate (ε_l, ε_t, and γ_{lt}).

To do so it is only necessary to change all the angles α in the matrix of Eq. (5-11) into $-\alpha$. It can be seen that all the trigonometric functions remain the same, except that $\sin(-\alpha) = -\sin\alpha$. By changing the sign of the four terms with $\sin\alpha$, we have

$$\begin{bmatrix} \varepsilon_l \\ \varepsilon_t \\ \gamma_{lt}/2 \end{bmatrix} = \begin{bmatrix} \cos^2\alpha & \sin^2\alpha & 2\sin\alpha\cos\alpha \\ \sin^2\alpha & \cos^2\alpha & -2\sin\alpha\cos\alpha \\ -\sin\alpha\cos\alpha & \sin\alpha\cos\alpha & (\cos^2\alpha - \sin^2\alpha) \end{bmatrix} \begin{bmatrix} \varepsilon_d \\ \varepsilon_r \\ \gamma_{dr}/2 \end{bmatrix} \quad (5\text{-}23)$$

This 3×3 matrix in Eq. (5-23) is the transformation matrix that transforms the strains in the *d-r* coordinate into the strains in the *l-t* coordinate. This transformation is based on the RC sign convention.

If we define the *d-r* axes as the principal axes, then γ_{dr} must vanish. Inserting $\gamma_{dr} = 0$ into Eq. (5-23) results in three equations as follows:

$$\varepsilon_l = \varepsilon_d \cos^2\alpha + \varepsilon_r \sin^2\alpha \quad (5\text{-}24)$$

$$\varepsilon_t = \varepsilon_d \sin^2\alpha + \varepsilon_r \cos^2\alpha \quad (5\text{-}25)$$

$$\frac{\gamma_{lt}}{2} = (-\varepsilon_d + \varepsilon_r)\sin\alpha\cos\alpha \quad (5\text{-}26)$$

It can be seen that these strain transformation equations are considerably simplified when the three strains in the *l-t* coordinate are expressed in terms of the principal strains ε_d and ε_r.

Equations (5-24) to (5-26) can be illustrated strictly by geometry in Figure 5.5. Figure 5.5 gives the geometric relationships between the three strain components ε_l, ε_t, and γ_{lt} in the *l-t* coordinate and the two normal strains ε_d and ε_r in the *d-r* coordinate. Take ε_l for example: It is contributed by both ε_d and ε_r. Taking a unit length *OA* on the *l* axis, the projections of this unit length on the *d* and *r* axes are $\cos\alpha$ and $-\sin\alpha$. The displacements along the *d* and *r* axes, *AB* and *BC*, are then $\varepsilon_d \cos\alpha$ and $(-\varepsilon_r)(-\sin\alpha)$. The projections of these two displacements on the *l* axis, *AD* and *DE*, are $\varepsilon_d \cos^2\alpha$ and $\varepsilon_r \sin^2\alpha$. The sum of these two displacement projections, *AE*, is the strain ε_l due to ε_d and ε_r, because it is measured from an original length *OA* of unity. This geometric relationship is expressed by Eq. (5-24). Similar geometric relationships can be demonstrated for Eqs. (5-25) and (5-26).

Equations (5-24) to (5-26) can also be looked at from a more general point of view. Because they represent the compatibility condition, they are also the compatibility equations. These three equations contain six variables, namely, ε_l, ε_t, γ_{lt}, ε_d, ε_r, and

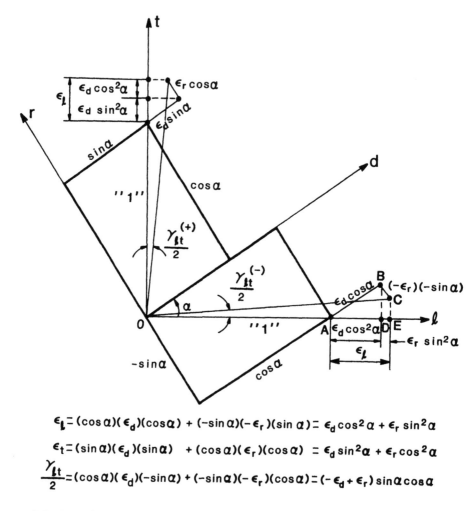

$$\epsilon_l = (\cos\alpha)(\epsilon_d)(\cos\alpha) + (-\sin\alpha)(-\epsilon_r)(\sin\alpha) = \epsilon_d\cos^2\alpha + \epsilon_r\sin^2\alpha$$

$$\epsilon_t = (\sin\alpha)(\epsilon_d)(\sin\alpha) + (\cos\alpha)(\epsilon_r)(\cos\alpha) = \epsilon_d\sin^2\alpha + \epsilon_r\cos^2\alpha$$

$$\frac{\gamma_{lt}}{2} = (\cos\alpha)(\epsilon_d)(-\sin\alpha) + (-\sin\alpha)(-\epsilon_r)(\cos\alpha) = (-\epsilon_d + \epsilon_r)\sin\alpha\cos\alpha$$

Figure 5.5 Transformation relationship in terms of principal strains (γ_{lt} is based on RC sign convention).

α. When any three of the six variables are given, the other three can be solved by these three compatibility equations.

It is interesting to note that the four normal strains ε_l, ε_t, ε_d, and ε_r have a simple relationship. Adding Eqs. (5-24) and (5-25) gives

$$\varepsilon_l + \varepsilon_t = \varepsilon_d + \varepsilon_r \tag{5-27}$$

This means that any normal stress can be determined when the other three are given. For example,

$$\varepsilon_r = \varepsilon_l + \varepsilon_t - \varepsilon_d \tag{5-28}$$

If we define the diameter of the Mohr strain circle as γ_m (Figure 5.6), then

$$\gamma_m = \varepsilon_r - \varepsilon_d = \varepsilon_l + \varepsilon_t - 2\varepsilon_d \tag{5-29}$$

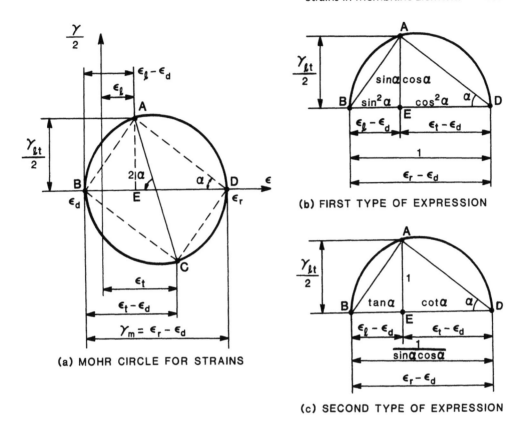

(a) MOHR CIRCLE FOR STRAINS

(b) FIRST TYPE OF EXPRESSION

(c) SECOND TYPE OF EXPRESSION

Figure 5.6   Geometric relationships in Mohr strain circle.

5.2.2   FIRST TYPE OF COMPATIBILITY EQUATIONS

In the case of a reinforced concrete element, ε_d is smaller by an order of magnitude when compared to the other four strains. Naturally, our interest will be focused on the relationship among the four strains ε_l, ε_t, ε_r, and γ_{lt}, while considering ε_d as a small secondary strain. With this aim in mind, the three transformation equations [Eqs. (5-24) to (5-26)] can be changed into three explicit relationships among the four strains ε_l, ε_t, ε_r, and γ_{lt}.

Inserting $\varepsilon_d \cos^2 \alpha = \varepsilon_d - \varepsilon_d \sin^2 \alpha$ into Eq. (5-24) gives

$$(\varepsilon_l - \varepsilon_d) = (\varepsilon_r - \varepsilon_d)\sin^2 \alpha \tag{5-30}$$

Similarly, inserting $\varepsilon_d \sin^2 \alpha = \varepsilon_d - \varepsilon_d \cos^2 \alpha$ into Eq. (5-25) gives

$$(\varepsilon_t - \varepsilon_d) = (\varepsilon_r - \varepsilon_d)\cos^2 \alpha \tag{5-31}$$

Equation (5-26) remains the same:

$$\frac{\gamma_{lt}}{2} = (\varepsilon_r - \varepsilon_d)\sin \alpha \cos \alpha \tag{5-32}$$

Equations (5-30) to (5-32) are the first type of expression for the compatibility condition. In this type of expression, the three strains ε_l, ε_t, and γ_{lt} are each related individually to the principal tensile strain ε_r. These three equations are convenient for the *analysis* of membrane elements.

Equations (5-30) to (5-32) represent the first type of geometric relationship in the Mohr circle as shown in Figure 5.6a and b. The geometric relationships of the half Mohr circle defined by *ABD* are illustrated in Figure 5.6b. If *BD* in Figure 5.6b is taken as unity, then $BE = \sin^2 \alpha$, $ED = \cos^2 \alpha$, and $AE = \sin \alpha \cos \alpha$. These three trigonometric values are actually the ratios of the three strains $(\varepsilon_l - \varepsilon_d)$, $(\varepsilon_t - \varepsilon_d)$, and $\gamma_{lt}/2$, respectively, divided by the absolute-value sum of the principal strains $(\varepsilon_r - \varepsilon_d)$.

Dividing Eq. (5-30) by Eq. (5-31) we have

$$\tan^2 \alpha = \frac{\varepsilon_l - \varepsilon_d}{\varepsilon_t - \varepsilon_d} \tag{5-33}$$

Equation (5-33) is the compatibility equation to determine the angle α. Its derivation heralded the arrival of the Mohr compatibility truss model. The equilibrium (plasticity) truss model in Chapter 3 is based on an α angle determined by Eq. (3-22) [or its generalized versions of Eq. (4-66) in Chapter 4], whereas the Mohr compatibility truss model to be introduced in Chapter 6 and the softened truss model to be introduced in Chapters 7 and 8 are based on an angle α determined by Eq. (5-33).

Multiplying Eqs. (5-30) and (5-31) gives

$$(\varepsilon_l - \varepsilon_d)(\varepsilon_t - \varepsilon_d) = (\varepsilon_r - \varepsilon_d)^2 \sin^2 \alpha \cos^2 \alpha \tag{5-34}$$

Squaring Eq. (5-32) gives

$$\left(\frac{\gamma_{lt}}{2}\right)^2 = (\varepsilon_r - \varepsilon_d)^2 \sin^2 \alpha \cos^2 \alpha \tag{5-35}$$

Equating Eqs. (5-34) and (5-35) results in

$$\frac{\gamma_{lt}}{2} = \pm \sqrt{(\varepsilon_l - \varepsilon_d)(\varepsilon_t - \varepsilon_d)} \tag{5-36}$$

Equation (5-36) shows that the shear strain can be calculated if the three normal strains are given.

5.2.3   SECOND TYPE OF COMPATIBILITY EQUATIONS

The three compatibility equations [Eqs. (5-30) to (5-32)] can also be expressed in another form. The three strains ε_l, ε_t, and ε_r can each be related individually to the shear strain γ_{lt}. Substituting $(\varepsilon_r - \varepsilon_d)$ from Eq. (5-32) into Eqs. (5-30) and (5-31) gives

$$(\varepsilon_l - \varepsilon_d) = \frac{\gamma_{lt}}{2} \tan \alpha \tag{5-37}$$

$$(\varepsilon_t - \varepsilon_d) = \frac{\gamma_{lt}}{2} \cot \alpha \tag{5-38}$$

$$(\varepsilon_r - \varepsilon_d) = \frac{\gamma_{lt}}{2} \frac{1}{\sin \alpha \cos \alpha} \tag{5-39}$$

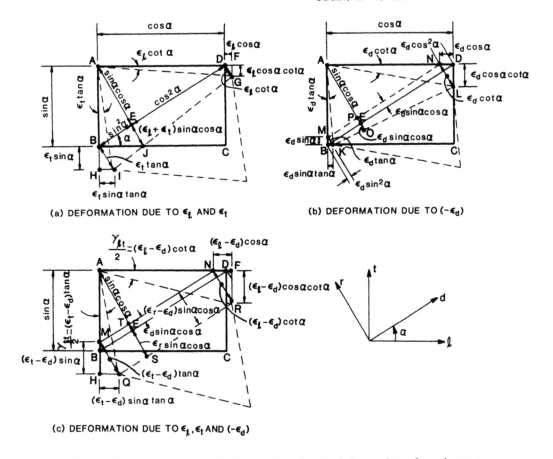

Figure 5.7 Strain compatibility relationship in deformation of an element.

Equations (5-37) to (5-39) are the second type of expression for the compatibility condition. They represent the second type of geometric relationship in the Mohr circle as shown in Figure 5.6c. When AE is taken as unity, then $BE = \tan\alpha$, $ED = \cot\alpha$, and $BD = 1/\sin\alpha\cos\alpha$. These three trigonometric values are actually the ratios of the three strains $(\varepsilon_l - \varepsilon_d)$, $(\varepsilon_t - \varepsilon_d)$, and $(\varepsilon_r - \varepsilon_d)$, respectively, divided by the shear strain $\gamma_{lt}/2$.

Dividing Eq. (5-37) by (5-38) gives Eq. (5-33), and multiplying Eq. (5-37) by Eq. (5-38) produces Eq. (5-36).

The relationships of $(\varepsilon_l - \varepsilon_d)$, $(\varepsilon_t - \varepsilon_d)$, $(\varepsilon_r - \varepsilon_d)$, and $\gamma_{lt}/2$ described by Eqs. (5-37) to (5-39) can also be illustrated directly by the strain compatibility relationships in the deformation of an element as shown in Figure 5.7. This rectangular element in the l-t coordinate has a diagonal in the d direction that is inclined at an angle α to the reference l axis. Because the length of the diagonal BD is taken as unity, the lengths AB, AD, BE, ED, and AE are $\sin\alpha$, $\cos\alpha$, $\sin^2\alpha$, $\cos^2\alpha$, and $\sin\alpha\cos\alpha$, respectively.

Figure 5.7a shows the deformations due to the normal tensile strains ε_l and ε_t. The strain ε_l causes a longitudinal displacement of DF equal to $\varepsilon_l\cos\alpha$. To maintain compatibility between the longitudinal length AF and the diagonal length BD, both lengths must rotate until they meet at point G. The two displacements due to the rotation, FG and DG, must be perpendicular to their respective lengths and should be equal to $\varepsilon_l\cos\alpha\cot\alpha$ and $\varepsilon_l\cot\alpha$, respectively. The shear strain due to the rotation of length AF is $(FG)/(AD)$ and is also equal to $\varepsilon_l \cot \alpha$. Similarly, the strain ε_t causes the point B to move first to point H and then to point I due to compatibility. The resulting displacement BI and the shear strain are both equal to $\varepsilon_t \tan \alpha$. From geometry of the trapezoid $BDGI$, it is easy to find that the length EJ is equal to $(\varepsilon_l + \varepsilon_t)\sin \alpha \cos \alpha$.

The deformation due to the compression strain $(-\varepsilon_d)$ is shown in Figure 5.7b. The negative signs for ε_d have been omitted in this particular figure. The displacements at both ends of the diagonal D and B are $\varepsilon_d \cos^2 \alpha$ and $\varepsilon_d \sin^2 \alpha$, respectively. To maintain compatibility, the diagonal BD must move to the position KL. Drawing two lines LN and KM perpendicular to the diagonal BD we obtain two triangles DLN and BKM. From the geometry of the two triangles, it is easy to see that the two lengths LN and KM are $\varepsilon_d \cot \alpha$ and $\varepsilon_d \tan \alpha$, respectively. Each of these two lengths has the same magnitude as its shear strain created. From geometry of the trapezoid $LNMK$ we can see that both EO and EP are equal to $\varepsilon_d \sin \alpha \cos \alpha$.

The combined deformations due to ε_l, ε_t, and $(-\varepsilon_d)$ are shown in Figure 5.7c. It can be seen that the diagonal BD has moved to QR in a parallel manner. The two lengths RN and QM are perpendicular to the diagonal BD and are equal to $(\varepsilon_l - \varepsilon_d)\cot \alpha$ and $(\varepsilon_t - \varepsilon_d)\tan \alpha$, respectively. Each of these two lengths also produce respective shear strain of equal magnitude. Because both the shear strains are $\gamma_{lt}/2$, we arrive at the two relationships described by Eqs. (5-37) and (5-38). In addition, the length ES is $\varepsilon_r \sin \alpha \cos \alpha$, because $\varepsilon_l + \varepsilon_t - \varepsilon_d = \varepsilon_r$. The length ST is then $(\varepsilon_r - \varepsilon_d)\sin \alpha \cos \alpha$. Because QR is parallel to MN, the length ST should also be equal to the shear strain $\gamma_{lt}/2$ and we obtain the relationship described by Eq. (5-39).

5.3 Design Based on Crack Control

5.3.1 STEEL STRAINS IN CRACK CONTROL

In the set of three compatibility equations, such as Eqs. (5-30) to (5-32) of the first type, we notice that they involve six variables, ε_l, ε_t, γ_{lt}, ε_r, ε_d, and α. Because γ_{lt} is involved only in the third equation, we can treat Eq. (5-32) as an independent equation for γ_{lt}. The remaining two equations [Eqs. (5-30) and (5-31)] involve only the four normal strains, ε_l, ε_t, ε_r, and ε_d plus the angle α.

Of the four normal strains, we should notice that the principal compressive strain ε_d is small by an order of magnitude when compared to the other three normal strains ε_l, ε_t, and ε_r. Also, the principal tensile strain ε_r is oriented in the cracking direction and can be taken as an indicator of the crack widths. In design, we can control the crack widths by specifying an allowable principal tensile strain ε_r.

A method of design has been developed on the basis of crack control. In this method, the two principal strains ε_r and ε_d are given, and the problem is to find the steel strains ε_l and ε_t as a function of the angle α. To do this, we nondimensionalize Eqs. (5-30) and (5-31) by dividing these two equations by the yield strain of steel, ε_y:

$$\frac{\varepsilon_l}{\varepsilon_y} = \frac{\varepsilon_d}{\varepsilon_y} + \left(\frac{\varepsilon_r}{\varepsilon_y} - \frac{\varepsilon_d}{\varepsilon_y}\right)\sin^2 \alpha \tag{5-40}$$

$$\frac{\varepsilon_t}{\varepsilon_y} = \frac{\varepsilon_d}{\varepsilon_y} + \left(\frac{\varepsilon_r}{\varepsilon_y} - \frac{\varepsilon_d}{\varepsilon_y}\right)\cos^2 \alpha \tag{5-41}$$

Equations (5-40) and (5-41) are plotted in Figure 5.8a, b, and c for $\varepsilon_d/\varepsilon_y = 0, -0.25$, and -0.5, respectively. In each figure, the two steel strain ratios $\varepsilon_l/\varepsilon_y$ and $\varepsilon_t/\varepsilon_y$ are plotted as a function of the angle α and using the crack strain ratio $\varepsilon_r/\varepsilon_y$ as a parameter. For a 60-ksi mild-steel bar, the yield strain $\varepsilon_y = 0.00207$.

The use of Figure 5.8 can be demonstrated by an example problem: Suppose $\varepsilon_d/\varepsilon_y = 0$ (i.e., very small) and the allowable crack strain ratio $\varepsilon_r/\varepsilon_y = 1.5$. Find the steel strain ratios $\varepsilon_l/\varepsilon_y$ and $\varepsilon_t/\varepsilon_y$ for an α angle of 30°. The solution can be easily found by examining Figure 5.8a, which is plotted for $\varepsilon_d/\varepsilon_y = 0$. The intersection of the vertical line at $\alpha = 30°$ and the $\varepsilon_l/\varepsilon_y$ curves for $\varepsilon_r/\varepsilon_y = 1.5$ gives $\varepsilon_l = 0.375\varepsilon_y$. The intersection of the same vertical line and the $\varepsilon_t/\varepsilon_y$ curve for $\varepsilon_r/\varepsilon_y = 1.5$ gives $\varepsilon_t = 1.125\varepsilon_y$.

It can be seen from Figure 5.8 that $\alpha = 45°$ gives the lowest strains for ε_l and ε_t. From a serviceability point of view, $\alpha = 45°$ provides the best design. For $\alpha = 45°$, we have

$$\varepsilon_l = \varepsilon_t = \tfrac{1}{2}(\varepsilon_r + \varepsilon_d) \tag{5-42}$$

For $\varepsilon_r = 1.5\varepsilon_y$ and $\varepsilon_d = 0$, both Eq. (5-42) and Figure 5.8a give the steel strains $\varepsilon_l = \varepsilon_t = 0.75\varepsilon_y$.

5.3.2   DESIGN OF REINFORCEMENT

For practical design of reinforcement in membrane elements, it is sufficiently accurate to assume $\varepsilon_d = 0$. Equations (5-40) and (5-41) can then be simplified to

$$\varepsilon_l = \varepsilon_y \left(\frac{\varepsilon_r}{\varepsilon_y} \sin^2 \alpha \right) \tag{5-43}$$

$$\varepsilon_t = \varepsilon_y \left(\frac{\varepsilon_r}{\varepsilon_y} \cos^2 \alpha \right) \tag{5-44}$$

In the design for crack control the steel should be working in the elastic range. Applying Hooke's law $f_l = E_s \varepsilon_l$ and $f_t = E_s \varepsilon_t$ to Eqs. (5-43) and (5-44), we have

$$f_l = f_y \left(\frac{\varepsilon_r}{\varepsilon_y} \sin^2 \alpha \right) \tag{5-45}$$

$$f_t = f_y \left(\frac{\varepsilon_r}{\varepsilon_y} \cos^2 \alpha \right) \tag{5-46}$$

The three equilibrium equations [Eqs. (4-79) to (4-81)] of the second type should be applicable to the design of reinforcement for membrane elements. Substituting f_l and f_t from Eqs. (5-45) and (5-46) into Eqs. (4-79) and (4-80) results in

$$\rho_l f_y = \frac{\sigma_l + \tau_{lt} \cot \alpha}{(\varepsilon_r/\varepsilon_y) \sin^2 \alpha} \tag{5-47}$$

$$\rho_t f_y = \frac{\sigma_t + \tau_{lt} \tan \alpha}{(\varepsilon_r/\varepsilon_y) \cos^2 \alpha} \tag{5-48}$$

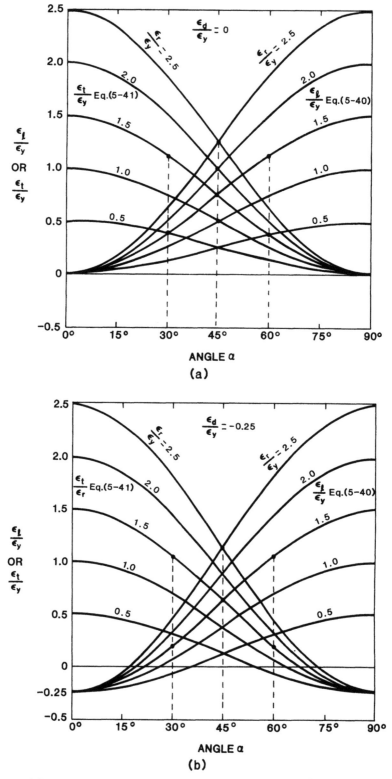

Figure 5.8 (a) Steel strains ε_l and ε_t as a function of cracking strain ε_r and angle $\alpha(\varepsilon_d/\varepsilon_y = 0)$. (b) Steel strains ε_l and ε_t as a function of cracking strain ε_r and angle $\alpha(\varepsilon_d/\varepsilon_y = -0.25)$. (c) Steel strains ε_l and ε_t as a function of cracking strain ε_r and angle $\alpha(\varepsilon_d/\varepsilon_y = -0.5)$.

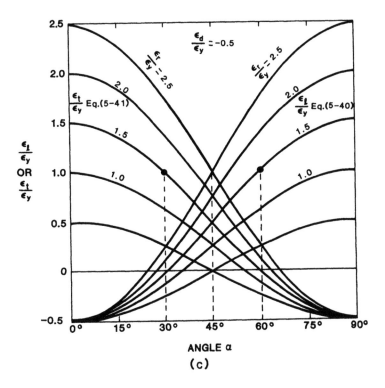

Figure 5.8 (*Continued*)

Equations (5-47) and (5-48) can be used to design the steel reinforcement in the l and t directions of a membrane element. In these equations, the stresses σ_l, σ_t, and τ_{lt} are obtained from the service loads. The amount of reinforcement required is inversely proportional to the allowable cracking strain ε_r. In other words, the smaller the specified ε_r, the larger the steel requirement.

There are two ways to select the angle α in Eqs. (5-47) and (5-48). The first and economical way is to find the angle α that will minimize the total steel requirement. This can be done by plotting a curve $\rho_l f_y + \rho_t f_y$ vs. α angle, and then graphically locating the α angle that corresponds to the minimum $\rho_l f_y + \rho_t f_y$.

The second and simple way is to use $\alpha = 45°$. This method should provide equal strains in the longitudinal and transverse steel. It should be noted, however, that the incorporation of $\sin^2 \alpha$ and $\cos^2 \alpha$ in the denominators of Eqs. (5-47) and (5-48) makes the amount of steel $\rho_l f_y$ and $\rho_t f_y$ very sensitive to the angle α. Therefore, significant benefit can be obtained by using the first way.

The method of design based on crack control was pioneered by Gupta (1981, 1984). This method satisfies the compatibility equations [Eqs. (5-30) to (5-32)] as well as the equilibrium equations [Eqs. (4-79) to (4-81)]. Unfortunately, the diagonal concrete stress σ_d in the equilibrium equations and the diagonal concrete strain ε_d in the compatibility equations do not have a realistic relationship based on the constitutive laws of concrete. In other words, the stress–strain relationship of concrete has not been satisfied. This design solution, therefore, is only an approximate solution, albeit a very good one. A rigorous solution that satisfies the equilibrium and compatibility conditions as well as the constitutive laws of materials will be given by the Mohr compatibility truss model in Chapter 6.

5.4   Cracking Condition at Yielding of Steel

The strain compatibility equations can also be used for the analysis of cracking condition at the yielding of steel. Two cases should be investigated, namely, the yielding of the longitudinal steel and the yielding of the transverse steel.

5.4.1   YIELDING OF LONGITUDINAL STEEL

As shown in Section 5.2.1, the first two compatibility equations [Eqs. (5-24) and (5-25)] involve four normal strains, ε_l, ε_t, ε_r, and ε_d, plus the angle α. In this analysis, the longitudinal steel strain is given, $\varepsilon_l = \varepsilon_y$, and the principal compressive strain ε_d is considered a small given value. Our aim is to express the transverse steel strain ε_t and the cracking strain ε_r as a function of ε_l, ε_d, and the angle α.

The transverse steel strain ε_t is related to the longitudinal steel strain ε_l by Eq. (5-33). Rearranging Eq. (5-33) to express ε_t gives

$$\varepsilon_t = \varepsilon_d + (\varepsilon_l - \varepsilon_d)\cot^2\alpha \tag{5-49}$$

Substituting ε_t from Eq. (5-49) into Eq. (5-28) provides the equation for ε_r:

$$\varepsilon_r = \varepsilon_l + (\varepsilon_l - \varepsilon_d)\cot^2\alpha \tag{5-50}$$

Dividing Eqs. (5-49) and (5-50) by ε_y and setting $\varepsilon_l = \varepsilon_y$ we obtain the nondimensionalized equations for ε_t and ε_r as

$$\frac{\varepsilon_t}{\varepsilon_y} = \frac{\varepsilon_d}{\varepsilon_y} + \left(1 - \frac{\varepsilon_d}{\varepsilon_y}\right)\cot^2\alpha \tag{5-51}$$

$$\frac{\varepsilon_r}{\varepsilon_y} = 1 + \left(1 - \frac{\varepsilon_d}{\varepsilon_y}\right)\cot^2\alpha \tag{5-52}$$

The cracking strain ratio $\varepsilon_r/\varepsilon_y$ and the transverse steel strain ratio $\varepsilon_t/\varepsilon_y$ are plotted in Figure 5.9 as a function of the angle α according to Eqs. (5-52) and (5-51), respectively. For each equation a range of $\varepsilon_d/\varepsilon_y$ ratios from 0 to -0.25 is given. It can be seen that the effect of the $\varepsilon_d/\varepsilon_y$ ratio is small. When $\alpha = 45°$, Figure 5.9 gives $\varepsilon_r = 2\varepsilon_y$ to $2.25\varepsilon_y$ and $\varepsilon_t = \varepsilon_y$. When α is reduced to $30°$, ε_r increases rapidly to the range of $4\varepsilon_y$ to $4.75\varepsilon_y$ and ε_t to the range of $3\varepsilon_y$ to $3.5\varepsilon_y$. These strains increase even faster when α is further reduced.

5.4.2   YIELDING OF TRANSVERSE STEEL

In this case of analysis, the transverse steel strain is given, $\varepsilon_t = \varepsilon_y$, and ε_d is considered a small given value. Our purpose is to express the longitudinal steel strain ε_l and the cracking strain ε_r as a function of ε_t, ε_d, and the angle α.

The longitudinal steel strain is related to the transverse steel strain ε_t by Eq. (5-33). Rearranging Eq. (5-33) to express ε_l gives

$$\varepsilon_l = \varepsilon_d + (\varepsilon_t - \varepsilon_d)\tan^2\alpha \tag{5-53}$$

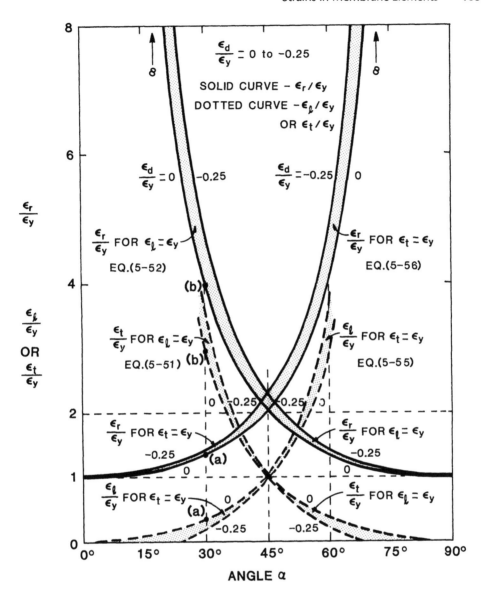

Figure 5.9   Cracking condition at yielding of steel.

Substituting ε_l from Eq. (5-53) into Eq. (5-28) provides the equation for ε_r:

$$\varepsilon_r = \varepsilon_t + (\varepsilon_t - \varepsilon_d)\tan^2\alpha \tag{5-54}$$

Dividing Eqs. (5-53) and (5-54) by ε_y and setting $\varepsilon_t = \varepsilon_y$ we obtain the nondimensionalized equations for ε_l and ε_r:

$$\frac{\varepsilon_l}{\varepsilon_y} = \frac{\varepsilon_d}{\varepsilon_y} + \left(1 - \frac{\varepsilon_d}{\varepsilon_y}\right)\tan^2\alpha \tag{5-55}$$

$$\frac{\varepsilon_r}{\varepsilon_y} = 1 + \left(1 - \frac{\varepsilon_d}{\varepsilon_y}\right)\tan^2\alpha \tag{5-56}$$

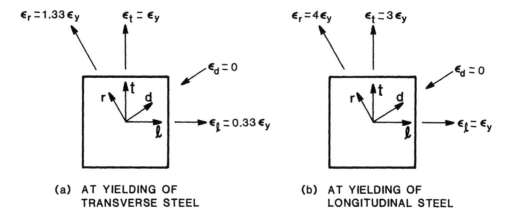

(a) AT YIELDING OF
TRANSVERSE STEEL

(b) AT YIELDING OF
LONGITUDINAL STEEL

Figure 5.10 Strain condition when $\alpha = 30°$ and $\varepsilon_d = 0$.

The cracking strain ratio $\varepsilon_r/\varepsilon_y$ and the longitudinal steel strain ratio $\varepsilon_l/\varepsilon_y$ are also plotted in Figure 5.9 as a function of the angle α according to Eqs. (5-56) and (5-55), respectively. For each of these two equations a range of $\varepsilon_d/\varepsilon_y$ ratios from 0 to -0.25 is given. Again, the effect of the $\varepsilon_d/\varepsilon_y$ ratio is shown to be small. When $\alpha = 45°$, Figure 5.9 gives $\varepsilon_r = 2\varepsilon_y$ to $2.25\varepsilon_y$ and $\varepsilon_l = \varepsilon_y$. When α is increased to $60°$, ε_r increases rapidly to the range of $4\varepsilon_y$ to $4.75\varepsilon_y$ and ε_l to the range of $3\varepsilon_y$ to $3.5\varepsilon_y$. These strains increase even faster when α is further increased.

A strain condition is shown in Figure 5.10 when $\alpha = 30°$ and $\varepsilon_d = 0$. At the first yielding of the transverse steel (Figure 5.10a), the cracking strain ε_r is $1.33\varepsilon_y$ and the longitudinal steel strain ε_l is $0.33\varepsilon_y$. These values can be obtained from the two points indicated by (a) in Figure 5.9. When straining increases to the stage of yielding in the longitudinal steel (Figure 5.10b), the cracking strain ε_r is $4\varepsilon_y$ and the transverse steel strain ε_t is $3\varepsilon_y$. These values can be obtained from the two points indicated by (b) in Figure 5.9.

The two solid curves for $\varepsilon_r/\varepsilon_y$ in the special case of $\varepsilon_d/\varepsilon_y = 0$ (Figure 5.9) was first presented by Thurlimann (1979). The trend of these two curves shows that the cracking strain ratio $\varepsilon_r/\varepsilon_y$ increases very rapidly after the first yield of steel when the angle α moves away from $45°$. It was, therefore, concluded that the CEB code limitation of $30.9° < \alpha < 59.1°$ is quite valid from the viewpoint of crack control.

6

Mohr Compatibility Truss Model

6.1   Membrane Elements

6.1.1   MOHR'S EQUILIBRIUM AND COMPATIBILITY CONDITIONS

Bernoulli compatibility truss model has been applied to flexural members in Chapter 2. This model rigorously satisfies the parallel stress equilibrium condition, the Bernoulli linear compatibility condition, and the uniaxial constitutive laws. In this chapter we will be dealing with two-dimensional membrane elements. A rigorous analysis of such elements should satisfy the two-dimensional equilibrium condition, Mohr's circular compatibility condition, and biaxial constitutive laws. If a uniaxial constitutive law, however, is adopted and is assumed to be linear (i.e., Hooke's law is satisfied), the theory is significantly simplified. This simplified theory is called the *Mohr compatibility truss model* and will be studied in this chapter. This model is applicable up to the service load stage, and could even be used up to the load stage when the steel begins to yield. If the constitutive laws are based on the biaxial, nonlinear, softened stress–strain relationships, then the theory will be more realistic, but more complex. This accurate theory is called the *softened truss model* and will be presented in Chapters 7 and 8. This later model should be applicable to the entire loading history including both the service load stage and the ultimate load stage.

In Chapter 4 we derived the three basic equilibrium equations for reinforced concrete membrane elements. These three equilibrium equations can be expressed graphically by Mohr's stress circles (Figures 4.11 and 4.12). They can also be expressed algebraically in three forms, namely, the transformation type (Section 4.3.1), the first type (Section 4.3.2), and the second type (Section 4.3.3). The basic transformation type of equations is given here:

$$\sigma_l = \sigma_d \cos^2 \alpha + \sigma_r \sin^2 \alpha + \rho_l f_l \qquad (6\text{-}1)$$

$$\sigma_t = \sigma_d \sin^2 \alpha + \sigma_r \cos^2 \alpha + \rho_t f_t \qquad (6\text{-}2)$$

$$\tau_{lt} = (-\sigma_d + \sigma_r)\sin \alpha \cos \alpha \qquad (6\text{-}3)$$

167

Assuming the yielding of both the longitudinal and the transverse steel ($f_l = f_y$, $f_t = f_y$), a design method based on the plasticity truss model was developed for membrane elements in Section 4.4. This method, of course, could not predict the actual deformation of the elements and the yielding of steel. In order to find the longitudinal and transverse steel strains at failure, the strain compatibility condition must be investigated.

In Chapter 5 the three basic compatibility equations for membrane elements were derived. These three compatibility equations can be expressed graphically by Mohr's strain circles (Figures 5.4 and 5.6). They can also be expressed algebraically by the same three types of equations used for stresses, namely, the transformation type (Section 5.2.1), the first type (Section 5.2.2), and the second type (Section 5.2.3). The basic transformation type of equations is given here:

$$\varepsilon_l = \varepsilon_d \cos^2 \alpha + \varepsilon_r \sin^2 \alpha \tag{6-4}$$

$$\varepsilon_t = \varepsilon_d \sin^2 \alpha + \varepsilon_r \cos^2 \alpha \tag{6-5}$$

$$\frac{\gamma_{lt}}{2} = (-\varepsilon_d + \varepsilon_r)\sin \alpha \cos \alpha \tag{6-6}$$

Utilizing the compatibility equations a design method for membrane elements had been developed based on crack control, Section 5.3. This design method satisfies both the compatibility condition and the equilibrium condition, but could not satisfy all the stress–strain relationships of materials.

6.1.2   HOOKE'S CONSTITUTIVE LAW

The Mohr compatibility truss model, which will now be introduced, should satisfy all the equilibrium and compatibility conditions, as well as Hooke's constitutive laws. To complete this rigorous method, we first introduce Hooke's laws for steel and concrete. Hooke's linear stress–strain relationship of steel is given in one-dimensional form as

$$\varepsilon_l = \frac{f_l}{E_s} \tag{6-7}$$

$$\varepsilon_t = \frac{f_t}{E_s} \tag{6-8}$$

The stress–strain relationship of concrete will first be expressed in a general two-dimensional form:

$$\varepsilon_d = f_d(\sigma_d, \sigma_r) \tag{6-9}$$

$$\varepsilon_r = f_r(\sigma_r, \sigma_d) \tag{6-10}$$

where f_d and f_r are functions to be specified. Using Hooke's law these stress–strain relationships of concrete can be categorized into three cases, which will be discussed in the following Sections 6.1.2.1 to 6.1.2.3.

6.1.2.1   *Simple Truss Model*

Similar to flexural members, the tensile stress of concrete is neglected in the analysis of cracked membrane elements, i.e.,

$$\sigma_r = 0 \qquad (6\text{-}11)$$

Also, ε_d is taken only as a linear function of σ_d:

$$\varepsilon_d = \frac{\sigma_d}{E_c} \qquad (6\text{-}12)$$

Because the tensile stress–strain relationship of concrete becomes irrelevant, we have 9 equations [Eqs. (6-1) to (6-8) and (6-12)] involving 14 variables. These variables include six stresses (σ_l, σ_t, τ_{lt}, σ_d, f_l, and f_t), five strains (ε_l, ε_t, γ_{lt}, ε_r, and ε_d), two cross-sectional properties (ρ_l and ρ_t), and one geometric parameter, α. Therefore, five variables must be given before the remaining nine variables can be solved by the nine equations.

In practice, the three stresses σ_l, σ_t, and τ_{lt} are obtained from the global analysis of a structure, and are considered given values in the design and analysis of a membrane element. Therefore, two additional variables must be given before the problem can be solved. Depending on the two additional given variables, the problem is divided into two types:

Types of Problem	Given Variables					Unknown Variables								
(a) Analysis	σ_l	σ_t	τ_{lt}	ρ_l	ρ_t	f_l	f_t	σ_d	ε_l	ε_t	γ_{lt}	ε_r	ε_d	α
(b) Design	σ_l	σ_t	τ_{lt}	f_l	f_t	ρ_l	ρ_t	σ_d	ε_l	ε_t	γ_{lt}	ε_r	ε_d	α

The problems in the analysis and design of membrane elements boil down to finding the most efficient way to solve the nine equations. To demonstrate the methodology of the solution process, the nine equations will first be applied to the analysis problem in Section 6.1.3. This will be followed by the application of the nine equations to the design problem in Section 6.1.4.

6.1.2.2   *Modified Truss Model*

The simple truss model may be improved if the tensile strength of concrete is considered. Because the tensile strength of concrete is only a small fraction of the compressive strength after cracking, it can be considered a small constant in the range of 100 to 200 psi for normal strength concrete:

$$\sigma_r = \text{small constant} \qquad (6\text{-}13)$$

The stress–strain relationship of concrete in compression remains linear:

$$\varepsilon_d = \frac{\sigma_d}{E_c} \qquad (6\text{-}14)$$

The solution procedures are identical to those for simple truss model, except that the value of σ_r must be incorporated. The reader should be able to make this modification, if desired, after studying the simple truss model in Sections 6.1.3 and 6.1.4.

6.1.2.3 *Uncracked Elastic Elements*

In the uncracked stage, the stress–strain relationship of concrete should obey Hooke's law in two-dimensional form, i.e.,

$$\varepsilon_d = \frac{1}{E_c}(\sigma_d - \mu\sigma_r) \tag{6-15}$$

$$\varepsilon_r = \frac{1}{E_c}(\sigma_r - \mu\sigma_d) \tag{6-16}$$

where μ is Poisson's ratio.

The behavior of uncracked elements will be treated in Section 6.1.5.

6.1.3   ANALYSIS OF CRACKED MEMBRANE ELEMENTS

In order to analyze the stresses and strains in the membrane element shown in Figure 6.1a, the five variables given are the three externally applied stresses, σ_l, σ_t, and τ_{lt}, and the two steel percentages, ρ_l and ρ_t. The nine equations [Eqs. (6-1) to (6-8) and (6-12)] will then be used to solve the remaining nine unknown variables, including the two steel stresses, f_l and f_t, the one concrete stress, σ_d, the five strains, ε_l, ε_t, γ_{lt}, ε_r, and ε_d, plus the angle α. The problem is summarized as

Given five variables: σ_l, σ_t, τ_{lt}, ρ_l, ρ_t

Find nine unknown variables: f_l, f_t, σ_d, ε_l, ε_t, γ_{lt}, ε_r, ε_d, α

Expressing the three equilibrium equations [Eqs. (6-1) to (6-3)] in terms of the second type of equations [Eqs. (4-76) to (4-78)] and taking $\sigma_r = 0$, result in Eqs. (4-79) to (4-81). From these three equations the stresses in the longitudinal steel, transverse steel, and concrete struts can be expressed as

$$f_l = \frac{\sigma_l + \tau_{lt}\cot\alpha}{\rho_l} \tag{6-17}$$

$$f_t = \frac{\sigma_t + \tau_{lt}\tan\alpha}{\rho_t} \tag{6-18}$$

$$\sigma_d = \frac{-\tau_{lt}}{\sin\alpha\cos\alpha} \tag{6-19}$$

By substituting the stress–strain relationships from Eqs. (6-7), (6-8), and (6-12) into the preceding three equations, we obtain the strains for the longitudinal steel, transverse steel, and concrete struts:

$$\varepsilon_l = \frac{\sigma_l + \tau_{lt}\cot\alpha}{\rho_l E_s} \tag{6-20}$$

$$\varepsilon_t = \frac{\sigma_t + \tau_{lt}\tan\alpha}{\rho_t E_s} \tag{6-21}$$

$$\varepsilon_d = \frac{1}{E_c}\left(\frac{-\tau_{lt}}{\sin\alpha\cos\alpha}\right) \tag{6-22}$$

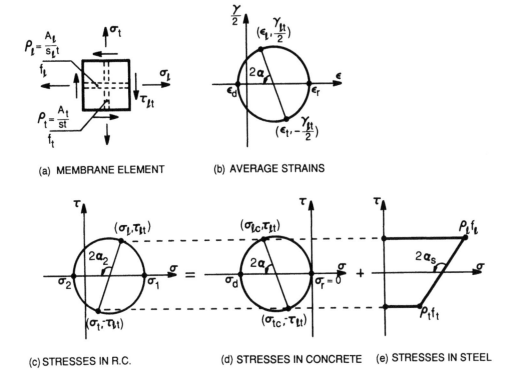

(a) MEMBRANE ELEMENT   (b) AVERAGE STRAINS

(c) STRESSES IN R.C.   (d) STRESSES IN CONCRETE   (e) STRESSES IN STEEL

Figure 6.1 Stresses and strains in analysis and design of membrane elements.

Equations (6-17) to (6-22) show that all six unknown stresses and strains in the steel reinforcement and concrete struts are expressed in terms of a single unknown variable α. This angle α can be determined by the compatibility Eq. (5-33):

$$\tan^2 \alpha = \frac{\varepsilon_l - \varepsilon_d}{\varepsilon_t - \varepsilon_d} \tag{6-23}$$

Substituting the strains ε_l, ε_t, and ε_d from Eqs. (6-20) to (6-22) into Eq. (6-23) and simplifying result in

$$\rho_l(1 + \rho_t n)\tan^4 \alpha + \frac{\sigma_t}{\tau_{lt}} \rho_l \tan^3 \alpha - \frac{\sigma_l}{\tau_{lt}} \rho_t \tan \alpha - \rho_t(1 + \rho_l n) = 0 \quad (6\text{-}24)$$

The solution of the angle α from Eq. (6-24) can best be obtained by a trial-and-error procedure. This procedure will be illustrated by Example Problem 6.1.

After the angle α is found, the three unknown stresses in the steel bars and concrete struts (f_l, f_t, σ_d) can be found directly from Eqs. (6-17) to (6-19), and the three corresponding strains $(\varepsilon_l, \varepsilon_t, \varepsilon_d)$ can be found from Hooke's laws. The last two unknown strains, ε_r and γ_{lt}, could then be obtained from two additional compatibility equations. The cracking strain ε_r can be calculated from Eq. (5-28):

$$\varepsilon_r = \varepsilon_l + \varepsilon_t - \varepsilon_d \tag{6-25}$$

and the shear strain γ_{lt} from Eq. (6-6).

Now that all the stresses and strains have been calculated, the two Mohr circles, one for stresses in the concrete struts and the other for average strains, can be plotted as shown in Figure 6.1e and b. It should be emphasized that the α angle is the same for these two Mohr circles, but is different from the angle α_2 in the Mohr circle for the whole reinforced concrete element, Figure 6.1c. The α_2 angle is calculated directly from the externally applied stresses σ_l, σ_t, and τ_{lt}.

EXAMPLE PROBLEM 6.1

The membrane element that has been designed using the plasticity truss model in Example Problem 4.1 will now be analyzed by the Mohr compatibility truss model. This membrane element is subjected to a set of membrane stresses $\sigma_l = 310$ psi (tension), $\sigma_t = -310$ psi (compression), and $\tau_{lt} = 537$ psi. For the sake of clarity, these applied stresses are again illustrated in Figure 6.2a.

Based on the plasticity truss model, the membrane element is reinforced with 1.033% of steel in both the longitudinal and the transverse directions. The steel is designed to have the same grade in both directions, $f_{ly} = f_{ty} = 60,000$ psi. In order to use the Mohr compatibility truss model, additional properties of the steel and the concrete are specified as follows: $E_s = 29,000$ ksi, $f_c' = 4000$ psi, $E_c = 3600$ ksi, and $n = 29,000/3600 = 8.06$. We are particularly interested in the stress and strain conditions at the first yield of steel.

SOLUTION

First of all, the angle α will be solved by Eq. (6-24). When $\rho_l = \rho_t = \rho$, Eq. (6-24) becomes

$$(1 + \rho n) \tan^4 \alpha + \frac{\sigma_t}{\tau_{lt}} \tan^3 \alpha - \frac{\sigma_l}{\tau_{lt}} \tan \alpha - (1 + \rho n) = 0$$

The given property, ρn, and the external stress ratios, σ_l/τ_{lt} and σ_t/τ_{lt}, in this equation are

$$\rho n = 0.01033(8.06) = 0.0833$$

$$\frac{\sigma_l}{\tau_{lt}} = \frac{310}{537} = 0.5773 \quad \text{and} \quad \frac{\sigma_t}{\tau_{lt}} = \frac{-310}{537} = -0.5773$$

Eq. (6-24) $1.0833 \tan^4 \alpha - 0.5773 \tan^3 \alpha - 0.5773 \tan \alpha - 1.0833 = 0$

This equation can be easily solved by a trial-and-error procedure using the tabulation in Table 6.1.

In the trial-and-error process we first assume an α value (first column) and calculate directly the values of $\tan \alpha$ (second column), $\tan^3 \alpha$ value (third column) and $\tan^4 \alpha$ (fourth column). Then the value of the left hand side of Eq. (6.-24) is computed (last column). If the value of the left hand side is equal to zero, a solution is found. If not, another cycle is performed until it is sufficiently close to zero. The convergence is usually quite rapid. It can be seen from Table 6.1 that we start with an assumption of $\alpha = 60°$, the α value obtained from the plasticity truss model (see Example 4.1). After six cycles of iteration a very accurate value of $\alpha = 52.46°$ is obtained for the Mohr compatibility truss model.

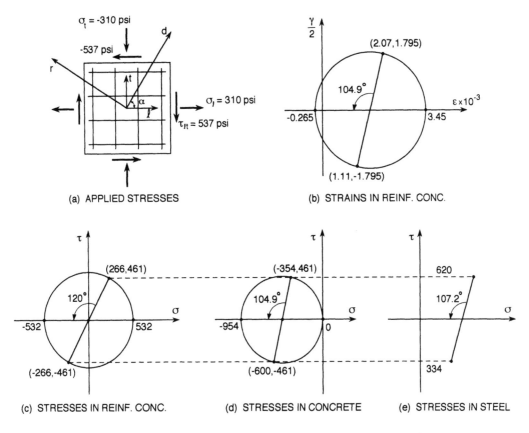

Figure 6.2 Stresses and strains in a membrane element (Example 6.1) at first yield.

TABLE 6.1 Trial-and-Error Method for Solution of Angle α

α (deg.)	Tan α	$\tan^3 \alpha$	$\tan^4 \alpha$	Left Hand Side of Eq. (6-24)
60°	1.7321	5.1961	9.0000	4.6668
55°	1.4281	2.9129	4.1599	0.9170
52.5°	1.3032	2.2134	2.8845	0.0113
52.4°	1.2985	2.1895	2.8431	−0.0170
52.45°	1.3009	2.2014	2.8638	−0.0028
52.46°	1.3013	2.2038	2.8678	−0.0001

For $\alpha = 52.46°$, $\tan \alpha = 1.3013$, $\cot \alpha = 0.7684$, and $\sin \alpha \cos \alpha = 0.4831$. The stresses are calculated as follows:

Eq. (6-19) $$\sigma_d = \frac{-\tau_{lt}}{\sin \alpha \cos \alpha} = \frac{-537}{0.4831} = -1112 \text{ psi}$$

Eq. (6-17) $$\rho_l f_l = \sigma_l + \tau_{lt} \cot \alpha = 310 + 537 (0.7684) = 722.6 \text{ psi}$$

$$f_l = \frac{722.6}{0.01033} = 69{,}954 \text{ psi} > 60{,}000 \text{ psi (yield)}$$

Eq. (6-18) $\rho_t f_t = \sigma_t + \tau_{lt} \tan \alpha = -310 + 537\,(1.3013) = 388.8$ psi

$$f_t = \frac{388.8}{0.01033} = 37{,}637 \text{ psi} < 60{,}000 \text{ psi (yield)}$$

It can be seen that the longitudinal steel stress f_l has exceeded the yield point, whereas the transverse steel stress f_t is still in the elastic range.

First Yield of Steel

The first yield of the longitudinal steel should occur when $f_l = f_y = 60{,}000$ psi. It will occur when the applied stresses are reduced by a factor of $60{,}000/69{,}954 = 0.8577$, i.e.,

$$\sigma_l = 0.8577\,(310) = 265.9 \approx 266 \text{ psi}$$

$$\sigma_t = 0.8577\,(-310) = -265.9 \approx -266 \text{ psi}$$

$$\tau_{lt} = 0.8577\,(537) = 460.6 \approx 461 \text{ psi}$$

These applied stresses at first yield are recorded in terms of Mohr circle in Figure 6.2c. Accordingly, the stresses in the steel are

$$f_t = 0.8577\,(37{,}637) = 32{,}281 \text{ psi}$$

$$\rho_t f_t = 0.8577\,(388.8) = 333.5 \approx 334 \text{ psi}$$

$$\rho_l f_l = 0.8577\,(722.6) = 619.8 \approx 620 \text{ psi}$$

These stresses for the steel are recorded in Figure 6.2e. From the applied stresses and the stresses in the steel we can calculate the stresses in the concrete struts at first yield:

$$\sigma_l - \rho_l f_l = 266 - 620 = -354 \text{ psi}$$

$$\sigma_t - \rho_t f_t = -266 - 334 = -600 \text{ psi}$$

$$\sigma_d = 0.8577\,(-1112) = -953.7 \approx -954 \text{ psi}$$

These stresses for the concrete are used to plot the Mohr circle in Figure 6.2d. Finally, the strains are calculated as follows:

Eq. (6-12) $\varepsilon_d = \dfrac{\sigma_d}{E_c} = \dfrac{-953.7}{3{,}600{,}000} = -0.265 \times 10^{-3}$

Eq. (6-7) $\varepsilon_l = \dfrac{f_l}{E_s} = \dfrac{60{,}000}{29{,}000{,}000} = 2.07 \times 10^{-3}$

Eq. (6-8) $\varepsilon_t = \dfrac{f_t}{E_s} = \dfrac{32{,}281}{29{,}000{,}000} = 1.11 \times 10^{-3}$

Eq. (6-25) $\varepsilon_r = \varepsilon_l + \varepsilon_t - \varepsilon_d = 2.07 + 1.11 + 0.265 = 3.45 \times 10^{-3}$

Eq. (6-6) $\dfrac{\gamma_{lt}}{2} = (-\varepsilon_d + \varepsilon_r)\sin \alpha \cos \alpha = (0.265 + 3.45)(0.4831)$

$$= 1.795 \times 10^{-3}$$

These strains are given in terms of Mohr's strain circle in Figure 6.2b. It should be observed that the α value of 52.46° for the strain condition is identical to that for the stress condition in the concrete. This α value of 52.46° at first yield of longitudinal steel (Mohr compatibility truss model) will rotate to reach a value of 60° when the load is increased until the yielding of both the longitudinal and the transverse steel (plasticity truss model in Example 4.1).

6.1.4   DESIGN OF REINFORCEMENT

We will now study the design of reinforcement in the membrane element shown in Figure 6.1a. The five variables given in this case are the three applied stresses, σ_l, σ_t, and τ_{lt}, and the two steel stresses, f_l and f_t. For design purpose, f_l and f_t will be specified as the allowable stresses f_{la} and f_{ta}. According to the 1983 ACI Code Commentary, f_{la} and f_{ta} are 20 ksi for Grades 40 and 50 steel and 24 ksi for Grade 60 and higher strength steel. The nine equations [Eqs. (6-1) to (6-8) and (6-12)] will then be used to solve the remaining nine unknown variables, including the two steel ratios ρ_l and ρ_t, the one concrete stress σ_d, the five strains ε_l, ε_t, γ_{lt}, ε_r, and ε_d, plus the angle α. The problem is summarized as

Given five variables: σ_l, σ_t, τ_{lt}, $f_l = f_{la}$, $f_t = f_{ta}$
Find nine unknown variables: ρ_l, ρ_t, σ_d, ε_l, ε_t, γ_{lt}, ε_r, ε_d, α

Similar to the analysis problem in Section 6.1.3, the solution of the nine equations starts with expressing the three equilibrium equations [Eqs. (6-1) to (6-3)] by the second type of equations and setting $\sigma_r = 0$. From Eqs. (4-79) to (4-81) we have

$$\rho_l = \frac{\sigma_l + \tau_{lt} \cot \alpha}{f_{la}} \tag{6-26}$$

$$\rho_t = \frac{\sigma_t + \tau_{lt} \tan \alpha}{f_{ta}} \tag{6-27}$$

$$\sigma_d = \frac{-\tau_{lt}}{\sin \alpha \cos \alpha} \tag{6-28}$$

From the linear stress–strain relationships of steel [Eqs. (6-7) and (6-8)] we have

$$\varepsilon_l = \frac{f_{la}}{E_s} \tag{6-29}$$

$$\varepsilon_t = \frac{f_{ta}}{E_s} \tag{6-30}$$

Substituting Eq. (6-28) into the stress–strain equation of compression concrete [Eq. (6-12)] gives

$$\varepsilon_d = \frac{1}{E_c} \left(\frac{-\tau_{lt}}{\sin \alpha \cos \alpha} \right) \tag{6-31}$$

It can be seen that ε_l and ε_t are constants and ρ_l, ρ_t, σ_d, and ε_d are all expressed in terms of a single unknown variable α. Substituting the strains ε_l, ε_t, and ε_d from Eqs. (6-29) to (6-31) into Eq. (6-23) results in

$$\tan^2 \alpha = \frac{f_{la} \sin \alpha \cos \alpha + n\tau_{lt}}{f_{ta} \sin \alpha \cos \alpha + n\tau_{lt}} \tag{6-32}$$

The angle α can be solved by Eq. (6-32) using a trial-and-error method.

Examination of Eq. (6-32) reveals two observations. First, when the allowable stresses in the two directions are equal, $f_{la} = f_{ta}$, then the Mohr compatibility truss model recommends the use of an angle $\alpha = 45°$. Second, the second terms in both the numerator and denominator are smaller than the first terms by an order of magnitude. If the two $n\tau_{lt}$ terms are neglected in Eq. (6-32), then

$$\tan^2 \alpha = \frac{f_{la}}{f_{ta}} \tag{6-33}$$

The angle α obtained from Eq. (6-33) should be a close approximate value of that obtained from Eq. (6-32) and becomes exact when $f_{la} = f_{ta}$.

6.1.5 ANALYSIS OF UNCRACKED MEMBRANE ELEMENTS

As pointed out in Section 6.1.1, the stress–strain relationships of concrete should be expressed by Eqs. (6-15) and (6-16) for uncracked sections. In this case, we have ten equations [Eqs. (6-1) to (6-8) plus (6-15) and (6-16)] including three for equilibrium, three for compatibility, and four for stress–strain relationships. These 10 equations involve 15 variables, including 7 stresses (σ_l, σ_t, τ_{lt}, σ_d, σ_r, f_l, and f_t), 5 strains (ε_l, ε_t, γ_{lt}, ε_r, and ε_d), 2 cross-sectional properties (ρ_l and ρ_t), and 1 geometric parameter, α. Therefore, five variables must be given before the remaining ten unknown variables can be solved by the ten equations. The problem posed is

Given five variables: σ_l, σ_t, τ_{lt}, ρ_l, ρ_t
Find ten unknown variables: f_l, f_t, σ_d, σ_r, ε_l, ε_t, γ_{lt}, ε_r, ε_d, α

In view of the linear stress–strain relationships of the materials, the solution of the ten equations is greatly simplified. First, the two compatibility equations [Eqs. (6-4) and (6-5)] will be expressed in terms of stresses. Substituting the stress–strain relationships of steel and concrete [Eqs. (6-7), (6-8), (6-15), and (6-16)] into Eqs. (6-4) and (6-5) gives

$$f_l = m\left[(\sigma_d \cos^2 \alpha + \sigma_r \sin^2 \alpha) - \mu(\sigma_d \sin^2 \alpha + \sigma_r \cos^2 \alpha) \right] \tag{6-34}$$

$$f_t = m\left[(\sigma_d \sin^2 \alpha + \sigma_r \cos^2 \alpha) - \mu(\sigma_d \cos^2 \alpha + \sigma_r \sin^2 \alpha) \right] \tag{6-35}$$

where $m = E_s/E_c - 1$. The subtraction of 1 in the m expression is required because the original steel area has been included in the concrete area for convenience of calculation. The Poisson ratio for concrete μ can be taken as 0.2. This first step reduces a set of ten equations with ten unknown variables to a set of five equations involving five unknowns. The five equations are Eqs. (6-1) to (6-3), (6-34), and (6-35). The five unknown variables are the four stresses $(f_l, f_t, \sigma_d, \sigma_r)$ and the angle α.

The second step is to eliminate f_l and f_t from the five equations. Substituting f_l and f_t from Eqs. (6-34) and (6-35) into Eqs. (6-1) and (6-2) and rearranging the terms give

$$\frac{\sigma_l}{1 + \rho_l m} = (\sigma_d \cos^2 \alpha + \sigma_r \sin^2 \alpha) - \frac{\mu \rho_l m}{1 + \rho_l m}(\sigma_d \sin^2 \alpha + \sigma_r \cos^2 \alpha) \quad (6\text{-}36)$$

$$\frac{\sigma_t}{1 + \rho_t m} = (\sigma_d \sin^2 \alpha + \sigma_r \cos^2 \alpha) - \frac{\mu \rho_t m}{1 + \rho_t m}(\sigma_d \cos^2 \alpha + \sigma_r \sin^2 \alpha) \quad (6\text{-}37)$$

Let

$$\frac{\mu \rho_l m}{1 + \rho_l m} = \mu_l^*$$

$$\frac{\mu \rho_t m}{1 + \rho_t m} = \mu_t^*$$

Then solving Eqs. (6-36) and (6-37) gives

$$\frac{1}{1 - \mu_l^* \mu_t^*}\left(\frac{\sigma_l}{1 + \rho_l m} + \mu_l^* \frac{\sigma_t}{1 + \rho_t m}\right) = \sigma_d \cos^2 \alpha + \sigma_r \sin^2 \alpha \quad (6\text{-}38)$$

$$\frac{1}{1 - \mu_l^* \mu_t^*}\left(\frac{\sigma_t}{1 + \rho_t m} + \mu_t^* \frac{\sigma_l}{1 + \rho_l m}\right) = \sigma_d \sin^2 \alpha + \sigma_r \cos^2 \alpha \quad (6\text{-}39)$$

Note that μ_l^* and μ_t^* are small values on the order of 2%. Consequently, $1/(1 - \mu_l^* \mu_t^*)$ is extremely close to unity. With an accuracy of about 2% we can also neglect the second terms on the left-hand sides of Eqs. (6-38) and (6-39), resulting in

$$\frac{\sigma_l}{1 + \rho_l m} = \sigma_d \cos^2 \alpha + \sigma_r \sin^2 \alpha \quad (6\text{-}40)$$

$$\frac{\sigma_t}{1 + \rho_t m} = \sigma_d \sin^2 \alpha + \sigma_r \cos^2 \alpha \quad (6\text{-}41)$$

Equation (6-3), of course, can be written as

$$-2\tau_{lt} = (\sigma_d - \sigma_r)\sin 2\alpha \quad (6\text{-}42)$$

Now we have three equations [Eqs. (6-40) to (6-42)] with three unknown variables σ_d, σ_r, and α.

The third step is to solve the angle α from the preceding three equations. Subtracting Eq. (6-41) from Eq. (6-40) and noticing that $\cos^2 \alpha - \sin^2 \alpha = \cos 2\alpha$, we have

$$\frac{\sigma_l}{1 + \rho_l m} - \frac{\sigma_t}{1 + \rho_t m} = (\sigma_d - \sigma_r)\cos 2\alpha \qquad (6\text{-}43)$$

Dividing Eq. (6-43) by Eq. (6-42) gives

$$\cot 2\alpha = \frac{\dfrac{\sigma_l}{1 + \rho_l m} - \dfrac{\sigma_t}{1 + \rho_t m}}{-2\tau_{lt}} \qquad (6\text{-}44)$$

If the steel is neglected, i.e., $\rho_l = \rho_t = 0$, then Eq. (6-44) becomes

$$\cot 2\alpha = \frac{\sigma_l - \sigma_t}{-2\tau_{lt}} \qquad (6\text{-}45)$$

Equation (6-45) is identical to Eq. (4-31), which had been derived for uncracked homogeneous elements. Equation (6-44) is also derived for uncracked sections but includes the effect of steel in the form of transformed area percentages $\rho_l m$ and $\rho_t m$.

Now that the angle α is obtained, we can trace backward to solve σ_d and σ_r from Eqs. (6-40) to (6-42). Multiplying Eq. (6-42) by $\tan \alpha$ and eliminating σ_r from Eqs. (6-40) and (6-42) give

$$\sigma_d = \frac{\sigma_l}{1 + \rho_l m} - \tau_{lt} \tan \alpha \qquad (6\text{-}46)$$

Similarly, multiplying Eq. (6-42) by $\cot \alpha$ and eliminating σ_d from Eqs. (6-40) and (6-42) give

$$\sigma_r = \frac{\sigma_l}{1 + \rho_l m} + \tau_{lt} \cot \alpha \qquad (6\text{-}47)$$

Now that σ_d, σ_r, and α are obtained, f_l and f_t can be calculated from Eqs. (6-34) and (6-35), the strains ε_l, ε_t, ε_d, and ε_r can be calculated from the stress–strain relationships of Eqs. (6-7), (6-8), (6-15), and (6-16), and, finally, γ_{lt} can be calculated from the compatibility Eq. (6-6).

The principal stresses σ_d and σ_r and the angle α in this section are originally used in the equilibrium and compatibility equations [Eqs. (6-1) to (6-6)], assuming the concrete sections to be cracked. However, it is now proven that these equations are also applicable to uncracked sections when the appropriate constitutive laws are adopted. According to the symbols defined in Section 4.3.1 and Figure 4.10, the principal stresses for uncracked reinforced concrete should be σ_2 and σ_1, and the angle α should be α_2. Therefore, σ_d, σ_r, and α in Eqs. (6-34) to (6-47) should be replaced by σ_2, σ_1, and α_2, respectively.

In conclusion, the ten equations for Mohr compatibility truss model [Eqs. (6-1) to (6-10)] are very general and powerful. They are applicable to cracked or uncracked sections. For cracked sections the tensile stress of concrete can be included if desired. For the uncracked sections the transformed steel areas can be taken into account if necessary.

6.2   **Beams Under Uniformly Distributed Load**

6.2.1   ANALYSIS OF BEAM ELEMENTS

In Section 3.3.2 of Chapter 3 we studied a beam subjected to uniformly distributed load according to the equilibrium (plasticity) truss model (Figure 3.17). By assuming an infinite plasticity for the materials, we obtained a discontinuous banded stress field as shown in Figure 3.17b. From this stress field a staggered shear diagram was derived as shown in Figure 3.17d.

In this section we will apply the Mohr compatibility truss model to the same beam (Figure 6.3a) subjected to the same uniformly distributed load w. In this model a continuous stress field will be introduced (Hsu, 1982). This continuous stress field will satisfy not only the equilibrium condition, but also the compatibility condition and the stress–strain relationship of materials according to Hooke's law (Hsu, 1983).

The continuous stress field can be illustrated by describing the variation of the stresses in the stirrups and longitudinal bars in a typical element B isolated from the beam as shown in Figure 6.3a. The length of the element B is taken as $d_v \cot \alpha$. Because $\cot \alpha$ is assumed to be $5/3$ and $d_v = 3s$ in the model beam, the length of the beam element B is equal to $5s$. This means the element contains five uniformly distributed stirrups.

The beam element B shown in Figure 6.3a is subjected to a shear force V and a bending moment M on the left face. On the right face the shear force receives an additional increment of $wd_v \cot \alpha$ from the uniform load w. Similarly, the bending moment receives two additional increments: $Vd_v \cot \alpha$ from the shear force V and $-(1/2)wd_v^2 \cot^2 \alpha$ from the uniform load w.

According to the truss model concept, the bending resistance of a beam element will be supplied by the stringers and the shear resistance by the main body. The forces resisted by the stringers and by the main body are shown separately in Figure 6.3b. It can be seen that the main body is subjected to shear stresses caused by both the shear force V and the uniform load w. The shear force on the right face of the main body is $V - wd_v \cot \alpha$ and the shear force on the top and bottom faces is $V \cot \alpha - (1/2)wd_v \cot^2 \alpha$.

To find a continuous stress field that satisfies both the equilibrium and the compatibility conditions in the main body of element B, we make two assumptions. First, the shear stresses are assumed to be *uniformly* distributed over the depth of the beam element B at each cross section. Second, in view of the uniformly distributed load w, it is logical to assume that the cross-sectional shear force due to w varies *linearly* along the length of beam element B from zero at the left face to $-wd_v \cot \alpha$ at the right face. As a result, the part of the shear force due to w, i.e., $-(1/2)wd_v \cot^2 \alpha$ on the top and bottom faces should also be distributed in a linear fashion from zero at the left face of element B to a maximum negative value at the right face.

With these two assumptions we can now analyze the stresses in the stirrups and in the longitudinal web bars within the main body. The variation of the stirrup forces and the longitudinal web steel forces will be studied in Sections 6.2.2 and 6.2.3, respectively. The two assumptions we have made can be proven by the theory of elasticity to satisfy the compatibility condition. This proof will be given in Section 6.2.4. The analysis by theory of elasticity will also provide the formulas for the forces in the stirrups and in the longitudinal steel. These formulas will be given in Section 6.2.5.

Assume $\cot\alpha = \dfrac{5}{3}$ and $s = \dfrac{d_v}{3}$

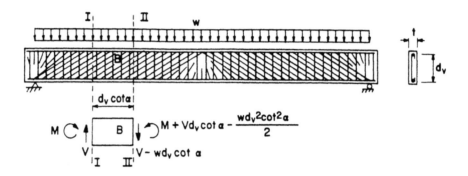

(a) FORCES ON BEAM ELEMENT

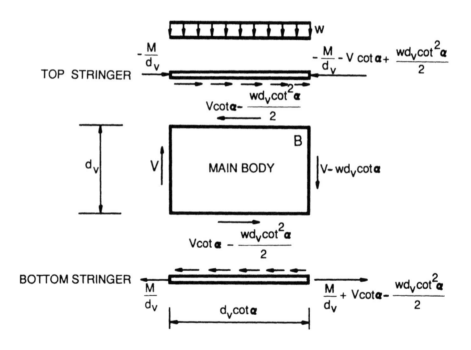

(b) FORCES ON MAIN BODY AND STRINGERS

Figure 6.3 Forces in beam under uniform load.

6.2.2 STIRRUP FORCES

The main body of the element as shown in Figure 6.3b is subjected to two sets of self-equilibrating forces: one caused by the shear force V and the other by the uniform load w. These two sets of forces due to V and w will produce two distinct stress fields. The stress field in the main body is the sum of these two stress fields.

The stress field due to V and the stress field due to w are shown separately in Figure 6.4a and b. The stress field due to V is one of pure shear, identical to that

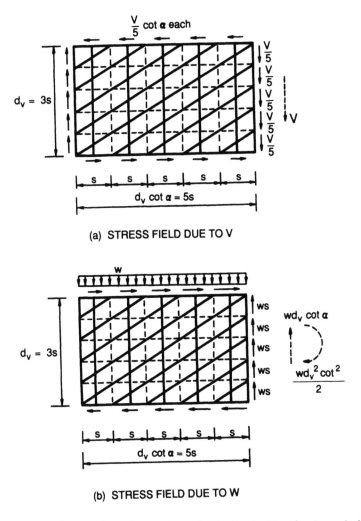

(a) STRESS FIELD DUE TO V

(b) STRESS FIELD DUE TO W

Figure 6.4 Separation of two stress fields due to *V* and *w* in main body.

given in Figure 3.5b (Section 3.1.3) or Figure 3.16 (Section 3.3.1) for simple beams subjected to midspan concentrated load. In this stress field the stirrup forces F_t are uniformly distributed and are equal to a constant $V/5$ as given by Eq. (3-100).

The stress field due to *w* (Figure 6.4b) is the same as that of a cantilever beam under a uniform load. The stirrup forces are nonuniformly distributed, i.e., F_t is not a constant. To illustrate this stress field, the beam element *B* is divided and separated into 25 equal subelements as shown in Figure 6.5. Each subelement contains a vertical steel bar at its center. Forces acting on each subelement are indicated and vary from subelement to subelement in accordance with the two assumptions.

Take, for example, the subelement at the upper right corner. It is subjected to a shear force of *ws* on the right face and 0.8*ws* on the left face. This change is in accordance with the second assumption. On the top face we have vertical force *ws* due to external load acting along the centerline of the subelement. (This force is drawn to the right of the centerline for clarity.) To maintain vertical equilibrium, the vertical force on the bottom face must be 0.8*ws* acting along the centerline (also drawn to the right of the centerline). This compressive force must be carried by the vertical stirrup, because vertical stresses cannot pass through the diagonal cracks

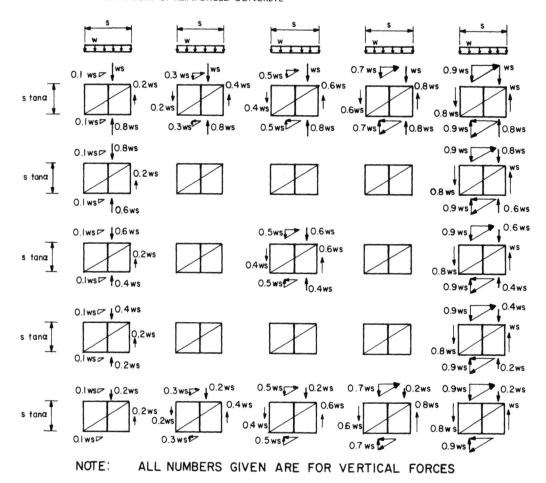

NOTE: ALL NUMBERS GIVEN ARE FOR VERTICAL FORCES

Figure 6.5 Equilibrium of subelements within a beam element *B* (stress field due to uniformly distributed load *w* on top surface).

except through connecting steel bars. To maintain moment equilibrium and horizontal force equilibrium, shear forces on the top and bottom faces must both be $(0.9ws)\cot\alpha$ and act in the opposite direction. The equality of the top and bottom shear forces is in accordance with the first assumption. Each of these two shear forces can be resolved into a diagonal tensile force in the concrete struts and a vertical compression force of $0.9ws$ in the stirrup. Summing the two vertical compressive forces in the stirrup on the top face gives $1.9ws$. On the bottom face, however, the sum of the two vertical forces is $1.7ws$, which is different from the top face.

If we observe the equilibrium of the five subelement of the extreme right column in Figure 6.5, it can be seen that the compressive stirrup forces on the top faces decrease linearly downward in the sequence $1.9ws$, $1.7ws$, $1.5ws$, $1.3ws$, and $1.1ws$. If we also look at the top faces of the top five subelements, the compressive stirrup forces decrease in this same sequence from right to left. Similarly, the sequence of compressive stirrup forces, $0.9ws$, $0.7ws$, $0.5ws$, $0.3ws$, and $0.1ws$, can be observed on the bottom faces of the five extreme left subelements from top to bottom as well as on the bottom faces of the five bottom subelements from right to left.

The compressive stress field of nonuniform stirrup forces as illustrated in Figure 6.5 can now be subtracted from the uniform tensile stress field of Figure 6.4a, where

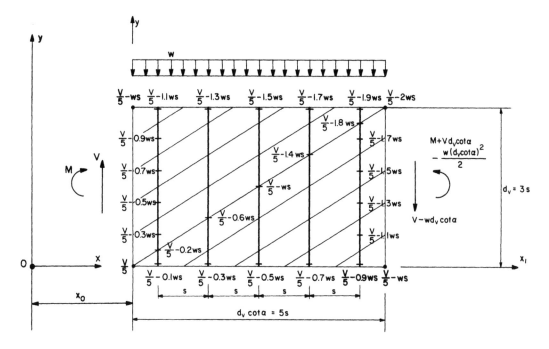

Figure 6.6 Variation of stirrup forces in compatibility truss model.

the stirrup force is a constant of $V/5$. The algebraic summation of the two stress fields is shown in Figure 6.6. Figure 6.6 clearly shows that the tensile forces in the stirrups increase linearly downward and that the maximum forces are located at the bottom of the bars. Stirrup forces at these lowest locations vary linearly along the beam length according to the conventional triangular shear diagram. This pattern of stress distribution in the stirrups has been verified by tests (Belarbi and Hsu, 1990).

In summary, the nonuniform stirrup forces of Figure 6.6 are based on the two assumptions that satisfy not only the equilibrium condition, but also the compatibility conditions and the linear stress–strain relationship of materials. This nonuniform continuous stress field derived from the Mohr compatibility truss model is quite different from the discontinuous banded stress field assumed in the equilibrium (plasticity) truss model of Figure 3.17b. The difference between these two models has two significant consequences. First, the equilibrium (plasticity) truss model gives the upper bound solution as far as material is concerned, whereas the Mohr compatibility truss model provides the lower bound solution. Second, the equilibrium (plasticity) truss model suggests a staggered shear diagram for design of stirrups, whereas the Mohr compatibility truss model requires the conventional triangular shear diagram.

6.2.3   LONGITUDINAL WEB STEEL FORCES

Distribution of horizontal forces in the main body of the beam element can also be derived from the two stress fields in Figure 6.4. For the stress field due to V (Figure 6.4a), the shear force on the vertical face of each subelement is $V/5$. This vertical force can be resolved into a diagonal compressive force and a longitudinal tensile force of $V/3$. This longitudinal tensile force should be taken by one longitudinal web bar at the center of each subelement. This longitudinal tensile force of $V/3$ should be uniform throughout the main body of the beam element.

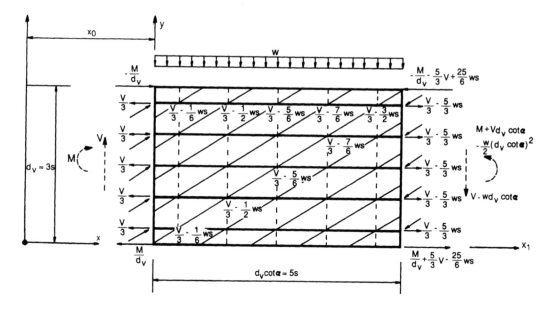

Figure 6.7 Variation of longitudinal steel forces (compatibility truss model).

For the stress field due to w (Figures 6.4b and 6.5), an upward shear force ws is shown to act on the right face of each of the five subelements in the far right column. Each of these five forces can be resolved into a diagonal tensile force and a longitudinal compressive force of $ws \cot \alpha = (5/3)ws$. This longitudinal compressive force should be taken by each of the longitudinal web bars on the right face of the element, and should decrease linearly from $(5/3)ws$ at the right face to zero at the left face of the element.

The summation of the longitudinal steel forces in the two stress fields due to V and w is given in Figure 6.7. It can be seen that the tensile force in each longitudinal web bar $N_{wl} = V/3$ on the left face of the element. This tensile force decreases linearly from left to right until it becomes $N_{wl} = V/3 - (5/3)ws$ on the right face of the element. The tensile forces at the centers of the five subelements in each row decrease linearly in the sequence $N_{wl} = V/3 - (1/6)ws$, $V/3 - (1/2)ws$, $V/3 - (5/6)ws$, $V/3 - (7/6)ws$, and $V/3 - (3/2)ws$.

Now that the forces in the stirrups and in the longitudinal bars of a beam element are clarified, we can observe the variation of these forces along a diagonal crack. A triangular free body of the beam element is shown in Figure 6.8 and all the forces acting on the free body are indicated. It can be seen that the stirrup forces decrease from left to right and the longitudinal web steel forces decrease from bottom to top, both in a linear manner. It would be educational for the reader to verify that the given force system satisfies the force equilibrium equations in both the longitudinal and transverse directions. The moment equilibrium equation about point O, however, is not exactly satisfied, because the centers of the subelements where the steel bars are placed are not located exactly at the centroids of the contributing stress diagrams for w. This slight error can be observed from the two triangular stress diagrams for w in Figure 6.8, one for stirrups and one for longitudinal steel.

If the five longitudinal web rebars are not available, a redistribution of the forces will occur. The longitudinal forces in the web rebars can be replaced by the longitudinal forces in the top and bottom stringers to ensure the equilibrium of the total longitudinal forces. This replacement can easily take care of the uniform

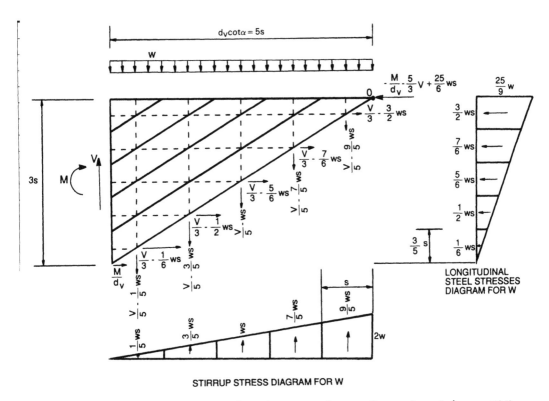

Figure 6.8 Stirrup and longitudinal steel stresses along a diagonal crack (compatibility truss model).

longitudinal forces ($V/3$ in each web rebar) produced by the shear force V. The equilibrium of the nonuniform longitudinal forces produced by the distributed load w is more difficult. Equilibrium of these local areas at the intersections of vertical stirrups and concrete struts will involve the dowel action of the vertical stirrups, the shear resistance of the concrete struts, and the angle change of the concrete struts.

6.2.4 SOLUTION BY THE THEORY OF ELASTICITY

In Sections 6.2.1 to 6.2.3 we derived a nonuniform stress field for the main body of the beam element subjected to uniformly distributed load. From this stress field it was concluded that the forces in the stirrups decrease linearly from the left to the right and from the bottom face to the top face. Design of the stirrups should be based conservatively on the conventional triangular shear diagram.

The derivation of the stress field was based on two assumptions. First, shear stresses are uniformly distributed over the depth of a cross section, and, second, shear stresses are linearly distributed along the length of the member. A question naturally arises: Does the stress field derived from these two assumptions really satisfy the compatibility condition and Hooke's constitutive law, in addition to the equilibrium condition? To answer this question in a rigorous way, we will resort to the inverse method of the theory of elasticity to verify the stress field.

In Figure 6.9 an x-y coordinate is imposed on one half span of a beam with the origin located at the bottom left corner of the beam. The simple beam is subjected to a uniformly distributed load w. The length of the beam is l and the depth is d_v. The stresses on an element in the x-y coordinate are σ_x, σ_y, and τ_{xy}. The corresponding

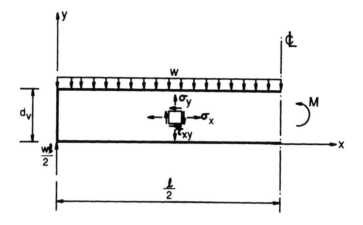

Figure 6.9 Notations and coordinates for continuous stress field.

strains are ε_x, ε_y, and γ_{xy}, and the displacements in the x and y directions are u and v, respectively.

In order to satisfy all the equilibrium conditions, compatibility condition, and Hooke's law, a stress field (σ_x, σ_y, τ_{xy}) and its corresponding strain field (ε_x, ε_y, γ_{xy}) and displacement field (u, v) must satisfy the following eight equations:

Equilibrium equations:

$$\frac{\partial \sigma_x}{\partial x} + \frac{\partial \tau_{xy}}{\partial y} = 0 \qquad (6\text{-}48)$$

$$\frac{\partial \sigma_y}{\partial y} + \frac{\partial \tau_{xy}}{\partial x} = 0 \qquad (6\text{-}49)$$

Strain-displacement relationships:

$$\varepsilon_x = \frac{\partial u}{\partial x} \qquad (6\text{-}50)$$

$$\varepsilon_y = \frac{\partial v}{\partial y} \qquad (6\text{-}51)$$

$$\gamma_{xy} = \frac{\partial v}{\partial x} + \frac{\partial u}{\partial y} \qquad (6\text{-}52)$$

Stress–strain relationships:

$$\sigma_x = \frac{E}{1 - \mu^2}(\varepsilon_x + \mu\varepsilon_y) \qquad (6\text{-}53)$$

$$\sigma_y = \frac{E}{1 - \mu^2}(\varepsilon_y + \mu\varepsilon_x) \qquad (6\text{-}54)$$

$$\tau_{xy} = G\gamma_{xy} \qquad (6\text{-}55)$$

where

E = Young's modulus of elasticity

G = shear modulus

μ = Poisson's ratio

In addition, the displacements u and v must be continuous functions, and the boundary conditions must be satisfied.

To define the boundary conditions, the beam in Figure 6.9 is redrawn in Figure 6.10a, but Figure 6.10a also includes the linear shear diagram and the parabolic moment diagram. Similar to the analysis of the beam element in Section 6.2.2, the force system on the half-beam can be resolved into two self-equilibrating force systems as shown in Figure 6.10b and c. Figure 6.10b is the force system caused by the support reaction $wl/2$ (or V), which is identical to the force system of a simply supported beam subjected to a midspan concentrated load. Figure 6.10c is the force system induced by the uniform load w, which is identical to the force system in a cantilever beam. The shear and moment diagrams corresponding to these two force systems are also shown. It can be easily checked that the shear and moment diagrams in Figure 6.10a are the sum of the corresponding diagrams in b and c.

The force systems in Figure 6.10a, b, and c can be idealized by truss models in Figure 6.11a, b, and c according to the fundamental assumption of the truss model. It is assumed that the bending moment is resisted by the stringers and the shear force is resisted by the main body. In Figure 6.11 the stringers are separated from the main body so that the boundary stresses on the main body can be clearly identified.

The boundary stresses on the main body shown in Figure 6.11a are determined from the two assumptions. The first assumption requires that the support reaction force $wl/2$ should be distributed uniformly over the depth with a force per unit length of $wl/2d_v$ at the left support. The second assumption requires that this shear stress decreases linearly from the left support to zero at the midspan according to the triangular shear diagram. To be consistent, the bond stresses between the stringer and the main body should also vary linearly from $wl/2d_v$ at the left support to zero at the midspan.

The boundary stresses so obtained can be separated into pure shear boundary stresses of $wl/2d_v$, shown in Figure 6.11b, and a boundary stress system created by the uniform load w, shown in Figure 6.11c. In Figure 6.11c the boundary stresses consist of: (1) the uniform vertical load w at the top boundary, (2) the vertical shear stress of zero and $wl/2d_v$ at the left support and midspan, respectively, and (3) the horizontal shear stresses at the top and bottom boundaries, which vary linearly from zero at the left support to $wl/2d_v$ at the midspan.

Now that the two sets of boundary stresses are defined, we will first search for a continuous displacement field that will satisfy the eight equations [Eqs. (6-48) to (6-55)] as well as the pure shear boundary condition of the main body in Figure 6.11b. This continuous displacement field is quite simple:

$$u = \frac{wl}{4d_vG}y \tag{6-56}$$

$$v = \frac{wl}{4d_vG}x \tag{6-57}$$

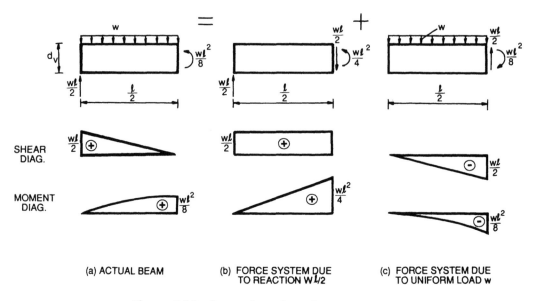

(a) ACTUAL BEAM

(b) FORCE SYSTEM DUE TO REACTION $w\ell/2$

(c) FORCE SYSTEM DUE TO UNIFORM LOAD w

Figure 6.10 Separation of two force systems.

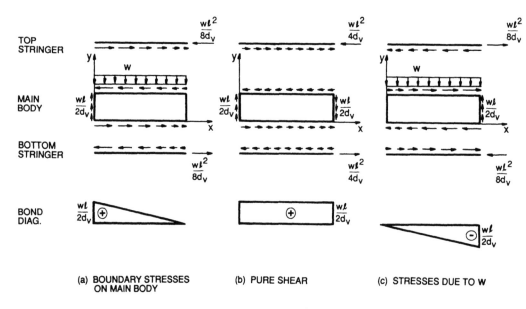

(a) BOUNDARY STRESSES ON MAIN BODY

(b) PURE SHEAR

(c) STRESSES DUE TO W

Figure 6.11 Truss model idealization (isolation of main body).

Substituting u and v from Eqs. (6-56) and (6-57) into Eqs. (6-50) through (6-52), we obtain the strain field:

$$\varepsilon_x = 0 \tag{6-58}$$

$$\varepsilon_y = 0 \tag{6-59}$$

$$\gamma_{xy} = \frac{wl}{2d_v G} \tag{6-60}$$

Substituting ε_x, ε_y, and γ_{xy} from Eqs. (6-58) to (6-60) into Eqs. (6-53) through (6-55) we arrive at the stress field for the case of pure shear:

$$\sigma_x = 0 \tag{6-61}$$

$$\sigma_y = 0 \tag{6-62}$$

$$\tau_{xy} = \frac{wl}{2d_v} \tag{6-63}$$

It can easily be seen that this stress field satisfies equilibrium [Eqs. (6-48) and (6-49)] as well as the pure shear boundary condition in Figure 6.11b.

The search for the continuous displacement field that satisfies the boundary condition due to w (Figure 6.11c) was not as simple. Fortunately, it was found after much trial and error:

$$u = -\frac{\mu w}{Ed_v}xy \tag{6-64}$$

$$v = -\frac{w}{2d_v}\left(\frac{1}{G} - \frac{\mu}{E}\right)x^2 + \frac{w}{2Ed_v}y^2 \tag{6-65}$$

Substituting u and v into Eqs. (6-50) through (6-52), the strain field was found:

$$\varepsilon_x = -\frac{\mu w}{Ed_v}y \tag{6-66}$$

$$\varepsilon_y = \frac{w}{Ed_v}y \tag{6-67}$$

$$\gamma_{xy} = -\frac{w}{Gd_v}x \tag{6-68}$$

Again, substituting ε_x, ε_y, and γ_{xy} into Eqs. (6-53) through (6-55), the stress field is obtained:

$$\sigma_x = 0 \tag{6-69}$$

$$\sigma_y = \frac{w}{d_v}y \tag{6-70}$$

$$\tau_{xy} = -\frac{w}{d_v}x \tag{6-71}$$

It is not difficult to show that this stress field satisfies the equilibrium equations [Eqs. (6-48) and (6-49)] as well as the boundary stresses due to w (Figure 6.11c).

Adding the stress field due to w in Eqs. (6-69) through (6-71) to the stress field of pure shear in Eqs. (6-61) through (6-63) gives the combined stress field for the actual boundary stresses in Figure 6.11a:

$$\sigma_x = 0 \tag{6-72}$$

$$\sigma_y = \frac{w}{d_v}y \tag{6-73}$$

$$\tau_{xy} = \frac{w}{d_v}\left(\frac{l}{2} - x\right) \tag{6-74}$$

Based on the principle of superposition in the theory of elasticity, this continuous stress field represented by Eqs. (6-72) to (6-74) and its corresponding strain and displacement fields should satisfy all eight equations of the theory of elasticity as well as the boundary conditions.

Equation (6-74) states that the shear stress τ_{xy} is $wl/2d_v$ at the left support ($x = 0$) and varies linearly to zero at the midspan ($x = l/2$). Because τ_{xy} is not a function of y, it is constant over the depth of the beam. These two observations verify the two assumptions we have made previously for this stress field. In other words, the stress field based on the two assumptions does satisfy all the equilibrium equations, the compatibility equations, and Hooke's linear constitutive law.

6.2.5   FORMULAS FOR STRESSES IN SIMPLE TRUSS MODEL

6.2.5.1   Stresses in a Half-Beam (Figure 6.9)

The stress field in a half-beam expressed by Eqs. (6-72) to (6-74) will now be applied to a simple truss model with the angle of diagonal concrete struts defined as α and the tensile stress (σ_r) taken as zero. Taking the l-t coordinate to be the same as the x-y coordinate, the stresses in the longitudinal web steel, transverse steel, and diagonal concrete struts are obtained from the second type of equilibrium expression for reinforced concrete [Eqs. (4-79) to (4-81)]:

$$\rho_l f_l = \tau_{xy} \cot \alpha + \sigma_x \tag{6-75}$$

$$\rho_t f_t = \tau_{xy} \tan \alpha + \sigma_y \tag{6-76}$$

$$\sigma_d = -\tau_{xy}\frac{1}{\sin \alpha \cos \alpha} \tag{6-77}$$

It should be noted that the stress field of Eqs. (6-72) to (6-74) was derived based on the basic sign convention (see Figure 4.1a). When a uniformly distributed load w that induced compression on the top surface of the beam is given, w should be taken as negative in Eqs. (6-73) and (6-74). When these equations are applied to the truss model using the RC sign convention, however, two changes of sign must be pointed out. First, the distributed load w was taken as an absolute value in the truss model of Sections 6.2.1 to 6.2.3. Consequently, a negative sign will be required in Eqs. (6-73) and (6-74) when these two equations are related to the stress distributions given in these Sections. Second, when the RC sign convention is used (see Figure 4.5a), the sign of the shear stress τ_{xy} must be reversed. In other words, an additional negative sign must be included in Eq. (6-74), restoring the right hand side of this equation to a positive value.

Considering these two changes of sign, we substitute $\sigma_x = 0$, $\sigma_y = -(w/d_v)y$ and $\tau_{xy} = (w/d_v)(l/2 - x)$ into Eqs. (6-75) to (6-77):

$$\rho_l f_l = \frac{w}{d_v}\left(\frac{l}{2} - x\right)\cot\alpha \tag{6-78}$$

$$\rho_t f_t = \frac{w}{d_v}\left(\frac{l}{2} - x\right)\tan\alpha - w\frac{y}{d_v} \tag{6-79}$$

$$\sigma_d = -\frac{w}{d_v}\left(\frac{l}{2} - x\right)\frac{1}{\sin\alpha\cos\alpha} \tag{6-80}$$

The forces in the stirrups and in the longitudinal web bars can be calculated from Eqs. (6-79) and (6-78):

$$F_t = \rho_t f_t(s)(1) = w\left(\frac{l}{2} - x\right)\frac{s}{d_v}\tan\alpha - ws\frac{y}{d_v} \tag{6-81}$$

$$N_{wl} = \rho_l f_l(s)(\tan\alpha) = w\left(\frac{l}{2} - x\right)\frac{s}{d_v} \tag{6-82}$$

The distributed load w now has a unit of force per unit length. The forces in the top and bottom stringers are easily found from the free body of the stringers in Figure 6.11a:

$$-N_{tl} = N_{bl} = \int_0^x \tau_{xy}\, dx = \int_0^x \frac{w}{d_v}\left(\frac{l}{2} - x\right)dx = \frac{w}{2d_v}x(l - x) \tag{6-83}$$

6.2.5.2   Stresses in a Beam Element (Figures 6.6 and 6.7)

We will now apply the preceding formulas to a beam element with $\tan\alpha = 3/5$ and $d_v = 3s$. Such a beam element has been extensively studied in Sections 6.2.2 and 6.2.3 and the stress fields have been plotted in Figures 6.6 and 6.7. According to Eqs. (6-81) and (6-82) the forces in the stirrups and in the longitudinal web steel are

$$F_t = \frac{1}{5}w\left(\frac{l}{2} - x\right) - ws\frac{y}{d_v} \tag{6-84}$$

$$N_{wl} = \frac{1}{3}w\left(\frac{l}{2} - x\right) \tag{6-85}$$

Supposing that the left face of the beam element is located at a distance x_0 from the left support of the beam, and the shear force at this distance x_0 is V as shown in Figure 6.6, then

$$V = w\left(\frac{l}{2} - x_0\right) \tag{6-86}$$

Supposing that the distance from a given section in the element to the left face of the element is x_1, then the horizontal coordinate of the given section x is equal to $x_0 + x_1$. Substituting $x = x_0 + x_1$ into Eqs. (6-84) and (6-85), and using Eq. (6-86) we have

$$F_t = \frac{V}{5} - \frac{wx_1}{5} - ws\frac{y}{d_v} \qquad (6\text{-}87)$$

$$N_{wl} = \frac{V}{3} - \frac{wx_1}{3} \qquad (6\text{-}88)$$

Equations (6-87) and (6-88) predict precisely the variation of stresses in the stirrups (Figure 6.6) and in the longitudinal web bars (Figure 6.7). Take, for example, the center point of the element, where $x_1 = 2.5s$ and $y = 0.5d_v$. Equation (6-87) gives $F_t = (V/5) - (ws)$ and Eq. (6-88) gives $N_{wl} = (V/3) - (5/6)ws$. These are exactly the values indicated in Figures 6.6 and 6.7.

Forces in the top and bottom stringers of the beam element can also be found. Inserting $x = x_0 + x_1$ into Eq. (6-83) results in

$$-N_{tl} = N_{bl} = \left[\frac{w}{2}x_0(l - x_0)\right]\frac{1}{d_v} + w\left(\frac{l}{2} - x_0\right)\frac{x_1}{d_v} - \frac{wd_v}{2}\left(\frac{x_1}{d_v}\right)^2 \qquad (6\text{-}89)$$

Noticing that $[(w/2)x_0(l - x_0)] = M$ and $w(l/2 - x_0) = V$, Eq. (6-89) becomes

$$-N_{tl} = N_{bl} = \frac{M}{d_v} + V\frac{x_1}{d_v} - \frac{wd_v}{2}\left(\frac{x_1}{d_v}\right)^2 \qquad (6\text{-}90)$$

At the right face of the beam element, $x_1 = (5/3)d_v$. Recalling $d_v = 3s$, Eq. (6-90) becomes

$$-N_{tl} = N_{bl} = \frac{M}{d_v} + \frac{5}{3}V - \frac{25}{6}ws \qquad (6\text{-}91)$$

Again, these top and bottom stringer forces are shown in Figures 6.7 and 6.8.

In conclusion, it is clear that the Mohr compatibility truss model, which utilizes Hooke's linear law for the materials, is consistent with the theory of elasticity. Therefore, the Mohr compatibility truss model and the theory of elasticity could be used to solve other important practical problems, in addition to the case of beams under uniformly distributed load shown in this section.

7

Softened Truss Model for Membrane Elements

7.1   Basic Equations for Membrane Elements

In Chapter 6 we studied the Mohr compatibility truss model, which utilizes ten equations to analyze the behavior of membrane elements subjected to two-dimensional loadings. The ten equations include three for equilibrium, three for compatibility, and four for the constitutive laws of materials. Because the constitutive equations for both concrete and steel are based on Hooke's law, the predicted behavior of the membrane element is linear. The Mohr compatibility truss model, therefore, provides a method of linear analysis.

In this chapter we will introduce a method of nonlinear analysis for membrane elements. In this model the constitutive equations are based on the actually observed stress–strain relationships of concrete and steel. The stress–strain curve of concrete must reflect two characteristics. The first is the nonlinear relationship between stress and strain. The second, and perhaps more important, is the softening of concrete in compression caused by cracking due to tension in the perpendicular direction. Consequently, a softening coefficient will be incorporated in the equation for the compressive stress–strain relationship of concrete.

Because of the crucial importance of the softening effect on the biaxial constitutive laws of reinforced concrete, this model has been named the *softened truss model*. The word "softened" in this name implies two characteristics: First, the analysis must be nonlinear and second, the softening of concrete must be taken into account.

From Chapters 4 and 5 the stresses and strains in a membrane element should satisfy the following equilibrium and compatibility equations:

Equilibrium Equations

$$\sigma_l = \sigma_d \cos^2 \alpha + \sigma_r \sin^2 \alpha + \rho_l f_l + \rho_{lp} f_{lp} \qquad \text{(7-1) or } \boxed{1}$$

$$\sigma_t = \sigma_d \sin^2 \alpha + \sigma_r \cos^2 \alpha + \rho_t f_t + \rho_{tp} f_{tp} \qquad \text{(7-2) or } \boxed{2}$$

$$\tau_{lt} = (-\sigma_d + \sigma_r)\sin \alpha \cos \alpha \qquad \text{(7-3) or } \boxed{3}$$

Compatibility Equations

$$\varepsilon_l = \varepsilon_d \cos^2 \alpha + \varepsilon_r \sin^2 \alpha \qquad\qquad \text{(7-4) or } \boxed{4}$$

$$\varepsilon_t = \varepsilon_d \sin^2 \alpha + \varepsilon_r \cos^2 \alpha \qquad\qquad \text{(7-5) or } \boxed{5}$$

$$\frac{\gamma_{lt}}{2} = (-\varepsilon_d + \varepsilon_r)\sin \alpha \cos \alpha \qquad\qquad \text{(7-6) or } \boxed{6}$$

where

ρ_{lp}, ρ_{tp} = prestressing steel ratio in the l and t directions, respectively

f_{lp}, f_{tp} = stresses in prestressing steel in the l and t directions, respectively

It should be pointed out that a new sequence of numbers is introduced for the fundamental equations that will be referred to frequently. Each of these equation numbers is enclosed in a *box* without the chapter designation.

It should also be noted that an improvement has been made in the preceding equations. Equations $\boxed{1}$ and $\boxed{2}$ include the stresses due to the prestressing steel, $\rho_{lp}f_{lp}$ and $\rho_{tp}f_{tp}$. These equations have now been generalized to include prestressed concrete membrane elements. The validity of adding these new terms can be verified easily by examining the equilibrium of a prestressed concrete membrane element as shown in Figure 7.1. Detailed derivation of these equations can proceed in the same manner as for reinforced concrete in Section 4.3.1. If the prestressing steel is absent, i.e., $\rho_{lp} = \rho_{tp} = 0$, these equations, of course, degenerate into those for reinforced concrete.

The solution of the preceding six equilibrium and compatibility equations requires six stress–strain relationships for (1) concrete in compression, relating σ_d to ε_d, (2) concrete in tension, relating σ_r to ε_r, (3) mild steel in longitudinal direction, relating f_l to ε_l, (4) mild steel in transverse direction, relating f_t to ε_t, (5) prestressing steel in longitudinal direction, relating f_{lp} to ε_{lp}, and (6) prestressing steel in transverse direction, relating f_{tp} to ε_{tp}. The first two stress–strain relationships for concrete are sketched in Figure 7.2a and b, respectively. The third and fourth stress–strain relationships for mild steel are given in Figure 7.2c, and the last two for prestressing steel in Figure 7.2d.

In general, each stress–strain relationship can be described by one equation, except in the case of concrete in compression. As shown in Figure 7.2a, the compressive stress–strain curve of concrete is a function of the peak softening coefficient ζ, which, in turn, is a function of the tensile and compression strains of concrete, ε_r and ε_d. Consequently, an additional equation will be required for the softening coefficient ζ. The seven equations for the constitutive laws of materials are summarized as follows:

Constitutive Laws of Materials

Concrete

$$\sigma_d = f_1(\varepsilon_d, \zeta) \qquad\qquad \text{(7-7) or } \boxed{7}$$

$$\zeta = f_2(\varepsilon_r, \varepsilon_d) \qquad\qquad \text{(7-8) or } \boxed{8}$$

$$\sigma_r = f_3(\varepsilon_r) \qquad\qquad \text{(7-9) or } \boxed{9}$$

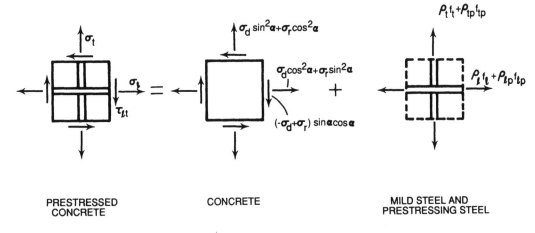

Figure 7.1 Stress condition in prestressed concrete.

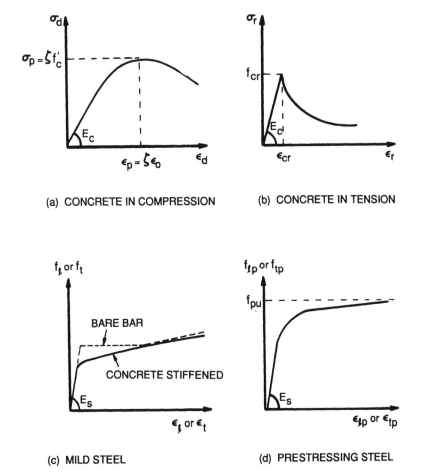

Figure 7.2 Stress –strain relationships in membrane elements.

Mild steel

$$f_l = f_4(\varepsilon_l) \qquad \text{(7-10) or } \boxed{10}$$

$$f_t = f_5(\varepsilon_t) \qquad \text{(7-11) or } \boxed{11}$$

Prestressing steel

$$f_{lp} = f_6(\varepsilon_{lp}) = f_6(\varepsilon_{\text{dec}} + \varepsilon_l) \qquad \text{(7-12) or } \boxed{12}$$

$$f_{tp} = f_7(\varepsilon_{lp}) = f_7(\varepsilon_{\text{dec}} + \varepsilon_t) \qquad \text{(7-13) or } \boxed{13}$$

where ε_{dec} is the strain in prestressing steel at decompression of concrete. This strain ε_{dec} depends on the amount of prestressing and is considered a given constant value in the analysis. The seven functions f_1 to f_7 in Eqs. $\boxed{7}$ to $\boxed{13}$ are to be determined.

Details of the constitutive laws will be presented in the next section.

7.2   Constitutive Laws in Membrane Elements

7.2.1   SOFTENED COMPRESSION STRESS–STRAIN RELATIONSHIP OF CONCRETE

The truss model has been applied to treat shear and torsion of reinforced concrete since the turn of the 20th century. However, the prediction based on the truss model consistently overestimated the shear and torsional strengths of tested specimens. The overestimation might exceed 50% in the case of low-rise shear walls and 30% in the case of torsional members. This nagging mystery has plagued researchers for over half a century. The source of this difficulty was first understood by Robinson and Demorieux (1972). They realized that a reinforced concrete membrane element subjected to shear stresses is actually subjected to biaxial compression–tension stresses in the 45° direction. Viewing the shear action as a two-dimensional problem, they discovered that the compressive strength in one direction was reduced by cracking due to tension in the perpendicular direction. Applying this softened effect of concrete struts to the thin webs of eight test beams with I-section, they were able to explain the equilibrium of stresses in the webs according to the truss model. Apparently, the mistake in applying the truss model theory before 1972 was the use of the compressive stress–strain relationship of concrete obtained from the uniaxial tests of standard cylinders without considering this two-dimensional softening effect.

The tests of Robinson and Demorieux, unfortunately, could not delineate the variables that govern the softening coefficient, because of the technical difficulties in the biaxial testing of large panels. The quantification of the softening phenomenon, therefore, had to wait for a decade until a unique "shear rig" test facility was built in 1981 by Vecchio and Collins (Vecchio and Collins, 1981). Based on their tests of 17 panels of 89 cm square and 7 cm thick, they proposed a softening coefficient that was a function of the ratio of the tensile principal strains to the compression principal strain, $\varepsilon_r / \varepsilon_d$.

The discovery and the quantification of this softening phenomenon during the last two decades has provided the major breakthrough in understanding of the shear and torsion problem in reinforced concrete. At present (1991), three giant panel testing

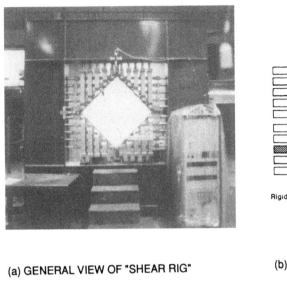

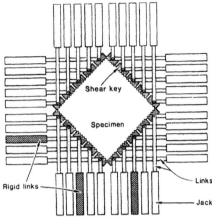

(a) GENERAL VIEW OF "SHEAR RIG"

(b) JACKS-AND-LINKS ASSEMBLY

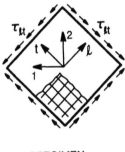

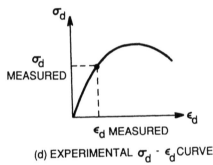

(c) TEST SPECIMEN

(d) EXPERIMENTAL σ_d - ϵ_d CURVE

Figure 7.3   Vecchio –Collins' experiment.

facilities with capacity higher than the "shear rig" by an order of magnitude have been constructed in Japan, Canada, and the United States. Using these facilities to test full-size panels, extensive research is still being carried out to define the softened compressive stress–strain curve of concrete and the softening coefficient.

7.2.1.1 *Experimental Stress–Strain Curve*

Vecchio and Collins' shear rig and their testing technique will be described briefly. As shown in Figure 7.3a and b, the shear rig consists of a steel frame housing 3 rigid links and 37 double-acting hydraulic jacks, each with a 10-ton capacity (in tension). The test specimen was oriented at 45° to the direction of the jacks with the coordinate as shown in Figure 7.3c. By compressing the panel specimen in the vertical direction (along axis 2) and tensioning with equal magnitude in the horizontal direction (along axis 1), a shear stress τ_{lt} was created at the four edges of the panel as shown in Figure 7.3c. This shear stress τ_{lt} was increased in load stages until the failure of the test panel.

The described testing produces a softened compressive stress–strain curve for concrete as shown in Figure 7.3d. The plotting of this σ_d versus ε_d curve from the

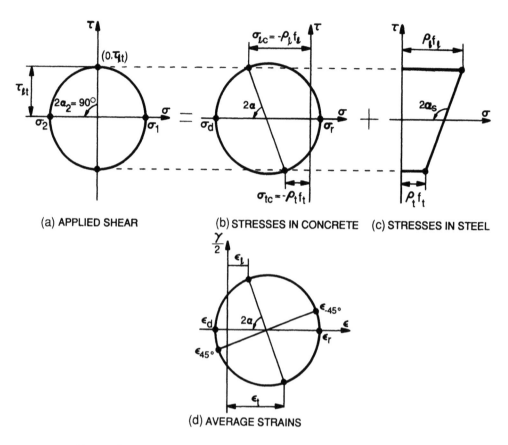

Figure 7.4 Mohr circles for stresses and strains in test panel.

test results was based on the following methodology:

1. At each load stage in the test panel, the stress condition due to the applied shear stress τ_{lt} was shown by a Mohr circle in Figure 7.4a.

2. At each load stage, the average strains were measured on the surfaces of the panel in all four directions (ε_l, ε_t, ε_1, and ε_2). From these four measured strains the Mohr circle for strains was established as shown in Figure 7.4d. (Actually only three strains are required to establish the Mohr circle; the fourth is an added redundant strain to increase the accuracy). From this Mohr strain circle, the strain in the d direction, ε_d, and the angle α were determined.

3. From the two measured steel strains ε_l and ε_t, the smeared steel stresses $\rho_l f_l$ and $\rho_t f_t$ were calculated from the constitutive laws of steel and were recorded in Figure 7.4c. From the equilibrium of stresses in the longitudinal and transverse direction the concrete stresses were computed: $\sigma_{lc} = -\rho_l f_l$ and $\sigma_{tc} = -\rho_t f_t$. Using the concrete stresses σ_{lc} and σ_{tc}, and the applied shear stress τ_{lt}, the Mohr circle for concrete stresses were completed in Figure 7.4b. From this Mohr circle for concrete stresses, the stress in the d direction, σ_d, and the angle α were determined.

4. From the measured strain ε_d and the measured stress σ_d, a point was obtained on the softened compressive stress–strain curve of concrete as shown in Figure

7.3d. By repeating this procedure for all the load stages, a complete stress–strain curve was established.

7.2.1.2 Mathematical Expressions for Stress–Strain Curves

Nonsoftened Curve

The stress–strain curve of a standard concrete cylinder subjected to a uniaxial compression is usually expressed mathematically by a parabolic curve:

$$\sigma_d = f_c' \left[2 \left(\frac{\varepsilon_d}{\varepsilon_0} \right) - \left(\frac{\varepsilon_d}{\varepsilon_0} \right)^2 \right] \qquad (7\text{-}14)$$

where ε_0 is the strain at the peak stress f_c' and is usually taken as 0.002.

Equation (7-14) is plotted graphically in Figure 7.5a and will be referred to as the nonsoftened stress–strain curve. The right side of the equation consists of two terms. The first is a linear term shown by the dotted straight line OP; the second represents the vertical lengths in the shaded area. Note also that the initial slope of the curve is $E_c = 2f_c'/\varepsilon_0$.

It should be noted that in an actual stress–strain curve of concrete the stress does not become zero when the strain reaches $2\varepsilon_0$. Because the last $1/8$ portion of the curve is not valid, the usefulness of Eq. (7-14) extends only up to about $1.75\varepsilon_0$, or about a strain of 0.0035.

Curve Softened in Stress Only

When the peak stress is softened linearly by a softened coefficient ζ, which varies from zero to unity, then the softened peak stress σ_p is

$$\sigma_p = \zeta f_c' \qquad (7\text{-}15)$$

and the stress-softened curve becomes

$$\sigma_d = \zeta f_c' \left[2 \left(\frac{\varepsilon_d}{\varepsilon_0} \right) - \left(\frac{\varepsilon_d}{\varepsilon_0} \right)^2 \right] \qquad (7\text{-}16)$$

Equation (7-16) is plotted as the solid curve in Figure 7.5b. The initial slope of this stress-softened curve is $E_c = 2\zeta f_c'/\varepsilon_0$.

Curve Softened in Both Stress and Strain

When both the stress and the strain at the peak point of the stress–strain curve are softened, then two softened coefficients must be defined, one ζ_σ for the stress and another ζ_ε for the strain:

$$\sigma_p = \zeta_\sigma f_c' \qquad (7\text{-}17)$$

$$\varepsilon_p = \zeta_\varepsilon \varepsilon_0 \qquad (7\text{-}18)$$

As shown in Figure 7.5c, the complete stress–strain curve must now be expressed by two equations: one for the ascending branch and another for the descending branch:

Ascending Branch $(\varepsilon_d / \zeta_\varepsilon \varepsilon_0 \leq 1)$

$$\sigma_d = \zeta_\sigma f_c' \left[2 \left(\frac{\varepsilon_d}{\zeta_\varepsilon \varepsilon_0} \right) - \left(\frac{\varepsilon_d}{\zeta_\varepsilon \varepsilon_0} \right)^2 \right] \qquad (7\text{-}19)$$

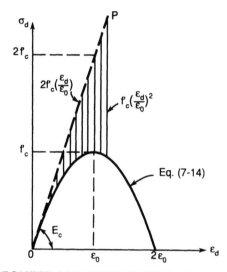

(a) GEOMETRY OF NONSOFTENED CURVE

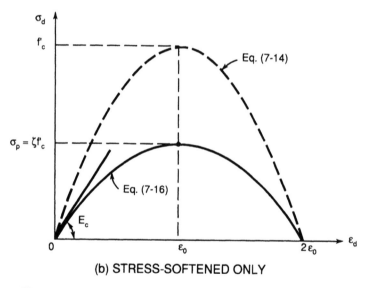

(b) STRESS-SOFTENED ONLY

Figure 7.5 Compression stress–strain curve of concrete.

Equation (7-19) is the ascending portion of the solid curve in Figure 7.5c with the initial slope $E_c = 2\zeta_\sigma f'_c / \zeta_\varepsilon \varepsilon_0$.

Descending Branch $(\varepsilon_d / \zeta_\varepsilon \varepsilon_0 > 1)$

The descending portion of the solid curve in Figure 7.5c is also assumed to be a parabolic curve from the peak point to the point of $2\varepsilon_0$ on the horizontal axis. The vertical distance from the parabolic curve to the peak stress level is designated as Δ. This vertical distance Δ is located at a horizontal distance $\varepsilon_d - \varepsilon_p$ from the peak point. At a horizontal distance of $2\varepsilon_0 - \varepsilon_p$ from the peak point, however, the vertical distance from the parabolic curve to the peak stress level is σ_p. The ratio of Δ / σ_p can

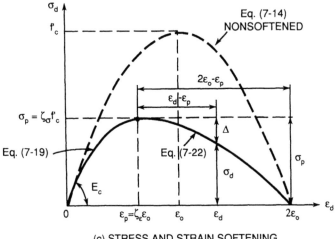

(c) STRESS AND STRAIN SOFTENING

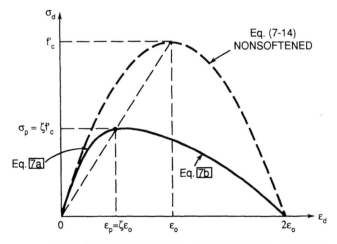

(d) PROPORTIONAL SOFTENING OF STRESS AND STRAIN

Figure 7.5 (*Continued*)

be obtained from the geometry of parabolic shape:

$$\frac{\Delta}{\sigma_p} = \left(\frac{\varepsilon_d - \varepsilon_p}{2\varepsilon_0 - \varepsilon_p}\right)^2 \tag{7-20}$$

Then, the stress σ_d at the location of ε_d is

$$\sigma_d = \sigma_p - \Delta = \sigma_p\left[1 - \left(\frac{\varepsilon_d - \varepsilon_p}{2\varepsilon_0 - \varepsilon_p}\right)^2\right] \tag{7-21}$$

or, more convenient for calculation,

$$\sigma_d = \zeta_\sigma f_c'\left[1 - \left(\frac{\varepsilon_d/\zeta_\varepsilon\varepsilon_0 - 1}{2/\zeta_\varepsilon - 1}\right)^2\right] \tag{7-22}$$

Proportional Softening of Stress and Strain

If the two softened coefficients are close, it is often convenient to assume a proportional softening of stress and strain, i.e., to set $\zeta_\sigma = \zeta_\varepsilon = \zeta$. In this case, Eqs. (7-19) and (7-22) are simplified to:

Ascending Branch $(\varepsilon_d / \zeta\varepsilon_0 \leq 1)$

$$\sigma_d = \zeta f_c' \left[2\left(\frac{\varepsilon_d}{\zeta\varepsilon_0} \right) - \left(\frac{\varepsilon_d}{\zeta\varepsilon_0} \right)^2 \right]$$ (7-23) or 7a

Descending Branch $(\varepsilon_d / \zeta\varepsilon_0 > 1)$

$$\sigma_d = \zeta f_c' \left[1 - \left(\frac{\varepsilon_d / \zeta\varepsilon_0 - 1}{2/\zeta - 1} \right)^2 \right]$$ (7-24) or 7b

A stress–strain curve with proportional softening is shown in Figure 7.5d. For a family of such curves with decreasing softening coefficient ζ, the locus of the peak points traces a straight line passing through the origin.

7.2.1.3   *Peak Softening Coefficient* ζ

The softened effects of concrete are apparently caused by the diagonal shear cracking of concrete. Consequently, the softening coefficient ζ must be a function of a parameter that measures the severity of cracking. The most important parameter to measure the severity of cracking is the tensile strain in the r direction, ε_r. To a lesser extent, ζ is also a function of the compressive strain in the d direction, ε_d, and other variables.

University of Toronto

The stress–strain curve proposed by Vecchio and Collins (1981) was Eqs. 7a and 7b with proportional softening for stress and strain. The corresponding softening coefficient was

$$\zeta = \frac{1}{\sqrt{0.7 - (\varepsilon_r / \epsilon_d)}}$$ (7-25)

In 1986, however, they proposed (Vecchio and Collins, 1986) the use of Eq. (7-16) for stress–strain curves with stress softening only. The softening coefficient was also simplified as

$$\zeta = \frac{1}{0.8 + 170\varepsilon_r}$$ (7-26)

Equation (7-26) is considerably more conservative than Eq. (7-25) when ε_r exceeds 1%.

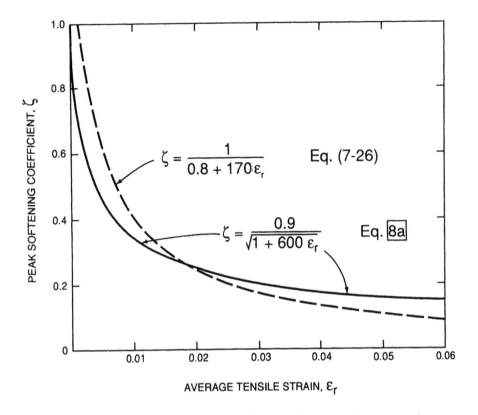

Figure 7.6 Peak softening coefficient of concrete in compression.

University of Houston

In 1991, formulas for the softening effect were proposed by the author and his colleagues at the University of Houston, based on the tests of 35 full-size panels 55 in. square and 7 in. thick. The tests were sponsored by the National Science Foundation (Belarbi and Hsu, 1991; Pang and Hsu, 1992). Two important conclusions were reached. First, both the stress-softening and the strain-softening effects were observed in the compressive stress–strain curves of concrete. Therefore, Eqs. 7a and 7b, which consider a proportional softening of stress and strain, are more realistic than Eq. (7-16), which takes into account only the stress softening. Second, the softening coefficient is a function of four parameters, namely the tensile strain ε_r, the compressive strain ε_d, the α_2 angle between the longitudinal bars and the principal compressive load, and the load path between the two principal strains. Among these parameters, the first one is the most important.

For simplicity, a softening coefficient ζ was proposed, which is only a function of the tensile strain ε_r:

$$\zeta = \frac{0.9}{\sqrt{1 + 600\varepsilon_r}} \qquad \text{(7-27) or } \boxed{8a}$$

The constant 0.9 in the numerator of Eq. 8a takes into account the size effect, the shape effect, and the loading rate effect between the testing of standard cylinders and the testing of panels. Equation 8a is a conservative, but practical, lower boundary of all the 35 test points.

Equations (7-26) and 8a are compared in Figure 7.6.

7.2.2   TENSILE STRESS–STRAIN RELATIONSHIP OF CONCRETE

From the tests of panels subjected to shear, it was clear that the tensile stress of concrete, σ_r, is not zero as assumed in the simple truss model. This can be verified by the measured Mohr circle for concrete stresses in Figure 7.4. The stress σ_r in the Mohr circle is an average uniform tensile stress of concrete, representing the stiffening of the steel bars by concrete in tension.

Figure 7.7 shows a typical tensile stress–strain curve of concrete. The curve consists of two distinct branches. Before cracking the stress–strain relationship is essentially linear. Upon cracking, however, a drastic drop of strength occurs and the descending branch of the curve becomes concave. In the descending branch, the concrete is cracked and the concept of concrete tensile stress σ_r and concrete tensile strain ε_r are quite different from those before cracking. σ_r is defined as the *average* concrete tensile stress and ε_r the *average* concrete tensile strain. These terms will be defined in Section 7.2.3.2 in conjunction with the stress and strain of mild steel.

University of Toronto

Formulas for the tensile stress–strain curve of concrete was given by Vecchio and Collins (1981) based on their panel tests at the University of Toronto:

Ascending branch $(\varepsilon_r \leq \varepsilon_{cr})$

$$\sigma_r = E_c \varepsilon_r \tag{7-28}$$

where

$E_c = 2f_c'/\varepsilon_0 =$ same initial slope as nonsoftened compression stress–strain curve

$\varepsilon_{cr} =$ strain at cracking of concrete $= f_{cr}/E_c$

$f_{cr} =$ stress at cracking of concrete $= 4\sqrt{f_c'}$, where f_c' and $\sqrt{f_c'}$ are in pounds per square inch

Descending branch $(\varepsilon_r > \varepsilon_{cr})$

The relationship of the average stress σ_r and the average strain ε_r is found from tests to be

$$\sigma_r = \frac{f_{cr}}{1 + \sqrt{\dfrac{\varepsilon_r - \varepsilon_{cr}}{0.005}}} \tag{7-29}$$

University of Houston

A set of formulas were recommended by the author and his colleagues at the University of Houston, based on the tests of 35 full-size panels, sponsored by the National Science Foundation (Belarbi and Hsu, 1991; Pang and Hsu, 1992):

Ascending branch $(\varepsilon_r \leq \varepsilon_{cr})$

$$\sigma_r = E_c \varepsilon_r \tag{7-30 or \boxed{9a}}$$

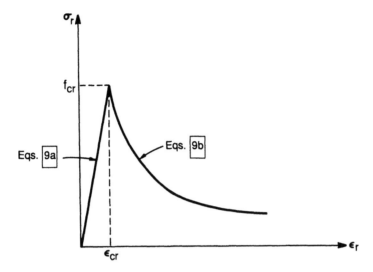

Figure 7.7 Tensile stress–strain curve of concrete.

where

$$E_c = 47{,}000 \sqrt{f_c'}, \text{ where } f_c' \text{ and } \sqrt{f_c'} \text{ are in pounds per square inch}$$

$$\varepsilon_{cr} = 0.00008 \text{ in./in.}$$

$$f_{cr} = 3.75\sqrt{f_c'}$$

Descending branch $(\varepsilon_r > \varepsilon_{cr})$

$$\sigma_r = f_{cr}\left(\frac{\varepsilon_{cr}}{\varepsilon_r}\right)^{0.4} \qquad\qquad \text{(7-31) or } \boxed{9b}$$

Equation $\boxed{9b}$ was first proposed by Tamai et al. (1987) at the University of Tokyo based on tension members. This equation was also found to be applicable to panels at the University of Houston as explained in Section 7.2.3.2. Equations $\boxed{9a}$ and $\boxed{9b}$ are plotted in Figure 7.7.

7.2.3   STRESS–STRAIN RELATIONSHIP OF MILD STEEL

7.2.3.1   *Bare Bars*

The stress–strain curve of mild steel is usually assumed to be elastic–perfectly plastic, i.e.,

$$\text{when } \varepsilon_l < \varepsilon_{ly}, \qquad f_l = E_s \varepsilon_l \qquad\qquad (7\text{-}32)$$

$$\text{when } \varepsilon_l \geq \varepsilon_{ly}, \qquad f_l = f_{ly} \qquad\qquad (7\text{-}33)$$

$$\text{when } \varepsilon_t < \varepsilon_{ty}, \qquad f_t = E_s \varepsilon_t \qquad\qquad (7\text{-}34)$$

$$\text{when } \varepsilon_t \geq \varepsilon_{ty}, \qquad f_t = f_{ty} \qquad\qquad (7\text{-}35)$$

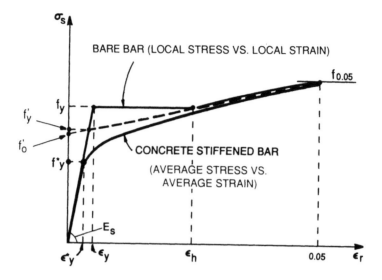

Figure 7.8   Stress–strain curve of mild steel.

where

E_s = modulus of elasticity of steel bars

f_{ly}, f_{ty} = yield stresses of longitudinal and transverse steel bars, respectively

$\varepsilon_{ly}, \varepsilon_{ty}$ = yield strains of longitudinal and transverse steel bars, respectively

Equations (7-32) and (7-33) [or Eqs. (7-34) and (7-35)], obtained from the testing of bare, mild steel bars, are plotted in Figure 7.8. Also shown in the figure is the stress–strain curve of steel bars embedded in concrete. It can be seen that these two stress–strain curves are quite different. The stress–strain curve of a steel bar in concrete relates the *average stress* to the *average strain* of a large length of bar crossing several cracks, whereas the stress–strain curve of a bare bar relates the stress to the strain at a local point.

7.2.3.2   Steel Bars Embedded in Concrete

Average Stress σ_r and Average Strain ε_r of Concrete in Tension

A steel bar surrounded by concrete and subjected to a tensile force P is shown in Figure 7.9. At the two cracks indicated, the steel stress will be designated σ_{s0} as indicated in Figure 7.9b. Between these two cracks, however, the steel stress $\sigma_s(r)$ at any section r will be less than σ_{s0}, and the difference will be carried by the concrete in tension, $\sigma_c(r)$. From the longitudinal equilibrium of forces at any section r between the two cracks (Figure 7.9a), we can write

$$P = A_s \sigma_{s0} = A_s \sigma_s(r) + A_c \sigma_c(r) \tag{7-36}$$

Expressing Eq. (7-36) in terms of stresses,

$$\sigma_{s0} = \sigma_s(r) + \frac{1}{\rho} \sigma_c(r) \tag{7-37}$$

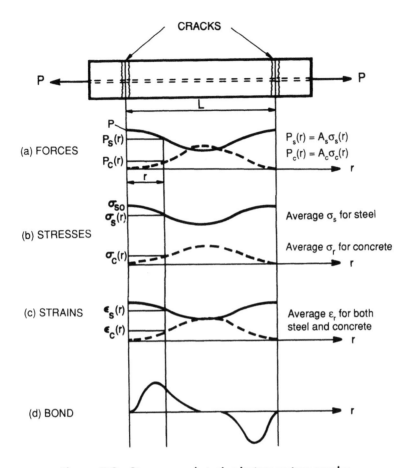

Figure 7.9   **Stresses and strains between two cracks.**

where $\rho = A_s/A_c$, the percentage of steel based on the net concrete section A_c. Equation (7-37) states that at any section between the two cracks, the sum of the steel stress $\sigma_s(r)$ and the concrete stress $\sigma_c(r)$ divided by ρ must be equal to the steel stress at the cracked section σ_{s0}.

The steel strain $\varepsilon_s(r)$ and the concrete strain $\varepsilon_c(r)$ are also sketched in Figure 7.9c. The steel strain $\varepsilon_s(r)$ decreases from a maximum at the crack to a minimum at the midpoint between the two cracks. In contrast, the concrete strain $\varepsilon_c(r)$ should be zero at the crack and increases to a maximum at the midpoint. The difference between $\varepsilon_s(r)$ and $\varepsilon_c(r)$ is caused by the slip between the steel bar and the surrounding concrete. The slip results in the bond stresses sketched in Figure 7.9d and the gaps that constitute the cracks.

Let us recall that the solution of the equilibrium and compatibility equations [Eqs. $\boxed{1}$ to $\boxed{6}$] requires the constitutive relationship between the *average tensile stress of concrete* σ_r and the *average tensile strain of concrete* ε_r in the r direction. This average strain of concrete ε_r should be measured along a length that crosses several cracks. This average strain ε_r is not the average value of the concrete strain $\varepsilon_c(r)$ shown in Figure 7.9c, because the average strain ε_r includes not only the strain of the concrete itself, but also the strain contributed by the crack widths. Hence, the average strain of concrete ε_r must be obtained from averaging the steel strain $\varepsilon_s(r)$. In other words, the average strain ε_r represents both the average strain of steel and the average strain of concrete including the crack widths. With this understanding in mind the average tensile strain ε_r is defined as

$$\varepsilon_r = \frac{1}{L} \int_0^L \varepsilon_s(r)\, dr \qquad (7\text{-}38)$$

Before the first yielding of the steel, the linear relationship $\varepsilon_s(r) = \sigma_s(r)/E_s$ is valid at any cross section r. Substituting $\varepsilon_s(r)$ into Eq. (7-38) gives

$$\varepsilon_r = \frac{1}{E_s} \left(\frac{1}{L} \int_0^L \sigma_s(r)\, dr \right) \qquad (7\text{-}39)$$

The term in the parentheses of Eq. (7-39) will be defined as the *average stress of steel* σ_s, i.e.,

$$\sigma_s = \frac{1}{L} \int_0^L \sigma_s(r)\, dr \qquad (7\text{-}40)$$

Then Eq. (7-39) gives a linear relationship between the average stress of steel σ_s and the average strain of steel ε_r:

$$\varepsilon_r = \frac{1}{E_s} \sigma_s \qquad (7\text{-}41)$$

Using this concept of averaging, we can now average the steel stresses $\sigma_s(r)$ and the concrete stresses $\sigma_c(r)$ in Eq. (7-37) by integrating these stresses along r and divided by the length L:

$$\sigma_{s0} = \frac{1}{L} \int_0^L \sigma_s(r)\, dr + \frac{1}{\rho} \left(\frac{1}{L} \int_0^L \sigma_c(r)\, dr \right) \qquad (7\text{-}42)$$

The first term on the right-hand side of Eq. (7-42) is obviously the average stress of steel σ_s as defined in Eq. (7-40). The quantity enclosed by the parentheses in the second term is the average tensile stress of concrete σ_r:

$$\sigma_r = \frac{1}{L} \int_0^L \sigma_c(r)\, dr \qquad (7\text{-}43)$$

Then Eq. (7-42) becomes

$$\sigma_{s0} = \sigma_s + \frac{1}{\rho} \sigma_r \qquad (7\text{-}44)$$

Substituting $\sigma_s = E_s \varepsilon_r$ from Eq. (7-41) into Eq. (7-44) we derive the relationship between the average tensile stress of concrete σ_r and the average tensile strain of concrete ε_r:

$$\sigma_r = \rho(\sigma_{s0} - E_s \varepsilon_r) = \rho \left(\frac{P}{A_s} - E_s \varepsilon_r \right) = \frac{P}{A_c} - \rho E_s \varepsilon_r \qquad (7\text{-}45)$$

For each load stage in a panel test, the load P and the average tensile strain of concrete ε_r are measured. An average tensile stress of concrete σ_r can then be calculated by Eq. (7-45). Plotting σ_r versus ε_r consecutively for all load stages will give an experimental tensile stress–strain curve for concrete in the r direction. It has been found that the best mathematical form to fit the descending branch of the experimental stress–strain curve is

$$\sigma_r = f_{cr} \left(\frac{\varepsilon_{cr}}{\varepsilon_r} \right)^c \tag{7-46}$$

and the best experimental values for f_{cr}, ε_{cr}, and c are

f_{cr} = tensile cracking strength of concrete, taken as $3.75\sqrt{f_c'}$

ε_{cr} = strain at cracking of concrete, taken as 0.00008 in./in.

c = a constant, equal to 0.4

Equation (7-46) is, of course, identical to Eq. (7-31) and is plotted in Figure 7.7.

Apparent Yield Stress of Steel f_y^*

Inserting $\varepsilon_r = \sigma_s/E_s$ from Eq. (7-41) and $c = 0.4$ into Eq. (7-46) and, in turn, substituting σ_r from Eq. (7-46) into Eq. (7-44) we have

$$\sigma_{s0} = \sigma_s + \frac{f_{cr}}{\rho} \left(\frac{E_s \varepsilon_{cr}}{\sigma_s} \right)^{0.4} \tag{7-47}$$

Equation (7-47) relates directly the steel stress at the crack σ_{s0} to the average steel stress σ_s.

Yielding of a reinforced concrete panel occurs when the steel stress at the cracked section reaches the yield plateau, i.e., $\sigma_{s0} = f_y$. At the same time, the average steel stress reaches a level that we shall call the apparent yield stress of steel, f_y^*, i.e., $\sigma_s = f_y^*$. Substituting $\sigma_{s0} = f_y$ and $\sigma_s = f_y^*$ into Eq. (7-47), dividing by f_y, and rearranging the terms result in

$$\left(\frac{f_y^*}{f_y} \right)^{0.4} - \left(\frac{f_y^*}{f_y} \right)^{1.4} = \frac{f_{cr}}{\rho f_y} \left(\frac{E_s \varepsilon_{cr}}{f_y} \right)^{0.4} \tag{7-48}$$

Using the relationships $f_{cr} = E_c \varepsilon_{cr}$ and $E_s/E_c = n$, Eq. (7-48) becomes

$$\left(\frac{f_y^*}{f_y} \right)^{0.4} - \left(\frac{f_y^*}{f_y} \right)^{1.4} = \frac{n^{0.4}}{\rho} \left(\frac{f_{cr}}{f_y} \right)^{1.4} \tag{7-49}$$

The solution of Eq. (7-49) determines the apparent yield stress f_y^*. Equation (7-49) also shows that f_y^* depends on the four variables, ρ, f_{cr}, f_y, and n, but the modulus ratio n is a dependent variable of f_{cr}.

If the cracking strength of concrete f_{cr} and the yield stress of steel f_y are grouped as one nondimensionalized variable f_{cr}/f_y, then the solution of Eq. (7-49) can be expressed in terms of three nondimensionalized variables ρ, f_{cr}/f_y, and n. The primary variable is the steel percentage ρ. The lower the percentage of steel ρ, the lower the yield stress ratio f_y^*/f_y. This ratio f_y^*/f_y is particularly sensitive to ρ at low percentages of steel. The second important variable is the cracking stress ratio f_{cr}/f_y. For a higher value of f_{cr}, the contribution of concrete to resist the tensile stress is greater, resulting in a lower apparent yield stress. Because f_{cr} varies in a narrow range for concrete, its effect on the yield stress ratio f_y^*/f_y is limited. The third variable of modulus ratio n, which is dependent on f_{cr}, varies from 5 for 3000-psi concrete to 9 for 10,000-psi concrete.

The yield stress ratio f_y^*/f_y is plotted against the steel percentage ρ using the cracking stress ratio f_{cr}/f_y and the modulus ratio n as parameters in Figure 7.10. The parameter n was chosen to vary between 5 and 9 to give a band of solutions for Eq. (7-49). The solution for the apparent yield stress ratio f_y^*/f_y in Eq. (7-49) can be very closely expressed in the form

$$\frac{f_y^*}{f_y} = 1 - 1.314\left[\frac{n^{0.4}}{\rho}\left(\frac{f_{cr}}{f_y}\right)^{1.4}\right]^{1.0836} = 1 - 1.314\frac{n^{0.434}}{\rho^{1.084}}\left(\frac{f_{cr}}{f_y}\right)^{1.517} \quad (7\text{-}50)$$

Note that the form of solution is expressed in terms of the quantity given on the right-hand side of Eq. (7-49). As compared to Eq. (7-49), Eq. (7-50) is exact with a coefficient of determination of 99.8%.

To simplify Eq. (7-50) for design purposes, the powers of ρ, f_{cr}/f_y, and n will be rounded out to 1, 1.5, and 0.5, respectively. This rounding out requires the constant 1.314 to be adjusted to 1.5. Then Eq. (7-50) becomes

$$\frac{f_y^*}{f_y} = 1 - \frac{1.5\sqrt{n}}{\rho}\left(\frac{f_{cr}}{f_y}\right)^{1.5} \quad (7\text{-}51)$$

where $\rho > 0.5\%$. Figure 7.10 shows that Eq. (7-51) is quite accurate when ρ is greater than 0.5%. Because $\rho = 0.5\%$ is approximately the minimum percentage of reinforcement required to develop a truss action (see Section 3.1.2.5), the limitation of $\rho > 0.5\%$ does not present any difficulty for all practical purposes. As an average solution for Eq. (7-51), the parameter n is taken as a constant 7 and the simplified equation becomes

$$\frac{f_y^*}{f_y} = 1 - \frac{4}{\rho}\left(\frac{f_{cr}}{f_y}\right)^{1.5} \quad (7\text{-}52)$$

Equation (7-52) is a good average solution for the two bands of curves shown in Figure 7-10. One band is determined by Eq. (7-49) or (7-50), and the other band from Eq. (7-51).

Post-Yield Stress–Strain Curve of Steel

The average stress–strain relationship of mild-steel bars embedded in concrete is more difficult to determine after yielding, because the steel strain at the cracked sections increases rapidly to reach the strain hardening region of the stress–strain curve. The averaging of the steel strains and the corresponding steel stresses along the length L becomes mathematically more complex. The integration process

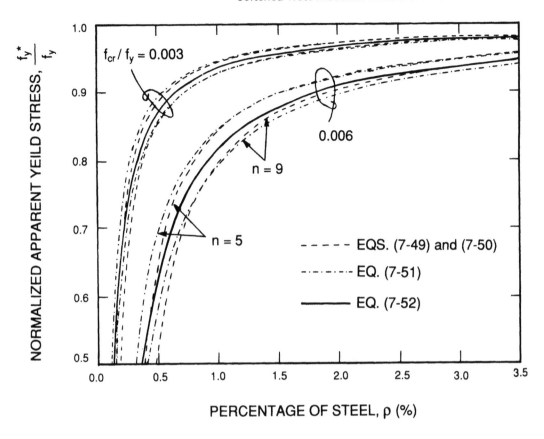

Figure 7.10 Comparison of the exact and the approximate solutions for the apparent yield stress of steel.

involved in the averaging requires numerical integration and the assistance of an electronic computer.

In order to simplify the averaging process, two assumptions are made by Tamai et al.:

a. The stress distribution in the steel between two adjacent cracks is assumed to follow a full cosine curve.

b. The average stress–strain relationship of concrete in tension [Eq.(7-46)] is valid both before and after yielding. That is to say, Eq. (7-46), which had been calibrated to fit the test results before yielding, remains valid after yielding.

From the first assumption we can write

$$\sigma_s(r) = \sigma_s + a_s \cos\frac{2\pi r}{L} \tag{7-53}$$

where a_s is the amplitude of the cosine curve. At the cracked sections, $\sigma_s(r) = \sigma_{s0}$ and $\cos(2\pi r/L) = 1$ (i.e., r equals 0 or L). Therefore

$$a_s = \sigma_{s0} - \sigma_s \tag{7-54}$$

Substituting $(\sigma_{s0} - \sigma_s)$ from Eq. (7-44) into Eq. (7-54) gives

$$a_s = \frac{\sigma_r}{\rho} \tag{7-55}$$

Substituting a_s from Eq. (7-55) into Eq. (7-53) gives

$$\sigma_s(r) = \sigma_s + \frac{\sigma_r}{\rho}\cos\frac{2\pi r}{L} \tag{7-56}$$

Now, using the second assumption and substituting σ_r from Eq. (7-46) into Eq. (7-56) gives

$$\sigma_s(r) = \sigma_s + \frac{1}{\rho}f_{cr}\left(\frac{\varepsilon_{cr}}{\varepsilon_r}\right)^{0.4}\cos\frac{2\pi r}{L} \tag{7-57}$$

With these two assumptions, the averaging process is summarized as follows:

Step 1: Select a value of the average steel stress σ_s.

Step 2: Assume an average strain ε_r.

Step 3: Calculate the distribution of steel stress $\sigma_s(r)$ from Eq. (7-57).

Step 4: Determine the corresponding distribution of steel strain $\varepsilon_s(r)$ according to the stress–strain curve of bare bars, including the strain hardening region.

Step 5: Calculate the average steel strain ε_r by numerical integration of the integral

$$\varepsilon_r = \frac{1}{L}\int_0^L \varepsilon_s(r)\,dr \tag{7-58}$$

Step 6: If ε_r calculated from Eq. (7-58) is not the same as that assumed, repeat Step 2 to 5 until the calculated ε_r is sufficiently close to the assumed value. The calculated ε_r and the selected value σ_s provide one point on the post-yield stress–strain curve.

Step 7: Select a series of σ_s values and find their corresponding ε_r values from Steps 2 to 6. Then the whole average stress–strain curve in the post-yielding range can be plotted.

A theoretical stress–strain curve of steel obtained using the preceding procedures is compared to an experimental stress–strain curve of steel in Figure 7.11. The agreement is acceptable.

The prediction of the preceding theoretical method can be improved if the stress distribution of the steel bar is expressed by a function closer to the actual condition [see Belarbi and Hsu (1991)]. The cosine function in Eq. (7-53) can be modified by adding more terms:

$$\sigma_s(r) = \sigma_s + a_s\cos\frac{2\pi r}{L} + b_s\left(\sin\frac{3\pi r}{L} - 0.6\sin\frac{5\pi r}{L} - 0.1358\right) \tag{7-59}$$

where b_s (ksi) $= 800\varepsilon_r \leq (5/7)a_s$. The constant 0.1358 is required, because the two sinusoidal terms $\sin(3\pi r/L)$ and $\sin(5\pi r/L)$ are not self-equilibrating within the length L. The original cosine term, the additional sinusoidal and constant terms, and

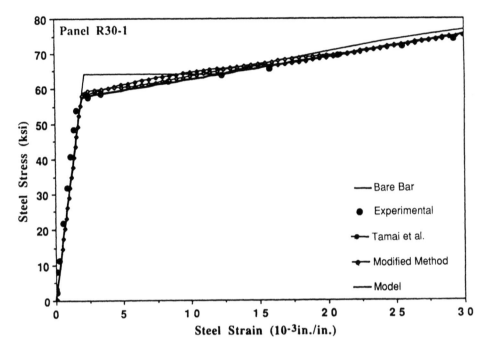

Figure 7.11 Average stress–strain curves of mild steel: theories and tests.

the resulting stress distribution are given in Figure 7.12. The modified theoretical curve based on the stress distribution of Eq. (7-59) is also shown in Figure 7.11. It can be seen that the modified method is closer to the test points. Notice also that the post-yield theoretical curve is very close to the experimental curve of the bare bar in the strain-hardening region. This is because the average tensile stress of concrete becomes negligible in this region of very large strains.

As shown in Eq. (7-52), the apparent yield stress f_y^* is a function of three variables, namely, the percentage of steel ρ, the cracking strength of concrete f_{cr}, and the yield stress of steel f_y. A parametric study using the modified method was carried out to examine the effect of each of these three variables on the post-yielding stress–strain curves of steel. When two variables are maintained constant and the third one is varied, a family of average stress–strain curves is obtained from the preceding post-yield calculation procedures. The three families of curves are shown in Figure 7.13a, b, and c.

Mathematical Modeling of Average Stress–Strain Curve

As shown in Figure 7.13a, b, and c, the shape of the average stress–strain curve of mild steel resembles two straight lines. These two straight lines will have a slope of E_s before yielding and a slope of E_p' after yielding, as illustrated in Figure 7.14. The plastic modulus E_p' after yielding is only a small fraction of the elastic modulus E_s before yielding. The stress level at which the two straight lines intersect is designated as f_y'. The equations of these two lines are then given as follows:

$$\text{when } f_s \le f_y' \qquad f_s = E_s \varepsilon_s \tag{7-60}$$

$$\text{when } f_s > f_y' \qquad f_s = f_0' + E_p' \varepsilon_s \tag{7-61}$$

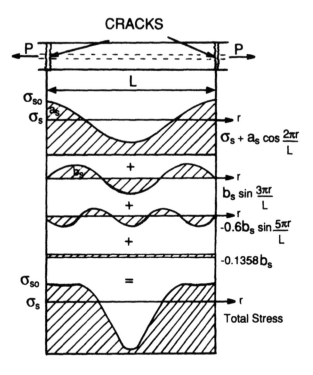

CRACKS

$\sigma_s + a_s \cos \dfrac{2\pi r}{L}$

$b_s \sin \dfrac{3\pi r}{L}$

$-0.6 b_s \sin \dfrac{5\pi r}{L}$

$-0.1358 b_s$

Total Stress

Figure 7.12 Stress distribution along a reinforcing bar in a cracked element.

where f_0' is the vertical intercept of the post-yield straight line. This vertical intercept f_0' can be calculated by

$$f_0' = \frac{E_s - E_p'}{E_s} f_y' \tag{7-62}$$

The intersection stress level f_y' and the plastic modulus E_p' depend mainly on the level of the apparent yield stress f_y^*. The lower the apparent yield stress, the lower the intersection level of the two lines and the higher the plastic modulus. Assuming that f_y' and E_p' are linear functions of f_y^*, these relationships are found to be

$$\frac{f_y'}{f_y} = 0.43 + 0.5 \frac{f_y^*}{f_y} \tag{7-63}$$

and

$$\frac{E_p'}{E_p} = 3.3 - 2.5 \frac{f_y^*}{f_y} \tag{7-64}$$

The modulus E_p in Eq. (7-64) is the slope for the strain-hardened region of the bare bar. In the case when the stress–strain curve of the bare bar is determined by tests, the plastic modulus E_p can be defined as

$$E_p = \frac{f_{0.05} - f_y}{0.005 - \varepsilon_h} \tag{7-65}$$

where

$f_{0.05}$ = stress of bare bar at a strain of 0.05 in the strain hardening region

ε_h = strain of a bare bar at the beginning of the strain hardening region

In practical applications, however, the plastic modulus E_p can be taken as 2.5% of the elastic modulus E_s or ten times the yield stress f_y. The elastic modulus E_s can be taken as 29,000 ksi.

A simple bilinear model of the average stress–strain relationship of mild steel embedded in concrete can now be derived.

Step 1: Assume $E_p = 0.025E_s$ in Eq. (7-64).

Step 2: Substitute f_y^* from Eq. (7-52) into Eqs. (7-63) and (7-64).

Step 3: Insert f_0', f_y', and E_p' from Eqs. (7-62) to (7-64) into Eqs. (7-60) and (7-61) and round up the term f_0'.

These three steps result in two simple straight lines given by

$$\text{when } f_s \le f_y' \qquad f_s = E_s \varepsilon_s \qquad\qquad (7\text{-}66) \text{ or } \boxed{10a} \text{ or } \boxed{11a}$$

$$\text{when } f_s > f_y' \qquad f_s = (0.91 - 2B)f_y + (0.02 + 0.25B)E_s \varepsilon_s \qquad (7\text{-}67)$$

where

$f_s = f_l$ or f_t when applied to longitudinal steel or transverse steel, respectively,

$\varepsilon_s = \varepsilon_l$ or ε_t when applied to longitudinal steel or transverse steel, respectively,

and

$$f_y' = (0.93 - 2B)f_y \qquad\qquad (7\text{-}68)$$

$$B = \frac{1}{\rho}\left(\frac{f_{cr}}{f_y}\right)^{1.5} \qquad\qquad (7\text{-}69)$$

In Eq. (7-69), f_{cr} is given as $3.75\sqrt{f_c'}$ (f_{cr} and f_c' in psi). Equations (7-66) and (7-67) are quite simple to use, because they are functions of only one parameter B.

Up to this point, Eqs. (7-66) and (7-67) have been derived for 90° reinforcement, or $\alpha_2 = 90°$. That means the principle compressive stress is perpendicular to the longitudinal reinforcement. It was also found from tests (Pang and Hsu, 1992), however, that the post-yield stress in the reinforcement is also a function of the reinforcement orientation. For 45° reinforcement, or $\alpha_2 = 45°$, the post-yield stress decreases by a factor of $(1 - 1/1000\rho)$. Incorporating this factor into Eqs. (7-67) and (7-68) and assuming a linear transition between 90° and 45°, we have:

$$\text{when } f_s > f_y' \qquad f_s = \left(1 - \frac{2 - \alpha_2/45°}{1000\rho}\right)\left[(0.91 - 2B)f_y + (0.02 + 0.25B)E_s \varepsilon_s\right]$$

$$\boxed{10b} \text{ or } \boxed{11b}$$

$$\text{and } f_y' = \left(1 - \frac{2 - \alpha_2/45°}{1000\rho}\right)(0.93 - 2B)f_y \qquad\qquad (7\text{-}70)$$

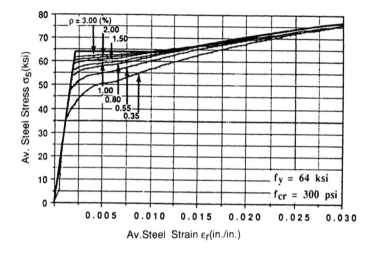

(a) PERCENTAGE OF STEEL AS PARAMETER

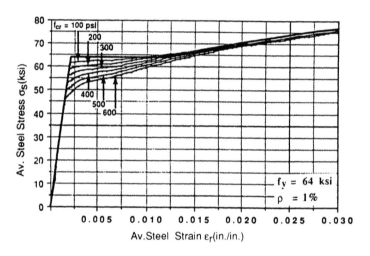

(b) CRACKING STRESS OF CONCRETE AS PARAMETER

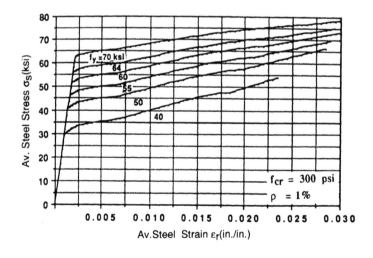

(c) YIELD STRESS OF STEEL AS PARAMETER

Figure 7.13 Average stress–strain curves of mild steel.

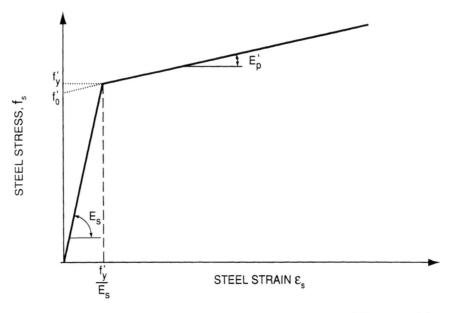

Figure 7.14 Stress–strain relationship for mild steel using bilinear model.

Equations $\boxed{10a}$ and $\boxed{10b}$ [or $\boxed{11a}$ and $\boxed{11b}$] are the average stress–strain relationships of mild-steel rebars embedded in concrete.

7.2.4   STRESS–STRAIN RELATIONSHIP OF PRESTRESSING STEEL

The stress–strain curve of bare prestressing steel consists essentially of two straight lines jointed by a curve knee (Figure 7.15). The asymptotic limit of the second straight line is a horizontal straight line having a zero slope and a vertical intercept of f_{pu}, where f_{pu} is the ultimate strength of the prestressing steel. The whole stress–strain curve of prestressing steel can be divided into two parts. The first part is a straight line up to $0.7f_{pu}$. The second part is expressed by Ramberg–Osgood equation that meets the first part at the stress level of $0.7f_{pu}$:

$$f_p \le 0.7f_{pu} \qquad f_p = E_{ps}(\varepsilon_{dec} + \varepsilon_s) \qquad\qquad \text{(7-71) or } \boxed{12a} \text{ or } \boxed{13a}$$

$$f_p > 0.7f_{pu} \qquad f_p = \frac{E'_{ps}(\varepsilon_{dec} + \varepsilon_s)}{\left[1 + \left\{(E'_{ps}(\varepsilon_{dec} + \varepsilon_s))/f_{pu}\right\}^m\right]^{1/m}} \qquad \text{(7-72) or } \boxed{12b} \text{ or } \boxed{13b}$$

where

f_p = stress in prestressing steel; f_p becomes f_{lp} or f_{tp} when applied to the longitudinal and transverse steel, respectively

ε_s = strain in the mild steel; ε_s becomes ε_l or ε_t, when applied to the longitudinal and transverse steel, respectively

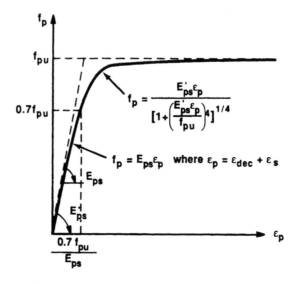

Figure 7.15 Stress–strain relationship of prestressing strands.

ε_{dec} = strain in prestressing steel at decompression of concrete

E_{ps} = elastic modulus of prestressed steel, taken as 29,000 ksi

E'_{ps} = tangential modulus of Ramberg–Osgood curve at zero load (taken as 31,060 ksi)

f_{pu} = ultimate strength of prestressing steel

m = shape parameter (taken as 4)

The strain in prestressing steel at decompression of concrete, ε_{dec}, is considered a known value and can be determined as

$$\varepsilon_{dec} = \varepsilon_{pi} + \varepsilon_i,$$

where

ε_{pi} = initial strain in prestressed steel after loss

ε_i = initial strain in mild steel after loss

ε_{dec} is approximately equal to 0.005 for 250- and 270-ksi prestressing strands.

It is, of course, possible to develop from tests the average stress–strain curve of bonded prestressing steel strands stiffened by concrete. However, the difference between such an average stress–strain curve and the stress–strain curve of a bare strand will not be significant, because the change of the strain ε_i along the length of the strand will be overwhelmed by the large and constant decompression strain ε_{dec}. Therefore, Eqs. [12a] and [12b], or [13a] and [13b], should be sufficiently accurate for prestressing strands embedded in concrete.

7.3   Solution of Equations for Membrane Elements

7.3.1   SUMMARY OF EQUATIONS

7.3.1.1   *Governing Equations*

The 13 governing equations for membrane elements are now summarized.

Equilibrium Equations

$$\sigma_l = \sigma_d \cos^2 \alpha + \sigma_r \sin^2 \alpha + \rho_l f_l + \rho_{lp} f_{lp} \qquad \boxed{1}$$

$$\sigma_t = \sigma_d \sin^2 \alpha + \sigma_r \cos^2 \alpha + \rho_t f_t + \rho_{tp} f_{tp} \qquad \boxed{2}$$

$$\tau_{lt} = (-\sigma_d + \sigma_r)\sin \alpha \cos \alpha \qquad \boxed{3}$$

Compatibility Equations

$$\varepsilon_l = \varepsilon_d \cos^2 \alpha + \varepsilon_r \sin^2 \alpha \qquad \boxed{4}$$

$$\varepsilon_t = \varepsilon_d \sin^2 \alpha + \varepsilon_r \cos^2 \alpha \qquad \boxed{5}$$

$$\frac{\gamma_{lt}}{2} = (-\varepsilon_d + \varepsilon_r)\sin \alpha \cos \alpha \qquad \boxed{6}$$

Constitutive Laws of Materials

Concrete in compression

$$\sigma_d = f_1(\varepsilon_d, \zeta) \qquad \boxed{7}$$

$$\zeta = f_2(\varepsilon_r, \varepsilon_d) \qquad \boxed{8}$$

Concrete in tension

$$\sigma_r = f_3(\varepsilon_r) \qquad \boxed{9}$$

Mild steel

$$f_l = f_4(\varepsilon_l) \qquad \boxed{10}$$

$$f_t = f_5(\varepsilon_t) \qquad \boxed{11}$$

Prestressing steel

$$f_{lp} = f_6(\varepsilon_{\text{dec}} + \varepsilon_l) \qquad \boxed{12}$$

$$f_{tp} = f_7(\varepsilon_{\text{dec}} + \varepsilon_t) \qquad \boxed{13}$$

The functions f_1 to f_7 in Eqs. $\boxed{7}$ to $\boxed{13}$ are determined by tests and can be selected for the desired degree of accuracy.

7.3.1.2 *Accurate Constitutive Equations*

If the deformation of the structure is important, such as in earthquake regions, the accurate constitutive equations summarized in this section should be used. In this set of equations, the tensile strength of concrete and the average stress–strain curve of mild steel stiffened by concrete are taken into account.

Concrete in Compression

$$\sigma_d = \zeta f_c' \left[2\left(\frac{\varepsilon_d}{\zeta \varepsilon_0} \right) - \left(\frac{\varepsilon_d}{\zeta \varepsilon_0} \right)^2 \right] \qquad \varepsilon_d / \zeta \varepsilon_0 \le 1 \qquad \boxed{7a}$$

$$\sigma_d = \zeta f_c' \left[1 - \left(\frac{\varepsilon_d / \zeta \varepsilon_0 - 1}{2/\zeta - 1} \right)^2 \right] \qquad \varepsilon_d / \zeta \varepsilon_0 > 1 \qquad \boxed{7b}$$

$$\zeta = \frac{0.9}{\sqrt{1 + 600 \varepsilon_r}} \qquad \boxed{8a}$$

Concrete in Tension

$$\sigma_r = E_c \varepsilon_r \qquad\qquad \varepsilon_r \le 0.00008 \qquad \boxed{9a}$$

$$\sigma_r = f_{cr} \left(\frac{0.00008}{\varepsilon_r} \right)^{0.4} \qquad \varepsilon_r > 0.00008 \qquad \boxed{9b}$$

where $E_c = 47{,}000\sqrt{f_c'}$ and $f_{cr} = 3.75\sqrt{f_c'}$. f_c' and $\sqrt{f_c'}$ are in pounds per square inch.

Mild Steel

$$f_s = E_s \varepsilon_s \qquad\qquad f_s \le f_y' \qquad \boxed{10a} \text{ or } \boxed{11a}$$

$$f_s = \left(1 - \frac{2 - \alpha_2/45°}{1000\rho} \right) [(0.91 - 2B)f_y$$

$$+ (0.02 + 0.25B) E_s \varepsilon_s] \qquad f_s > f_y' \qquad \boxed{10b} \text{ or } \boxed{11b}$$

where

B = a parameter defined as $(1/\rho)(f_{cr}/f_y)^{1.5}$

E_s = modulus of elasticity of steel bars, taken as 29,000 ksi

f_s = stress in mild steel; f_s becomes f_l or f_t when applied to longitudinal steel or transverse steel, respectively

$f_y' = [1 - (2 - \alpha_2/45°)/1000\rho](0.93 - 2B)f_y$

ε_s = strain in mild steel; ε_s becomes ε_l or ε_t, when applied to longitudinal and transverse steel, respectively

Prestressing Steel

$$f_p = E_{ps}(\varepsilon_{dec} + \varepsilon_s) \qquad\qquad f_p \le 0.7f_{pu} \quad \boxed{12a} \text{ or } \boxed{13a}$$

$$f_p = \frac{E'_{ps}(\varepsilon_{dec} + \varepsilon_s)}{\left[1 + \left\{(E'_{ps}(\varepsilon_{dec} + \varepsilon_s))/f_{pu}\right\}^m\right]^{1/m}} \qquad f_p > 0.7f_{pu} \quad \boxed{12b} \text{ or } \boxed{13b}$$

where

E_{ps} = modulus of elasticity of prestressed steel, take as 29,000 ksi

E'_{ps} = tangential modulus of Ramberg–Osgood curve at zero load (taken as 31,060 ksi)

f_p = stress in prestressing steel; f_p becomes f_{lp} or f_{tp} when applied to the longitudinal and transverse steel, respectively

f_{pu} = ultimate strength of prestressing steel

m = shape parameter (taken as 4 for 250- and 270-ksi prestressing strands)

ε_{dec} = strain in prestressing steel at decompression of concrete, usually taken as 0.005 for 250-ksi and 270-ksi prestressing strands

7.3.1.3 Simplified Constitutive Equations

If a structure is subjected to static loads and the deformation of the structure is not important, then a simplification of the constitutive equations could be made. In this simplification, two measures are taken simultaneously. First, the tensile strength of concrete is neglected, i.e.,

$$\sigma_r = 0 \qquad\qquad \boxed{9c}$$

and second, the stress–strain equations for bare mild-steel bars [Eqs. (7-32) to (7-35)] are used:

$$f_l = E_s\varepsilon_l \qquad \varepsilon_l < \varepsilon_{ly} \qquad\qquad \boxed{10c}$$

$$f_l = f_{ly} \qquad \varepsilon_l \ge \varepsilon_{ly} \qquad\qquad \boxed{10d}$$

$$f_t = E_s\varepsilon_t \qquad \varepsilon_t < \varepsilon_{ty} \qquad\qquad \boxed{11c}$$

$$f_t = f_{ty} \qquad \varepsilon_t \ge \varepsilon_{ty} \qquad\qquad \boxed{11d}$$

Because the first measure is conservative and the second is unconservative, the errors induced by these two measures should cancel each other in terms of strength. However, the deformations will be overestimated, because the stiffening effect of the steel bars due to concrete is neglected.

The constitutive equations for the concrete in compression ($\boxed{7a}$, $\boxed{7b}$, and $\boxed{8a}$), and for the prestressing steel ($\boxed{12a}$ and $\boxed{12b}$, or $\boxed{13a}$ and $\boxed{13b}$) remain the same.

It should be emphasized that the simultaneous employment of the stress–strain curve of bare steel bars ($\boxed{\text{10c}}$, $\boxed{\text{10d}}$, $\boxed{\text{11c}}$, and $\boxed{\text{11d}}$) and that of the concrete in tension ($\boxed{\text{9a}}$ and $\boxed{\text{9b}}$) will commit a conceptional error. Such a misleading treatment will result in "concrete strengthening," in addition to "concrete stiffening." In other words, it will lead to an unwarranted increase in strength, in addition to a correct reduction in deformation.

7.3.2   METHOD OF SOLUTION

The 13 governing equations for a membrane element (Eqs. $\boxed{1}$ to $\boxed{13}$) contain 16 unknown variables. These unknown variables include 9 stresses (σ_l, σ_t, τ_{lt}, σ_d, σ_r, f_l, f_t, f_{lp}, and f_{tp}) and 5 strains (ε_l, ε_t, γ_{lt}, ε_d, and ε_r), as well as the angle α and the material coefficient ζ. If 3 unknown variables are given, then the remaining 13 unknown variables can be solved using the 13 equations (Hsu, 1991a).

If the stresses and strains are required throughout the post-cracking loading history, then two variables must be given and a third variable selected. In general, ε_d is selected as the third variable because it varies monotonically from zero to maximum. For each given value of ε_d, the remaining 13 unknown variables can be solved. The series of solutions for various ε_d values allows us to trace the loading history.

In most of the structural applications, the two given variables come from the three externally applied stresses σ_l, σ_t, and τ_{lt}. They are generally given in two ways. First, the applied normal stresses σ_l and σ_t are given as constants, whereas the shear stress τ_{lt} is a variable to be solved. This type of problem occurs in nuclear containment structures. The stresses σ_l and σ_t in a wall element are induced by the internal pressure and are given as constants, whereas the shear stress τ_{lt} is caused by earthquake motion and is treated as an unknown variable. This first type of problem will be treated in this section. In the second type of relationship, the three stresses σ_l, σ_t, and τ_{lt} increase proportionally. They are related to the principal tensile stress σ_1 by given constants. In this way, the three variables are reduced to one variable. This type of problem occurs in all elastic structures where the three stresses on a wall element are produced simultaneously by an increasing load. This second type of problem will be treated in Section 7.4.

7.3.2.1   Characteristics of Equations

An efficient algorithm was discovered from a careful observation of the 4 characteristics of the 13 equations:

1. The three equilibrium equations (Eqs. $\boxed{1}$ to $\boxed{3}$) and the three compatibility equations (Eqs. $\boxed{4}$ to $\boxed{6}$) are transformation equations. In other words, the stresses and strains in the l-t coordinate (σ_l, σ_t, τ_{lt}, ε_l, ε_t, and γ_{lt}) are each expressed in terms of stresses and strains in the d-r direction (σ_d, σ_r, ε_d, and ε_r).

2. Equations $\boxed{7}$, $\boxed{8}$, and $\boxed{9}$ for the concrete material law involve only four unknown variables in the d-r coordinate (σ_d, σ_r, ε_d, and ε_r). If the strains ε_d and ε_r are given, then the stresses σ_d and σ_r can be calculated from these three equations.

3. Equations $\boxed{3}$ and $\boxed{6}$ are independent, because each contains one variable (τ_{lt} and γ_{lt}, respectively) that is not involved in any other equations.

4. The longitudinal steel stresses f_l and f_{lp} in the equilibrium equation Eq. $\boxed{1}$ are coupled to the longitudinal steel strain ε_l in the compatibility equation Eq. $\boxed{4}$ through the longitudinal steel stress–strain relationship of Eqs. $\boxed{10}$ and $\boxed{12}$. Similarly, the transverse steel stresses f_t and f_{tp} in the equilibrium equation Eq. $\boxed{2}$ are coupled to the transverse steel strain ε_t in compatibility equation Eq. $\boxed{5}$ through the transverse steel stress–strain relationships of Eqs. $\boxed{11}$ and $\boxed{13}$.

7.3.2.2 ε_l as a function of f_l and f_{lp}

In view of characteristic 4, the longitudinal steel strain ε_l can be expressed directly as a function of the longitudinal steel stresses f_l and f_{lp} by eliminating the angle α from Eqs. $\boxed{1}$ and $\boxed{4}$. From Eq. $\boxed{1}$, we use the first type of expression from Eq. (4-63) but including the prestressing terms, $\rho_{lp}f_{lp}$:

$$\cos^2 \alpha = \frac{-\sigma_l + \sigma_r + \rho_l f_l + \rho_{lp} f_{lp}}{\sigma_r - \sigma_d} \qquad (7\text{-}73)$$

Inserting $\varepsilon_r \sin^2 \alpha = \varepsilon_r - \varepsilon_r \cos^2 \alpha$ into Eq. $\boxed{4}$ gives

$$\cos^2 \alpha = \frac{-\varepsilon_l + \varepsilon_r}{\varepsilon_r - \varepsilon_d} \qquad (7\text{-}74)$$

Equating Eqs. (7-73) and (7-74) results in

$$\varepsilon_l = \varepsilon_r + \frac{\varepsilon_r - \varepsilon_d}{\sigma_r - \sigma_d}(\sigma_l - \sigma_r - \rho_l f_l - \rho_{lp} f_{lp}) \qquad (7\text{-}75) \text{ or } \boxed{14}$$

ε_l, f_l, and f_{lp} in Eq. $\boxed{14}$ can be solved simultaneously with the stress–strain relationships of Eqs. $\boxed{10}$ and $\boxed{12}$ for longitudinal steel.

7.3.2.3 ε_t as a function of f_t and f_{tp}

Similarly, the transverse steel strain ε_t can be expressed directly as a function of the transverse steel stresses f_t and f_{tp} by eliminating the angle α from Eqs. $\boxed{2}$ and $\boxed{5}$. From Eq. $\boxed{2}$, we use the first type of expression from Eq. (4-64) but include the prestressing terms, $\rho_{tp}f_{tp}$:

$$\sin^2 \alpha = \frac{-\sigma_t + \sigma_r + \rho_t f_t + \rho_{tp} f_{tp}}{\sigma_r - \sigma_d} \qquad (7\text{-}76)$$

Inserting $\varepsilon_r \cos^2 \alpha = \varepsilon_r - \varepsilon_r \sin^2 \alpha$ into Eq. $\boxed{5}$ gives

$$\sin^2 \alpha = \frac{-\varepsilon_t + \varepsilon_r}{\varepsilon_r - \varepsilon_d} \qquad (7\text{-}77)$$

Equating Eqs. (7-76) and (7-77) results in

$$\varepsilon_t = \varepsilon_r + \frac{\varepsilon_r - \varepsilon_d}{\sigma_r - \sigma_d}(\sigma_t - \sigma_r - \rho_t f_t - \rho_{tp} f_{tp}) \qquad (7\text{-}78) \text{ or } \boxed{15}$$

ε_t, f_t, and f_{tp} in Eq. $\boxed{15}$ can be solved simultaneously with the stress–strain relationships of Eqs. $\boxed{11}$ and $\boxed{13}$ for transverse steel.

7.3.2.4 ε_r and α as functions of ε_l, ε_t, and ε_d

Adding Eqs. $\boxed{4}$ and $\boxed{5}$ gives

$$\varepsilon_r = \varepsilon_l + \varepsilon_t - \varepsilon_d \qquad \text{(7-79) or } \boxed{16}$$

The angle α can also be derived from Eqs. $\boxed{4}$ and $\boxed{5}$. The derivation was made in Chapter 5, resulting in Eq. (5-33):

$$\tan^2 \alpha = \frac{\varepsilon_l - \varepsilon_d}{\varepsilon_t - \varepsilon_d} \qquad \text{(7-80) or } \boxed{17}$$

Equations $\boxed{14}$, $\boxed{15}$, $\boxed{16}$, and $\boxed{17}$ will be used in the proposed solution procedure.

7.3.2.5   Solution Procedures

A set of solution procedures is proposed as shown in the flow chart of Figure 7.16. The procedures are described as follows:

Step 1: Select a value of strain in the d direction, ε_d.

Step 2: Assume a value of strain in the r direction, ε_r.

Step 3: Calculate the softened coefficient ζ and the concrete stresses σ_d and σ_r from Eqs. $\boxed{8}$, $\boxed{7}$, and $\boxed{9}$, respectively.

Step 4: Solve the strains and stresses in the longitudinal steel (ε_l, f_l, and f_{lp}) from Eqs. $\boxed{14}$, $\boxed{10}$, and $\boxed{12}$, and those in the transverse steel (ε_t, f_t, and f_{tp}) from Eqs. $\boxed{15}$, $\boxed{11}$, and $\boxed{13}$.

Step 5: Calculate the strain $\varepsilon_r = \varepsilon_l + \varepsilon_t - \varepsilon_d$ from Eq. $\boxed{16}$. If ε_r is the same as assumed, the values obtained for all the strains are correct. If ε_r is not the same as assumed, then another value of ε_r is assumed and Steps 3 to 5 are repeated.

Step 6: Calculate the angle α, the shear stress τ_{lt}, and the shear strain γ_{lt} from Eqs. $\boxed{17}$, $\boxed{3}$, and $\boxed{6}$, respectively. This will provide one point on the τ_{lt} versus γ_{lt} curve.

Step 7: Select another value of ε_d and repeat Steps 2 to 6. Calculation for a series of ε_d values will provide the whole τ_{lt} versus γ_{lt} curve.

The preceding solution procedures have two distinct advantages. First, the variable angle α does not appear in the iteration process from Step 2 to Step 5. Second, the calculation of ε_l and ε_t in Step 4 can easily accommodate the nonlinear stress–strain relationships of reinforcing steel, including those for prestressing strands. These advantages were derived from an understanding of the four characteristics of the 13 governing equations. Steps 1 to 3 are proposed because of the second characteristic. Steps 4 and 5 are the results of the first and fourth characteristics, and Step 6 is possible based on the third characteristic.

EXAMPLE PROBLEM 7.1

A nuclear containment vessel is reinforced with 1.2% of mild steel in both directions and is post-tensioned only in the longitudinal direction by 0.3% of bonded

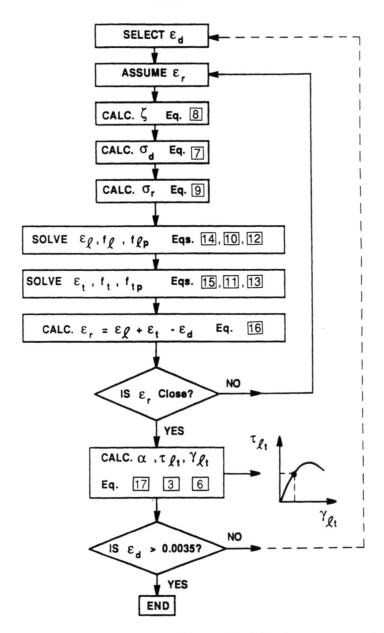

Figure 7.16 Flow chart showing the efficient algorithm (constant normal stresses).

prestressing steel. It is subjected to an internal pressure and an earthquake loading. An element taken from the vessel is shown in Figure 7.17. It is subjected to a longitudinal stress $\sigma_l = 500$ psi (3.44 MPa) and a transverse stress $\sigma_t = 250$ psi (1.72 MPa) due to an internal pressure. A shear stress τ_{lt} is induced by earthquake loading. Analyze the response of the element under the shear stress τ_{lt}, i.e., plot the τ_{lt} versus γ_{lt} curve. Use the compression stress–strain curve of softened concrete expressed by Eqs. ⟨7a⟩, ⟨7b⟩, and ⟨8a⟩, and the stress–strain relationships of prestressing steel expressed by ⟨12a⟩, ⟨12b⟩, ⟨13a⟩, and ⟨13b⟩. Also utilize the simplified constitutive laws (see Section 7.3.1.3) by neglecting the tensile strength of concrete (assume $\sigma_r = 0$, Eq. ⟨9c⟩) and by using the stress–strain curve of bare bars for mild steel (Eqs. ⟨10c⟩, ⟨10d⟩, ⟨11c⟩, and ⟨11d⟩).

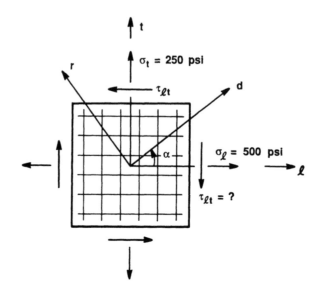

CONCRETE:	f'_c	=	6,000 psi
MILD STEEL:	f_y	=	60,000 psi
	E_s	=	29,000 ksi
	ρ_ℓ	=	ρ_t = 1.2%
PREST. STEEL:	f_{pu}	=	270 ksi
	E_{ps}	=	29,000 ksi
	E'_{ps}	=	31,060 ksi
	$\rho_{\ell p}$	=	0.3%
	ρ_{tp}	=	0
	ε_{dec}	=	0.005

Figure 7.17 Example Problem 7.1.

SOLUTION

(1) **Select** $\varepsilon_d = -0.0002$

After several cycles of trial-and-error process, assume

$$\varepsilon_r = 0.00385$$

Eq. 8a

$$\zeta = \frac{0.9}{\sqrt{1 + 600\varepsilon_r}} = \frac{0.9}{\sqrt{1 + 600(0.00385)}} = 0.495$$

$$\frac{\varepsilon_d}{\zeta\varepsilon_0} = \frac{-0.0002}{0.495(-0.002)} = 0.202 < 1 \quad \text{ascending branch}$$

Eq. 7a

$$\sigma_d = \zeta f'_c \left[2\left(\frac{\varepsilon_d}{\zeta\varepsilon_0} \right) - \left(\frac{\varepsilon_d}{\zeta\varepsilon_0} \right)^2 \right]$$

$$= 0.495(-6000)\left[2(0.202) - (0.202)^2 \right] = -1079 \text{ psi}$$

Eq. $\boxed{9c}$ $\sigma_r = 0$

$$\frac{\varepsilon_r - \varepsilon_d}{\sigma_r - \sigma_d} = \frac{0.00385 + 0.0002}{0 + 1079} = 3.754 \times 10^{-6} \text{ psi}^{-1}$$

Solve the longitudinal steel strain ε_l:

Eq. $\boxed{14}$ $\varepsilon_l = \varepsilon_r + \dfrac{\varepsilon_r - \varepsilon_d}{\sigma_r - \sigma_d}[\sigma_l - \sigma_r - \rho_l f_l - \rho_{lp} f_{lp}]$

$$= 0.00385 + 3.754 \times 10^{-6}[500 - 0 - 0.012 f_l - 0.003 f_{lp}]$$

Assume

Eq. $\boxed{10c}$ $f_l = 29 \times 10^6 \; \varepsilon_l$ before yielding

Eq. $\boxed{12a}$ $f_{lp} = 29 \times 10^6 (0.005 + \varepsilon_l)$ before elastic limit

Then

$$\varepsilon_l + 1.306 \varepsilon_l + 0.3266 \varepsilon_l = 0.00385 + 0.00188 - 0.00163$$

$$\varepsilon_l = \frac{0.00410}{2.633} = 0.00156 < 0.00207(\varepsilon_y) \quad \textbf{O.K.}$$

Because ε_{ps} at $0.7 f_{pu} - \varepsilon_{dec} = 0.00652 - 0.005 = 0.00152$,

$$\varepsilon_l = 0.00156 \approx 0.00152 \quad \text{say } \textbf{O.K.}$$

Solve the transverse steel strain ε_t:

Eq. $\boxed{15}$ $\varepsilon_t = \varepsilon_r + \dfrac{\varepsilon_r - \varepsilon_d}{\sigma_r - \sigma_d}[\sigma_t - \sigma_r - \rho_t f_t - \rho_{tp} f_{tp}]$

$$= 0.00385 + 3.754 \times 10^{-6}[250 - 0 - 0.012 f_t - 0]$$

Assume

Eq. $\boxed{11c}$ $f_t = 29 \times 10^6 \; \varepsilon_t$ before yielding

Then

$$\varepsilon_t + 1.306 \varepsilon_t = 0.00385 + 0.000939$$

$$\varepsilon_t = \frac{0.004789}{2.306} = 0.00208 \approx 0.00207(\varepsilon_y) \quad \text{say } \textbf{O.K.}$$

The tensile strain ε_r can now be checked:

Eq. $\boxed{16}$ $\varepsilon_r = \varepsilon_l + \varepsilon_t - \varepsilon_d = 0.00156 + 0.00208 + 0.0002$

$$= 0.00384 \approx 0.00385 \text{ (assumed)} \quad \textbf{O.K.}$$

Now that the strains ε_d, ε_r, ε_l, and ε_t and the stress σ_d are calculated from the trial-and-error process, we can determine the angle α, the shear stress τ_{lt}, and the shear strain γ_{lt}:

Eq. $\boxed{17}$ $\tan^2 \alpha = \dfrac{\varepsilon_l - \varepsilon_d}{\varepsilon_t - \varepsilon_d} = \dfrac{0.00156 + 0.0002}{0.00208 + 0.0002} = 0.772$

$\tan \alpha = 0.879$

$\alpha = 41.30°$ $2\alpha = 82.60°$

Eq. $\boxed{3}$ $\tau_{lt} = (-\sigma_d + \sigma_r)\sin \alpha \cos \alpha = (1079 + 0)(0.660)(0.751)$

$= 535$ psi

Eq. $\boxed{6}$ $\gamma_{lt} = 2(-\varepsilon_d + \varepsilon_r)\sin \alpha \cos \alpha$

$= 2(0.0002 + 0.00384)(0.660)(0.751) = 0.00400$

The stresses in the mild steel and prestressing steel can be calculated from the strains using the stress–strain relationships:

Eq. $\boxed{10c}$ $f_l = E_s \varepsilon_l = 29,000(0.00156) = 45.2$ ksi

Eq. $\boxed{11d}$ $f_t = 60.0$ ksi   (just yielded)

Eq. $\boxed{12a}$ $f_{lp} = E_{ps}(\varepsilon_{dec} + \varepsilon_l) = 29,000(0.005 + 0.00156) = 190.2$ ksi

Eq. $\boxed{13a}$ $f_{tp} = 0$

Mohr's circles for stresses on the concrete, in the steel, and on the reinforced concrete as a whole, as well as Mohr's circle for strains, are all plotted in Figure 7.18 for the selected strain of $\varepsilon_d = -0.0002$. In order to plot these Mohr circles, the following additional stresses are calculated:

$$\rho_l f_l + \rho_{lp} f_{lp} = 0.012(45,200) + 0.003(190,200) = 1114 \text{ psi}$$

$$\rho_t f_t = 0.012(60,000) = 720 \text{ psi}$$

$$\sigma_l - (\rho_l f_l + \rho_{lp} f_{lp}) = 500 - 1114 = -614 \text{ psi}$$

$$\sigma_t - \rho_t f_t = 250 - 720 = -470 \text{ psi}$$

$$\sigma_1 = \frac{\sigma_l + \sigma_t}{2} + \sqrt{\left(\frac{\sigma_l - \sigma_t}{2}\right)^2 + \tau_{lt}^2}$$

$$= \frac{500 + 250}{2} + \sqrt{\left(\frac{500 - 250}{2}\right)^2 + (535)^2}$$

$$= 375 + 549 = 924 \text{ psi}$$

$$\sigma_2 = 375 - 549 = -174 \text{ psi}$$

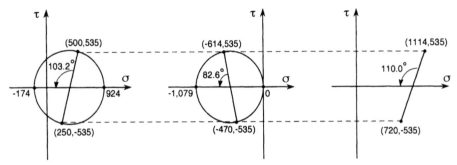

STRESSES IN REINF. CONC.   STRESSES IN CONCRETE   STRESSES IN STEEL

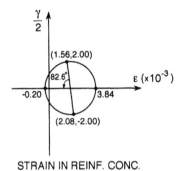

STRAIN IN REINF. CONC.

Figure 7.18 Stresses (psi) and strains for $\varepsilon_d = -0.0002$ (close to first yield in Example 7.1).

(2) **Select $\varepsilon_d = -0.0004$**

After several cycles of trial-and-error process, assume

$$\varepsilon_r = 0.0200$$

$$\zeta = \frac{0.9}{\sqrt{1 + 600(0.0200)}} = 0.250$$

$$\frac{\varepsilon_d}{\zeta\varepsilon_0} = \frac{-0.0004}{0.250(-0.002)} = 0.800 < 1 \quad \text{ascending branch}$$

$$\sigma_d = 0.250(-6000)\left[2(0.800) - (0.800)^2\right] = -1440 \text{ psi}$$

$$\sigma_r = 0$$

$$\frac{\varepsilon_r - \varepsilon_d}{\sigma_r - \sigma_d} = \frac{0.0200 + 0.0004}{0 + 1440} = 14.17 \times 10^{-6} \text{ psi}^{-1}$$

Solve the longitudinal steel strain ε_l:
Assume yielding of longitudinal steel $f_l = 60,000$ psi and assume the strain of the longitudinal steel to be $\varepsilon_l = 0.00630$:

$$E'_p(\varepsilon_{\text{dec}} + \varepsilon_l) = 31,060(0.005 + 0.00630) = 351.0 \text{ ksi}$$

Eq. $\boxed{12b}$

$$f_{lp} = \frac{E_p'(\varepsilon_{\text{dec}} + \varepsilon_l)}{\left[1 + \left((E_p'(\varepsilon_{\text{dec}} + \varepsilon_l))/f_{\text{pu}}\right)^4\right]^{1/4}} = \frac{351.0}{\left[1 + (351.0/270)^4\right]^{1/4}}$$

$$= 250.5 \text{ ksi}$$

$$\varepsilon_l = 0.0200 + 14.17 \times 10^{-6}[500 - 0 - 0.012(60,000) - 0.003(250,500)]$$

$$= 0.00623 \approx 0.00630 \quad \textbf{O.K.}$$

Solve the transverse steel strain ε_t:
Assume yielding of transverse steel $f_t = 60,000$ psi:

$$\varepsilon_t = 0.0200 + 14.17 \times 10^{-6}[250 - 0 - 0.012(60,000) - 0]$$

$$= 0.01334 > 0.00207 \quad \textbf{O.K.}$$

The tensile strain ε_r can now be checked:

$$\varepsilon_r = 0.00623 + 0.01334 + 0.0004 = 0.01997 \approx 0.0200 \quad \textbf{O.K.}$$

The angle α, the shear stress τ_{lt}, and the shear strain γ_{lt} are

$$\tan^2 \alpha = \frac{0.00623 + 0.0004}{0.01334 + 0.0004} = 0.4825$$

$$\tan \alpha = 0.6946$$

$$\alpha = 34.78° \qquad 2\alpha = 69.56°$$

$$\tau_{lt} = (1440 + 0)(0.570)(0.821) = 674 \text{ psi}$$

$$\gamma_{lt} = 2(0.0004 + 0.0200)(0.570)(0.821) = 0.0191$$

The stresses in the mild steel and prestressing steel are

$$f_l = 60,000 \text{ psi}$$

$$f_t = 60,000 \text{ psi}$$

$$f_{lp} = 250.5 \text{ ksi}$$

$$f_{tp} = 0 \text{ ksi}$$

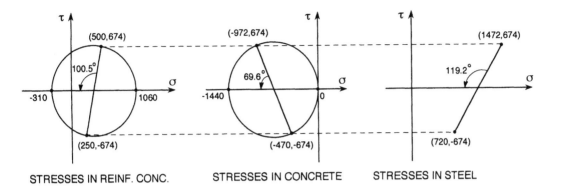

Figure 7.19 Stresses (psi) and strains for $\varepsilon_d = -0.0004$ (close to ultimate strength in Example 7.1).

Mohr's circles for stresses and strains are all plotted in Figure 7.19 for the selected strain of $\varepsilon_d = -0.0004$, using the following additional stresses:

$$\rho_l f_l + \rho_{lp} f_{lp} = 0.012(60,000) + 0.003(250.500) = 1472 \text{ psi}$$

$$\rho_t f_t = 0.012(60,000) = 720 \text{ psi}$$

$$\sigma_l - (\rho_l f_l + \rho_{lp} f_{lp}) = 500 - 1472 = -972 \text{ psi}$$

$$\sigma_t - \rho_t f_t = 250 - 720 = -470 \text{ psi}$$

$$\sigma_1 = \frac{500 + 250}{2} + \sqrt{\left(\frac{500 - 250}{2}\right)^2 + (674)^2} = 1060 \text{ psi}$$

$$\sigma_2 = 125 - 685 = -310 \text{ psi}$$

In addition to the strains $\varepsilon_d = -0.0002$ and -0.0004 selected previously, calculations were made for the strains $\varepsilon_d = -0.0003$ and -0.0005. The results of all these calculations are recorded in Table 7.1. Using the τ_{lt} and γ_{lt} values given in Table 7.1, the shear stress versus shear strain curve are plotted in Figure 7.20.

TABLE 7.1 Results of Calculations for Example Problem 7.1 ($\varepsilon_d = -0.2, -0.3, -0.4,$ and -0.5×10^{-3})

Variables	Eqs.	Calculated Values			
ε_d (10^{-3}) selected	—	-0.200	-0.300	-0.400	-0.500
ε_r (10^{-3}) last assumed	—	3.85	11.50	20.00	21.7
ζ	8	0.495	0.320	0.250	0.240
$\varepsilon_d / \zeta\varepsilon_0$	—	0.202	0.469	0.800	1.042
σ_d (psi)	7	-1079	-1379	-1440	-1440
σ_r (psi)	9	0	0	0	0
$(\varepsilon_r - \varepsilon_d)/(\sigma_r - \sigma_d)$ (10^{-6} psi^{-1})	—	3.754	8.557	14.17	15.42
ε_l (10^{-3})	14	1.56	3.75	6.23	6.63
ε_t (10^{-3})	15	2.08	7.48	13.34	14.45
ε_r (10^{-3}) checked	16	3.84	11.53	19.97	21.6
α (deg)	17	41.30	35.82	34.78	34.63
τ_{lt} (psi)	3	535	654	674	673
γ_{lt} (10^{-3})	6	4.00	11.2	19.1	20.6
f_l (ksi)	10	45.2	60.0	60.0	60.0
f_t (ksi)	11	60.0	60.0	60.0	60.0
f_{lp} (ksi)	12	190.2	227.8	250.5	252.3
f_{tp} (ksi)	13	0	0	0	0

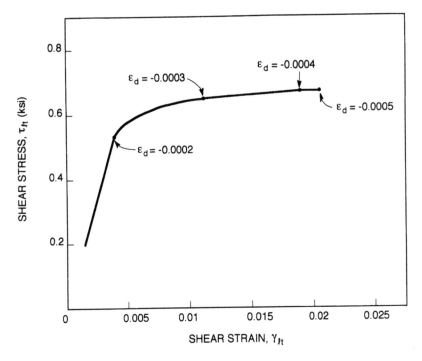

Figure 7.20 Shear stress versus shear strain curve for Example Problem 7.1.

7.3.3   SHEAR DUCTILITY

It can be seen from Table 7.1 and Figure 7.20 that the condition under the strain $\varepsilon_d = -0.0002$ represents very closely the *first yield condition*, because the transverse steel strain ε_t of 0.00208 just exceeds the yield stress of 0.00207, whereas the longitudinal steel strain is still within the elastic range. Under this condition the yield shear stress $(\tau_{lt})_y$ and the yield shear strain $(\gamma_{lt})_y$ are 535 psi and 0.00400, respectively.

The condition under the strain $\varepsilon_d = -0.0004$, on the other hand, represents very closely the *ultimate condition*. This can be observed as follows: When the strain ε_d is increased negatively from -0.0004 to -0.0005, the concrete struts increase from a level of $\varepsilon_d/\zeta\varepsilon_0 = 0.800$ to a level of 1.042, just exceeding the peak stress of unity. In the vicinity of the peak stress, the -1440-psi stress in the concrete struts levels off and the rapid increase of the various strains comes to a screeching halt. The shear strain τ_{lt} decreases very slightly from 674 to 673 psi, and the shear strain γ_{lt} increases only moderately from 0.0191 to 0.0206. Therefore, the ultimate shear stress $(\tau_{lt})_u$ and ultimate shear strain $(\gamma_{lt})_u$ can be closely taken as 674 psi and 0.0200, respectively.

If the shear ductility, μ_s is defined as the ratio of the ultimate shear strain $(\gamma_{lt})_u$ to the yield shear strain $(\gamma_{lt})_y$, i.e.,

$$\mu_s = \frac{(\gamma_{lt})_u}{(\gamma_{lt})_y} \tag{7-81}$$

then the shear ductility for the example membrane element is $0.0200/0.00400 = 5.0$.

7.4 Membrane Elements under Proportional Loadings

7.4.1 PROPORTIONAL MEMBRANE LOADINGS

The stress state described by the three applied stresses σ_l, σ_t, and τ_{lt} in an element is shown in Figure 7.21a. This stress state can also be expressed in terms of three principal stress variables σ_1, S, and α_2 as shown in Figure 7.21b. These three principal stress variables are defined as

σ_1 = larger principal stress, algebraically; always positive and in tension

$S = \sigma_2/\sigma_1$, ratio of smaller to larger principal stresses; positive when σ_2 is tension; negative when σ_2 is compression

α_2 = orientation angle, or angle between the direction of smaller principal stress, algebraically, and the longitudinal axis

When an element is subjected to a *proportional membrane loading*, the larger principal stress σ_1 increases while the other two variables, S and α_2, remain constant. Therefore, the set of applied stresses σ_l, σ_t, and τ_{lt} can be defined in terms of the principal stress σ_1 as

$$\sigma_l = m_l \sigma_1 \tag{7-82}$$

$$\sigma_t = m_t \sigma_1 \tag{7-83}$$

$$\tau_{lt} = m_{lt} \sigma_1 \tag{7-84}$$

The set of three coefficients m_l, m_t, and m_{lt} should also remain constant when the principal stress σ_1 increases under proportional loading.

The relationship between the set of three coefficients m_l, m_t, and m_{lt} and the set of two principal stress variables S and α_2 can be obtained through the transformation relationship. Substituting Eqs. (7-82) to (7-84), $\sigma_2 = S\sigma_1$, and $\alpha = \alpha_2$ into the transformation equations, (4-38) to (4-40), and then canceling the principal stress σ_1 result in

$$m_l = S \cos^2 \alpha_2 + \sin^2 \alpha_2 \tag{7-85}$$

$$m_t = S \sin^2 \alpha_2 + \cos^2 \alpha_2 \tag{7-86}$$

$$m_{lt} = (-S + 1)\sin \alpha_2 \cos \alpha_2 \tag{7-87}$$

Take, for example, a given set of applied stresses ($\sigma_l = -25$ psi, $\sigma_t = 125$ psi, and $\tau_{lt} = 129.9$ psi) shown in Figure 7.22a. Then the set of principal stress variables σ_1, S,

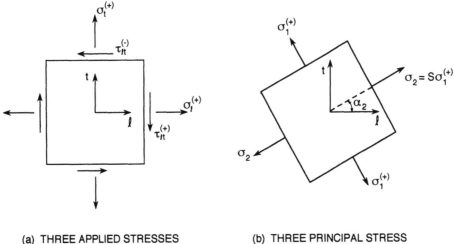

(a) THREE APPLIED STRESSES
σ_t, σ_l AND τ_{lt}

(b) THREE PRINCIPAL STRESS
VARIABLES σ_1, S AND α_2

Figure 7.21 Relationship between applied stresses and principal stress variables.

and α_2 can be calculated as

$$\sigma_1 = \frac{\sigma_l + \sigma_t}{2} + \sqrt{\left(\frac{\sigma_l - \sigma_t}{2}\right)^2 + \tau_{lt}^2}$$

$$= \frac{-25 + 125}{2} + \sqrt{\left(\frac{-25 - 125}{2}\right)^2 + (129.9)^2}$$

$$= 50 + 150 = 200 \text{ psi}$$

$$\sigma_2 = 50 - 150 = -100 \text{ psi}$$

$$S = \frac{\sigma_2}{\sigma_1} = \frac{-100}{200} = -0.5$$

$$\cot 2\alpha_2 = \frac{-\sigma_l + \sigma_t}{2\tau_{lt}} = \frac{25 + 125}{2(129.9)} = 0.5774$$

$$2\alpha_2 = 60° \quad \text{and} \quad \alpha_2 = 30°$$

The preceding calculated values of σ_1, S, and α_2 are recorded in Figure 7.22b. Both the given set of applied stresses and the calculated set of principal stress variables are shown in Figure 7.22c in terms of a Mohr circle. Points A and C give the values of the three applied stresses ($\sigma_l = -25$ psi, $\sigma_t = 125$ psi, and $\tau_{lt} = 129.9$ psi). Points D and B indicate the principal stresses ($\sigma_1 = 200$ psi and $\sigma_2 = -100$ psi), and the angle $2\alpha_2$ is 60°.

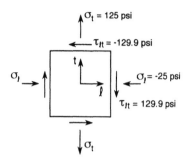

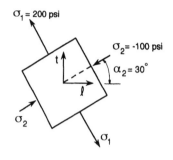

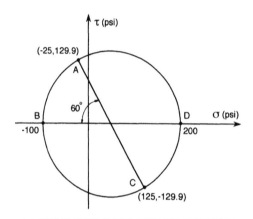

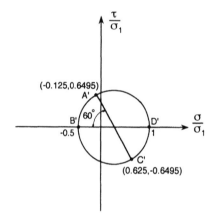

Figure 7.22 *m* coefficients and normalized Mohr circle.

The set of *m* coefficients is calculated by their definitions:

$$m_l = \frac{\sigma_l}{\sigma_1} = \frac{-25}{200} = -0.125$$

$$m_t = \frac{\sigma_t}{\sigma_1} = \frac{125}{200} = 0.625$$

$$m_{lt} = \frac{\tau_{lt}}{\sigma_1} = \frac{129.9}{200} = 0.6495$$

According to these definitions, the coefficients m_l, m_t, and m_{lt} have a physical meaning. They are simply the applied stresses σ_l, σ_t, and τ_{lt}, normalized by the principal stress σ_1. These normalized stresses can be represented by a normalized Mohr circle as shown in Figure 7.22d. The values of the *m* coefficients are given by points A' and C'. The normalized principal stress of σ_1 is, of course, equal to unity, and the normalized principal stress of σ_2 is the ratio $S = -0.5$. These two normalized principal stresses are indicated by the points D' and B', respectively. The $2\alpha_2$ angle, of course, remains 60°.

If the two principal stress variables S and α_2 are given, then the three *m* coefficients can be calculated from Eqs. (7-85) to (7-87). Take, for example, the case

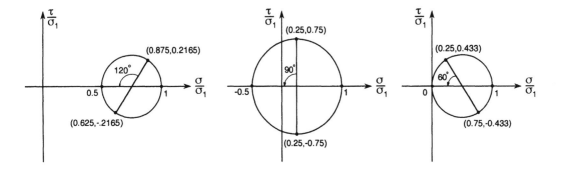

(a) S = 0.5, $\alpha_2 = 60°$
   m_l = 0.875
   m_t = 0.625
   m_{lt} = 0.2165

(b) S = -0.5, $\alpha_2 = 45°$
   m_l = 0.25
   m_t = 0.25
   m_{lt} = 0.75

(c) S = 0, $\alpha_2 = 30°$
   m_l = 0.25
   m_t = 0.75
   m_{lt} = 0.433

Figure 7.23   *m* coefficients for three pairs of *S* and α_2.

of $S = 0.5$ and $\alpha_2 = 60°$ as shown in Figure 7.23a:

$$\sin 60° = 0.866 \quad \cos 60° = 0.5$$

$$\sin^2 60° = 0.75 \quad \cos^2 60° = 0.25 \quad \sin 60° \cos 60° = 0.433$$

$$m_l = S \cos^2 \alpha_2 + \sin^2 \alpha_2 = 0.5(0.25) + (0.75) = 0.875$$

$$m_t = S \sin^2 \alpha_2 + \cos^2 \alpha_2 = 0.5(0.75) + (0.25) = 0.625$$

$$m_{lt} = (-S + 1)\sin \alpha_2 \cos \alpha_2 = (-0.5 + 1)(0.433) = 0.2165$$

These calculated *m* coefficients are also shown in Figure 7.23a, together with the normalized Mohr circles. In addition, Figure 7.23b and c gives two more cases of *S* and α_2, their corresponding *m* coefficients, and their normalized Mohr circles. Comparison of the three cases illustrates how the Mohr circles are affected by *S* and α_2.

7.4.2   STRENGTH OF MEMBRANE ELEMENTS

The basic equilibrium equations for membrane elements subjected to proportional loadings can be easily derived by substituting the definitions of the *m* coefficients [Eqs. (7-82) to (7-84)] into the equilibrium equations (Eqs. $\boxed{1}$ to $\boxed{3}$):

$$m_l \sigma_1 = \sigma_d \cos^2 \alpha + \sigma_r \sin^2 \alpha + \rho_l f_l + \rho_{lp} f_{lp} \qquad (7\text{-}88)$$

$$m_t \sigma_1 = \sigma_d \sin^2 \alpha + \sigma_r \cos^2 \alpha + \rho_t f_t + \rho_{tp} f_{tp} \qquad (7\text{-}89)$$

$$m_{lt} \sigma_1 = (-\sigma_d + \sigma_r)\sin \alpha \cos \alpha \qquad (7\text{-}90)$$

The strength of the element under proportional loadings can be represented by a single stress σ_1. This stress σ_1 can be solved from Eqs. (7-88) to (7-90) by eliminating the angle α. Utilizing the relationship $\sin^2\alpha + \cos^2\alpha = 1$ and rearranging the terms give

$$-m_l\sigma_1 + \sigma_r + \rho_l f_l + \rho_{lp} f_{lp} = (-\sigma_d + \sigma_r)\cos^2\alpha \tag{7-91}$$

$$-m_t\sigma_1 + \sigma_r + \rho_t f_t + \rho_{tp} f_{tp} = (-\sigma_d + \sigma_r)\sin^2\alpha \tag{7-92}$$

$$m_{lt}\sigma_1 = (-\sigma_d + \sigma_r)\sin\alpha\cos\alpha \tag{7-93}$$

Multiplying Eq. (7-91) by Eq. (7-92) gives

$$(-m_l\sigma_1 + \sigma_r + \rho_l f_l + \rho_{lp} f_{lp})(-m_t\sigma_1 + \sigma_r + \rho_t f_t + \rho_{tp} f_{tp})$$

$$= (-\sigma_d + \sigma_r)^2 \sin^2\alpha\cos^2\alpha \tag{7-94}$$

Squaring Eq. (7-93) gives

$$(m_{lt}\sigma_1)^2 = (-\sigma_d + \sigma_r)^2 \sin^2\alpha\cos^2\alpha \tag{7-95}$$

The angle α can now be eliminated by equating the left-hand sides of Eqs. (7-94) and (7-95). Multiplication and regrouping of terms will result in a quadratic equation for σ_1 as follows:

$$(m_l m_t - m_{lt}^2)\sigma_1^2 - \left[m_l(\sigma_r + \rho_t f_t + \rho_{tp} f_{tp}) + m_t(\sigma_r + \rho_l f_l + \rho_{lp} f_{lp})\right]\sigma_1$$

$$+ (\sigma_r + \rho_l f_l + \rho_{lp} f_{lp})(\sigma_r + \rho_t f_t + \rho_{tp} f_{tp}) = 0 \tag{7-96}$$

Define

$$S = (m_l m_t - m_{lt}^2) \tag{7-97}$$

$$B = \left[m_l(\sigma_r + \rho_t f_t + \rho_{tp} f_{tp}) + m_t(\sigma_r + \rho_l f_l + \rho_{lp} f_{lp})\right] \tag{7-98}$$

$$C = (\sigma_r + \rho_l f_l + \rho_{lp} f_{lp})(\sigma_r + \rho_t f_t + \rho_{tp} f_{tp}) \tag{7-99}$$

The solution of the quadratic equation is

$$\sigma_1 = \frac{1}{2S}\left(B \pm \sqrt{B^2 - 4SC}\right) \tag{7-100} \text{ or } \boxed{18}$$

It should be noted that the symbol S in Eq. (7-97) is the same as the definition of $S = \sigma_2/\sigma_1$. This can be easily proven by inserting Eqs. (7-85) to (7-87) into Eq. (7-97). Equation $\boxed{18}$ will be used in the proposed solution procedure.

In the case of a non-prestressed element ($\rho_{lp} = \rho_{tp} = 0$), zero tensile stress of concrete ($\sigma_r = 0$), and equal yield strengths of mild steel in both directions ($f_l = f_t =$

f_y), Eq. (7-100) can be simplified to the form

$$\sigma_1 = \frac{f_y}{2S}\left[(m_l\rho_t + m_t\rho_l) \pm \sqrt{(m_l\rho_t + m_t\rho_l)^2 - 4S\rho_l\rho_t}\right] \qquad (7\text{-}101)$$

Equation (7-101) was first derived by Han and Mau (1988). It is valid when both the longitudinal and the transverse steels are assumed to yield according to the plasticity truss model.

7.4.3   METHOD OF SOLUTION

The 13 equations (Eqs. [3] and [6] to [17]), which have been successfully used to solve the case of elements under constant normal stresses (Section 7.3.2), can also be applied to the case of elements subjected to proportional loadings. Minor modifications, however, need to be made in Eqs. [14] and [15]. Equation [14] for the longitudinal steel strains ε_l is still valid for proportional loading, except that the applied longitudinal stress σ_l should be replaced by $m_l\sigma_1$:

$$\varepsilon_l = \varepsilon_r + \frac{\varepsilon_r - \varepsilon_d}{\sigma_r - \sigma_d}(m_l\sigma_1 - \sigma_r - \rho_l f_l - \rho_{lp}f_{lp}) \qquad \boxed{14P}$$

The letter **P** in the equation number [14P] indicates that it is derived specifically for proportional loading. Similarly, Eq. [15] for the transverse steel strain ε_t is still valid, except that the applied transverse stresses σ_t should be replaced by $m_t\sigma_1$:

$$\varepsilon_t = \varepsilon_r + \frac{\varepsilon_r - \varepsilon_d}{\sigma_r - \sigma_d}(m_t\sigma_1 - \sigma_r - \rho_t f_t - \rho_{tp}f_{tp}) \qquad \boxed{15P}$$

In Eqs. [14P] and [15P] we notice that a new unknown variable σ_1 has been introduced. Therefore, an additional equation is required. This new equation is furnished by Eq. [18]. Based on the 14 equations (Eqs. [3] and [6] to [18]), a solution procedure is proposed as shown by the flow chart in Figure 7.24. This procedure is somewhat more complex than the flow chart shown in Figure 7.16 for constant normal stresses, because of the additional iteration cycles required by the variable σ_1 for the proportional loading. For this trial-and-error procedure with a nested do loop, it would be convenient to write a computer program and to take advantage of an electronic computer. However, the solution can still be obtained by hand calculations for some simple cases, such as the non-prestressed membrane element in Example Problem 7.2.

EXAMPLE PROBLEM 7.2

A non-prestressed membrane element was designed by the plasticity truss model in Example 4.1 (Figure 4.15) and then analyzed by the Mohr compatibility truss model in Example 6.1 (Figure 6.2). This membrane element is subjected to a set of membrane stresses $\sigma_l = 310$ psi (tension), $\sigma_t = -310$ psi (compression), and $\tau_{lt} = 537$ psi. Assuming that this set of stresses is applied in a proportional manner, we can now

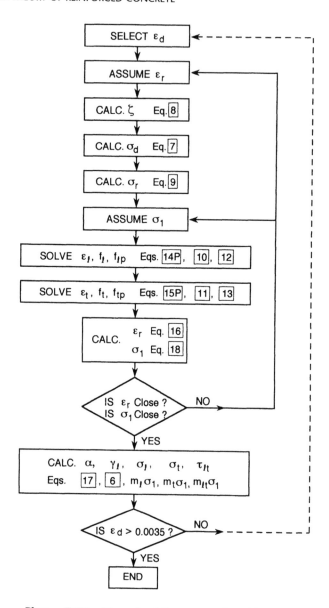

Figure 7.24 Flow chart for proportional loadings.

analyze the behavior of this element by the softened truss model. Particular interest is placed in the stress and strain conditions at the first yield of steel and at the ultimate load stage.

Based on the plasticity truss model, the membrane element is reinforced with 1.033% of steel in both the longitudinal and the transverse directions (Example 4.1). The steel is designed to have the same grade in both directions: $f_{ly} = f_{ty} = 60,000$ psi. In the Mohr compatibility truss model (Example 6.1) the strength of concrete is assumed to be $f'_c = 4000$ psi. In order to use the softened truss model, the constitutive laws of concrete and steel are specified as follows: the compression stress–strain curve of softened concrete is expressed by Eqs. $\boxed{7a}$, $\boxed{7b}$, and $\boxed{8a}$. The tensile strength of concrete is neglected (assume $\sigma_r = 0$; Eq. $\boxed{9c}$) and the stress–strain curve of bare bars for mild steel (Eqs. 10c , 10d , 11c , and 11d) is adopted.

SOLUTION

For the given set of applied stresses the principal stress variables σ_1, S, and α_2 are

$$\sigma_1 = \frac{\sigma_l + \sigma_t}{2} + \sqrt{\left(\frac{\sigma_l - \sigma_t}{2}\right)^2 + \tau_{lt}^2}$$

$$= \frac{310 - 310}{2} + \sqrt{\left(\frac{310 + 310}{2}\right)^2 + (537)^2}$$

$$= 0 + 620 = 620 \text{ psi}$$

$$\sigma_2 = 0 - 620 = -620 \text{ psi}$$

$$S = \frac{\sigma_2}{\sigma_1} = \frac{-620}{620} = -1$$

$$\cot 2\alpha_2 = \frac{-\sigma_l + \sigma_t}{2\tau_{lt}} = \frac{-310 - 310}{2(537)} = -0.5774$$

$$2\alpha_2 = 120° \quad \text{and} \quad \alpha_2 = 60°$$

The three m coefficients are

$$m_l = \frac{\sigma_l}{\sigma_1} = \frac{310}{620} = 0.5$$

$$m_t = \frac{\sigma_t}{\sigma_1} = \frac{-310}{620} = -0.5$$

$$m_{lt} = \frac{\tau_{lt}}{\sigma_1} = \frac{537}{620} = 0.866$$

The normalized Mohr circle for this case of $S = -1$ and $\alpha_2 = 60°$ is shown in Figure 7.25.

(1) Select $\varepsilon_d = -0.000275$ (for the first yield condition)

The stress and strain conditions at first yield have been given in Figure 6.2 (see Example 6.1) based on the Mohr compatibility truss model. At this load stage $\varepsilon_d = -0.000265$, $\varepsilon_r = 0.00345$, and $\sigma_1 = 532$ psi. In this softened truss model, however, ε_d should be slightly greater than -0.000265 (in an absolute sense), because of the slightly nonlinear stress–strain relationship of concrete at low level loading. After several cycles of trial-and-error process we select the values of $\varepsilon_d = -0.000275$, and assume

$$\varepsilon_r = 0.00345 \quad \text{and} \quad \sigma_1 = 532 \text{ psi}$$

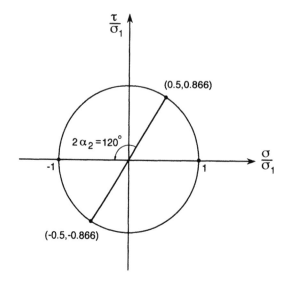

PRINCIPAL STRESS VARIABLES:

$$S = -1, \qquad \alpha_2 = 60°$$

m-COEFFICIENT:

$$m_l = 0.5, \quad m_t = -0.5, \quad m_{lt} = 0.866$$

Figure 7.25   Normalized Mohr circle for proportional loading in Example Problem 7.2.

Eq. 8a

$$\zeta = \frac{0.9}{\sqrt{1 + 600\varepsilon_r}} = \frac{0.9}{\sqrt{1 + 600(0.00345)}} = 0.514$$

$$\frac{\varepsilon_d}{\zeta\varepsilon_0} = \frac{-0.000275}{0.514(-0.002)} = 0.268 < 1 \quad \text{ascending branch}$$

Eq. 7a

$$\sigma_d = \zeta f_c' \left[2\left(\frac{\varepsilon_d}{\zeta\varepsilon_0}\right) - \left(\frac{\varepsilon_d}{\zeta\varepsilon_0}\right)^2 \right]$$

$$= 0.514(-4000)\left[2(0.268) - (0.268)^2\right] = -954 \text{ psi}$$

Eq. 9c

$$\sigma_r = 0$$

$$\frac{\varepsilon_r - \varepsilon_d}{\sigma_r - \sigma_d} = \frac{0.00345 + 0.000275}{0 + 954} = 3.905 \times 10^{-6} \text{ psi}^{-1}$$

Solve the longitudinal steel strain ε_l:

Eq. 14P

$$\varepsilon_l = \varepsilon_r + \frac{\varepsilon_r - \varepsilon_d}{\sigma_r - \sigma_d}[m_l\sigma_1 - \sigma_r - \rho_l f_l - \rho_{lp} f_{lp}]$$

$$= 0.00345 + 3.905 \times 10^{-6}[0.5(532) - 0 - 0.01033 f_l - 0]$$

Assume

Eq. $\boxed{10c}$ $f_l = 29 \times 10^6 \, \varepsilon_l$ before yielding

Then

$$\varepsilon_l + 1.170\varepsilon_l = 0.00345 + 0.001039$$

$$\varepsilon_l = \frac{0.004489}{2.170} = 0.00207 = 0.00207(\varepsilon_y) \quad \textbf{O.K.}$$

Solve the transverse steel strain ε_t:

Eq. $\boxed{15P}$ $\varepsilon_t = \varepsilon_r + \dfrac{\varepsilon_r - \varepsilon_d}{\sigma_r - \sigma_d}[m_t\sigma_1 - \sigma_r - \rho_t f_t - \rho_{tp} f_{tp}]$

$$= 0.00345 + 3.905 \times 10^{-6}[-0.5(532) - 0 - 0.01033 f_t - 0]$$

Assume

Eq. $\boxed{11c}$ $f_t = 29 \times 10^6 \, \varepsilon_t$ before yielding

Then

$$\varepsilon_t + 1.170\varepsilon_t = 0.00345 - 0.001039$$

$$\varepsilon_t = \frac{0.002411}{2.170} = 0.00111 < 0.00207(\varepsilon_y) \quad \textbf{O.K.}$$

The tensile strain ε_r can now be checked:

Eq. $\boxed{16}$ $\varepsilon_r = \varepsilon_l + \varepsilon_t - \varepsilon_d = 0.00207 + 0.00111 + 0.000275$

$$= 0.003455 \approx 0.00345 \text{ (assumed)} \quad \textbf{O.K.}$$

Check σ_1:

Eq. $\boxed{10c}$ $f_l = E_s \varepsilon_l = 0.00207(29,000,000) = 60,000$ psi   (just yielded)

Eq. $\boxed{11c}$ $f_t = E_s \varepsilon_t = 0.00111(29,000,000) = 32,200$ psi   (elastic range)

$$\rho_l f_l = 0.01033(60,000) = 620 \text{ psi}$$

$$\rho_t f_t = 0.01033(32,200) = 333 \text{ psi}$$

$$B = m_l(\rho_t f_t) + m_t(\rho_l f_l) = 0.5(333) - 0.5(620) = -144 \text{ psi}$$

$$C = (\rho_l f_l)(\rho_t f_t) = (620)(333) = 206,500 \text{ psi}$$

Eq. $\boxed{18}$ $\quad \sigma_1 = \dfrac{1}{2S}\left(B \pm \sqrt{B^2 - 4SC}\right)$

$$= \dfrac{1}{2(-1)}\left(-144 \pm \sqrt{(-144)^2 - 4(-1)206{,}500}\right)$$

$$= -0.5(-144 - 920) = 532 \text{ psi} = 532 \text{ psi (assumed)} \quad \textbf{O.K.}$$

Now that the strains ε_d, ε_r, ε_l, and ε_t and the stress σ_d are calculated from the trial-and-error process, we can determine the angle α, the shear stress τ_{lt}, and the shear strain γ_{lt}:

Eq. $\boxed{17}$ $\quad \tan^2 \alpha = \dfrac{\varepsilon_l - \varepsilon_d}{\varepsilon_t - \varepsilon_d} = \dfrac{0.00207 + 0.000275}{0.00111 + 0.000275} = 1.693$

$$\tan \alpha = 1.301$$

$$\alpha = 52.46° \qquad 2\alpha = 104.9°$$

Eq. $\boxed{3}$ $\quad \tau_{lt} = (-\sigma_d + \sigma_r)\sin \alpha \cos \alpha = (954 + 0)(0.793)(0.609)$

$$= 461 \text{ psi}$$

Eq. $\boxed{6}$ $\quad \gamma_{lt}/2 = (-\varepsilon_d + \varepsilon_r)\sin \alpha \cos \alpha$

$$= (0.000275 + 0.00345)(0.793)(0.609) = 0.00180$$

The applied stresses at the level of first yield are

$$\sigma_l = m_l \sigma_1 = 0.5(532) = 266 \text{ psi}$$

$$\sigma_t = m_t \sigma_1 = -0.5(532) = -266 \text{ psi}$$

$$\tau_{lt} = m_{lt} \sigma_1 = 0.866(532) = 461 \text{ psi}$$

The preceding calculations show that the stress and strain conditions analyzed by the softened truss model are almost identical to those obtained by the Mohr compatibility truss model in Figure 6.2 (Example 6.1). Therefore, the Mohr circles for the stresses and strains calculated by the softened truss model in this problem will not be given here.

(2) Select $\varepsilon_d = -0.0004$

This is a point after the yielding of longitudinal steel, but before the yielding of transverse steel. After several cycles of trial-and-error processes, assume

$$\varepsilon_r = 0.00790 \quad \text{and} \quad \sigma_1 = 603 \text{ psi}$$

Eq. $\boxed{8a}$ $\quad \zeta = \dfrac{0.9}{\sqrt{1 + 600(0.00790)}} = 0.376$

$$\dfrac{\varepsilon_d}{\zeta \varepsilon_0} = \dfrac{-0.0004}{0.376(-0.002)} = 0.532 < 1 \quad \text{ascending branch}$$

Eq. 7a $\sigma_d = 0.376(-4000)\left[2(0.532) - (0.532)^2\right] = -1175$ psi

Eq. 9c $\sigma_r = 0$

$$\frac{\varepsilon_r - \varepsilon_d}{\sigma_r - \sigma_d} = \frac{0.00790 + 0.0004}{0 + 1175} = 7.06 \times 10^{-6}\ \text{psi}^{-1}$$

Solve the longitudinal steel strain ε_l (assume yielding):

Eq. 14P $\varepsilon_l = 0.00790 + 7.06 \times 10^{-6}[0.5(603) - 0 - 620 - 0]$

$$= 0.00565 > 0.00207(\varepsilon_y)\quad \textbf{O.K.}$$

Solve the transverse steel strain ε_t:

Eq. 15P $\varepsilon_t = 0.00790 + 7.06 \times 10^{-6}[-0.5(603) - 0 - 0.01033f_t - 0]$

Assume

Eq. 11c $f_t = 29 \times 10^6\ \varepsilon_t$ before yielding

Then

$$\varepsilon_t + 2.115\varepsilon_t = 0.00790 - 0.00213$$

$$\varepsilon_t = \frac{0.00577}{3.115} = 0.00185 < 0.00207(\varepsilon_y)\quad \textbf{O.K.}$$

The tensile strain ε_r can now be checked:

Eq. 16 $\varepsilon_r = 0.00565 + 0.00185 + 0.0004$

$$= 0.00790 = 0.00790\ (\text{assumed})\quad \textbf{O.K.}$$

Check σ_1:

Eq. 10d $f_l = 60{,}000$ psi   (after yielding of longitudinal steel)

Eq. 11c $f_t = 0.00185(29{,}000{,}000) = 53{,}600$ psi   (elastic range)

$\rho_l f_l = 0.01033(60{,}000) = 620$ psi

$\rho_t f_t = 0.01033(53{,}600) = 554$ psi

$B = 0.5(554) - 0.5(620) = -33$ psi

$C = (620)(554) = 343{,}500$ psi

Eq. 18 $\sigma_1 = \dfrac{1}{2(-1)}\left(-33 \pm \sqrt{(-33)^2 - 4(-1)343{,}500}\right)$

$$= -0.5(-33 - 1173) = 603\ \text{psi} = 603\ \text{psi (assumed)}\quad \textbf{O.K.}$$

The angle α, the shear stress τ_{lt}, and the shear strain γ_{lt} are

Eq. $\boxed{17}$ $$\tan^2 \alpha = \frac{0.00565 + 0.0004}{0.00185 + 0.0004} = 2.689$$

$$\tan \alpha = 1.640$$

$$\alpha = 58.62° \qquad 2\alpha = 117.2°$$

Eq. $\boxed{3}$ $$\tau_{lt} = (1175 + 0)(0.854)(0.521) = 522.8 \text{ psi}$$

Eq. $\boxed{6}$ $$\gamma_{lt}/2 = (0.0004 + 0.00790)(0.854)(0.521) = 0.00369$$

The applied stresses at this level of $\varepsilon_l > \varepsilon_y$ and $\varepsilon_t < \varepsilon_y$ are

$$\sigma_l = 0.5(603) = 301.5 \text{ psi}$$

$$\sigma_t = -0.5(603) = -301.5 \text{ psi}$$

$$\tau_{lt} = 0.866(603) = 522.2 \text{ psi}$$

(3) Select $\varepsilon_d = -0.0005$

This is a point after the yielding of both the longitudinal and the transverse steel. After several cycles of trial-and-error process, assume:

$$\varepsilon_r = 0.0108 \quad \text{and} \quad \sigma_1 = 620 \text{ psi}$$

Eq. $\boxed{8a}$ $$\zeta = \frac{0.9}{\sqrt{1 + 600(0.0108)}} = 0.329$$

$$\frac{\varepsilon_d}{\zeta\varepsilon_0} = \frac{-0.0005}{0.329(-0.002)} = 0.760 < 1 \quad \text{ascending branch}$$

Eq. $\boxed{7a}$ $$\sigma_d = 0.329(-4000)\left[2(0.760) - (0.760)^2\right] = -1240 \text{ psi}$$

Eq. $\boxed{9c}$ $$\sigma_r = 0$$

$$\frac{\varepsilon_r - \varepsilon_d}{\sigma_r - \sigma_d} = \frac{0.0108 + 0.0005}{0 + 1240} = 9.11 \times 10^{-6} \text{ psi}^{-1}$$

Solve the longitudinal steel strain ε_l (assume yielding):

Eq. $\boxed{14P}$ $$\varepsilon_l = 0.0108 + 9.11 \times 10^{-6}[0.5(620) - 0 - 620 - 0]$$

$$= 0.00798 > 0.00207(\varepsilon_y) \quad \textbf{O.K.}$$

Solve the transverse steen strain ε_t (assume yielding):

Eq. $\boxed{15P}$ $$\varepsilon_t = 0.0108 + 9.11 \times 10^{-6}[-0.5(620) - 0 - 620 - 0]$$

$$= 0.00233 > 0.00207(\varepsilon_y) \quad \textbf{O.K.}$$

The tensile strain ε_r can now be checked:

Eq. $\boxed{16}$ $\qquad\qquad \varepsilon_r = 0.00798 + 0.00233 + 0.0005$

$$= 0.01081 \approx 0.0108 \text{ (assumed)} \quad \textbf{O.K.}$$

Check σ_1:

Eq. $\boxed{10d}$ $\qquad f_l = 60{,}000 \text{ psi} \quad$ (longitudinal steel yielded)

Eq. $\boxed{11d}$ $\qquad f_t = 60{,}000 \text{ psi} \quad$ (transverse steel yielded)

$$\rho_l f_l = \rho_t f_t = 0.01033(60{,}000) = 620 \text{ psi}$$

$$B = 0.5(620) - 0.5(620) = 0 \text{ psi}$$

$$C = (620)(620) \text{ psi}$$

Eq. $\boxed{18}$ $\qquad \sigma_1 = \dfrac{1}{2(-1)}\left(0 \pm \sqrt{(0)^2 - 4(-1)(620)^2}\right)$

$$= -0.5(0 - 1240) = 620 \text{ psi} = 620 \text{ psi (assumed)} \quad \textbf{O.K.}$$

The angle α, the shear stress τ_{lt}, and the shear strain γ_{lt} are

Eq. $\boxed{17}$ $\qquad \tan^2\alpha = \dfrac{0.00798 + 0.0005}{0.00233 + 0.0005} = 2.996$

$$\tan\alpha = 1.731$$

$$\alpha = 60.0° \qquad 2\alpha = 120°$$

Eq. $\boxed{3}$ $\qquad \tau_{lt} = (1240 + 0)(0.866)(0.500)$

$$= 537 \text{ psi}$$

Eq. $\boxed{6}$ $\qquad \gamma_{lt}/2 = (0.0005 + 0.0108)(0.866)(0.500) = 0.00489$

The applied stresses at this level of $\varepsilon_l \gg \varepsilon_y$ and $\varepsilon_t > \varepsilon_y$ are

$$\sigma_l = 0.5(620) = 310 \text{ psi}$$

$$\sigma_t = -0.5(620) = -310 \text{ psi}$$

$$\tau_{lt} = 0.866(620) = 537 \text{ psi}$$

The results of calculations shown for $\varepsilon_d = -0.000275$, -0.0004, and -0.0005 are summarized in Table 7.2, together with two additional cases of $\varepsilon_d = -0.0006$ and 0.00062. Comparison of these five cases illustrates clearly the trends of all the variables. It should be kept in mind that $\varepsilon_d = -0.000275$ is the point of first yield of the longitudinal steel; $\varepsilon_d = -0.0004$ represents a point after the yielding of longitudinal steel, but before the yielding of transverse steel; $\varepsilon_d = -0.0005$ gives a point after

TABLE 7.2 Results of Calculations for Example Problem 7.2 ($\varepsilon_d = -0.275, -0.4, -0.5, -0.6,$ and -0.62×10^{-3})

Variables	Eqs.	Calculated Values				
ε_d (10^{-3}) selected	—	-0.275	-0.400	-0.500	-0.600	-0.620
ε_r (10^{-3}) last assumed	—	3.45	7.90	10.80	12.35	12.40
σ_1 (psi) last assumed	—	532	603	620	620	620
ζ	[8]	0.514	0.376	0.329	0.3103	0.3098
$\varepsilon_d/\zeta\varepsilon_0$	—	0.268	0.532	0.760	0.967	1.000
σ_d (psi)	[7]	-954	-1175	-1240	-1240	-1239
σ_r (psi)	[9]	0	0	0	0	0
$(\varepsilon_r - \varepsilon_d)/(\sigma_r - \sigma_d)$ (10^{-6} psi^{-1})	—	3.905	7.06	9.11	10.44	10.51
ε_l (10^{-3})	[14]	2.07	5.65	7.98	9.11	9.14
ε_t (10^{-3})	[15]	1.11	1.85	2.33	2.64	2.63
ε_r (10^{-3}) checked	[16]	3.455	7.90	10.81	12.35	12.39
f_l (ksi)	[10]	60.0	60.0	60.0	60.0	60.0
f_t (ksi)	[11]	32.2	53.6	60.0	60.0	60.0
f_{lp} (ksi)	[12]	0	0	0	0	0
f_{tp} (ksi)	[13]	0	0	0	0	0
σ_1 (psi) checked	[18]	532	603	620	620	620
α (deg)	[17]	52.46	58.62	60.0	60.0	60.0
τ_{lt} (psi)	[3]	461	523	537	537	537
γ_{lt} (10^{-3})	[6]	3.60	7.38	9.78	11.22	11.28
σ_l (psi)	—	266	302	310	310	310
σ_t (psi)	—	-266	-302	-310	-310	-310
τ_{lt} (psi)	—	461	522	537	537	537

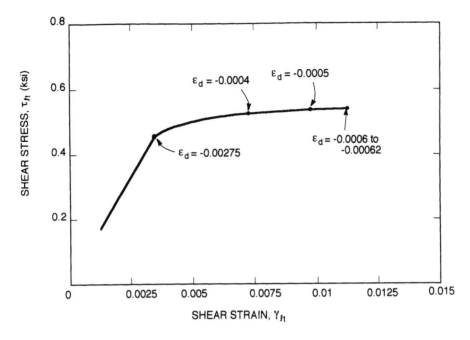

Figure 7.26 Shear stresses versus shear strain curve for Example Problem 7.2.

the yielding of both the longitudinal and the transverse steel. The element fails at $\varepsilon_d = -0.00062$.

The relationship between the shear stress τ_{lt} and shear strain γ_{lt} is shown in Figure 7.26. The shear strains at first yield and at ultimate are 0.00360 and 0.01128, respectively. The shear ductility μ_s [Eq. (7-81)] for this membrane element subjected to the given proportional loading is

$$\mu_s = \frac{0.01128}{0.00360} = 3.13$$

7.5   Failure Modes of Membrane Elements

A reinforced concrete membrane element subjected to a given set of biaxial forces and shear force may be designed in different ways. Depending on the selected thickness of the element, the steel may yield before the crushing of concrete or the concrete may crush before the yielding of steel. Depending on the ratio of the percentages of steel in the two directions, yielding of the steel may first occur in the longitudinal rebars or in the transverse rebars. An element designed in different ways will behave very differently.

Depending on the percentages of steel in the longitudinal and the transverse directions, a membrane element may fail in four modes:

1. *Underreinforced element:* Both the longitudinal steel and the transverse steel yield before the crushing of concrete.

2. *Element partially underreinforced in l direction:* Longitudinal steel yields before the crushing of concrete: Transverse steel does not yield.

3. *Element partially underreinforced in t direction:* Transverse steel yields before the crushing of concrete. Longitudinal steel does not yield.

4. *Overreinforced element:* Concrete crushes before the yielding of steel in both directions.

These four modes of failure can be divided graphically by the *failure modes diagrams* discussed in Section 7.5.3. To understand this diagram we must first study the *equal strain condition* for steel in Section 7.5.1 and the *balanced condition* of steel and concrete in Section 7.5.2.

7.5.1   EQUAL STRAIN CONDITION

Suppose that the longitudinal steel and transverse steel have the same stress, $f_l = f_t$, throughout the loading history. Then the strains in both directions must be equal, i.e.,

$$\varepsilon_l = \varepsilon_t \tag{7-102}$$

Inserting the compatibility equations (Eqs. $\boxed{4}$ and $\boxed{5}$) into Eq. (7-102) we have

$$\varepsilon_d \cos^2 \alpha + \varepsilon_r \sin^2 \alpha = \varepsilon_d \sin^2 \alpha + \varepsilon_r \cos^2 \alpha \tag{7-103}$$

Regrouping the terms results in

$$(\varepsilon_d - \varepsilon_r)(\cos^2 \alpha - \sin^2 \alpha) = 0 \tag{7-104}$$

Equation (7-104) is satisfied only if $\alpha = \pm 45°$, except when $\varepsilon_d = \varepsilon_r$, the special case of uniform tension in biaxial stresses.

Let us now limit our study to non-prestressed elements ($\rho_{lp} = \rho_{tp} = 0$) and neglect the tensile stress of concrete ($\sigma_r = 0$). The three equilibrium equations, Eqs. (7-88) to (7-90), can be further simplified by taking $\alpha = \pm 45°$:

$$m_l \sigma_1 = \frac{\sigma_d}{2} + \rho_l f_l \tag{7-105}$$

$$m_t \sigma_1 = \frac{\sigma_d}{2} + \rho_t f_t \tag{7-106}$$

$$m_{lt} \sigma_1 = -\frac{\sigma_d}{2} \tag{7-107}$$

Multiply Eq. (7-105) by m_{lt} and Eq. (7-107) by m_l. Then, we can subtract the latter from the former to eliminate the variable σ_1 and obtain

$$\rho_l f_l = \frac{-\sigma_d}{2}\left(\frac{m_l + m_{lt}}{m_{lt}}\right) \tag{7-108}$$

Similarly, multiplying Eq. (7-106) by m_{lt} and Eq. (7-107) by m_t, and then subtracting the latter from the former give

$$\rho_t f_t = \frac{-\sigma_d}{2} \left(\frac{m_t + m_{lt}}{m_{lt}} \right) \qquad (7\text{-}109)$$

Dividing Eq. (7-108) by Eq. (7-109) and taking $f_l = f_t$ result in the equal strain condition

$$\frac{\rho_l}{\rho_t} = \frac{m_l + m_{lt}}{m_t + m_{lt}} \qquad (5\text{-}110)$$

The equal strain condition in Eq. (5-110) was first derived by Fialkow (1985). This equation expresses the ratio of the percentage of longitudinal steel to the percentage of transverse steel in terms of the three m coefficients for proportional loadings. If a membrane element is designed according to this equal strain condition, the yielding of the steel is expected to occur simultaneously in the longitudinal and the transverse steel.

In Example 7.2, the three m coefficients were found to be $m_l = 0.5$, $m_t = -0.5$, and $m_{lt} = 0.866$. The ρ_l/ρ_t ratio under the equal strain condition is

$$\frac{\rho_l}{\rho_t} = \frac{0.5 + 0.866}{-0.5 + 0.866} = 3.73$$

Because the membrane element in Example 7.2 was designed to have a ρ_l/ρ_t ratio of unity, which is less than 3.73, the yielding of the steel is expected to first occur in the longitudinal rebars rather than in the transverse rebars.

Equation (5-110) is also applicable to the special case of uniform tension. In this special case of $S = 1$, the m coefficients are $m_l = m_t = 1$ and $m_{lt} = 0$ according to Eqs. (7-85) to (7-87). As a result, $\rho_l = \rho_t$ and α becomes indeterminate.

7.5.2   BALANCED CONDITION

Now that the steel in a membrane element can be designed to yield simultaneously under the equal strain condition, we can proceed to determine the balanced condition between the steel and the concrete. The balanced condition defines a mode of failure where both the longitudinal and the transverse steel yield simultaneously with the crushing of concrete. The balanced percentage of steel, therefore, divides the under-reinforced element from the overreinforced element under the equal strain condition.

Adding Eqs. (7-108) and (7-109), and taking $f_l = f_t = f_s$ results in

$$\rho_l + \rho_t = \frac{-\sigma_d}{2f_s} \left(\frac{m_l + 2m_{lt} + m_t}{m_{lt}} \right) \qquad (7\text{-}111)$$

Equation (7-111) is a parametric representation of a point on the equal strain line with σ_d/f_s as an unknown parameter. The relationship between the concrete stress σ_d and the steel stress f_s will have to be determined from the strain compatibility condition and the stress–strain relationships of concrete and steel. The balanced condition is obtained when the normalized concrete stress σ_d/f_c' peaks at the same time the steel stress f_s reaches the yield point.

Assuming the yielding of steel under the equal strain condition, i.e., $\varepsilon_l = \varepsilon_t = \varepsilon_y$ and $\alpha = 45°$, both the strain compatibility equations (Eqs. [4] and [5]) are reduced to

$$\varepsilon_r = 2\varepsilon_y - \varepsilon_d \tag{7-112}$$

Inserting ε_r from Eq. (7-112) into Eq. [8a] for the softening coefficient of concrete,

$$\zeta = \frac{0.9}{\sqrt{1 + 600(2\varepsilon_y - \varepsilon_d)}} \tag{7-113}$$

The yield strain ε_y in Eq. (7-113) is equal to 0.00207 for a mild steel with $f_y = 60,000$ psi.

Using the softened stress–strain relationship of concrete represented by Eqs. [7a], [7b], and (7-113), we can now trace the normalized concrete stress σ_d/f_c' with the increase of the concrete strain ε_d (in an absolute sense) as shown in Table 7.3. In Table 7.3 the peak strain ε_0 of nonsoftened concrete is taken as -0.002 in the calculation of the ratio $\varepsilon_d/\zeta\varepsilon_0$. It should be recalled that the concrete strain ε_d is in the ascending branch when $\varepsilon_d/\zeta\varepsilon_0 < 1$ and is in the descending branch when $\varepsilon_d/\zeta\varepsilon_0 > 1$. Table 7.3 illustrates that the normalized concrete stress σ_d/f_c' peaks at a value of $0.44918 \approx 0.4492$ at a concrete strain ε_d of 0.00087. The corresponding $\varepsilon_d/\zeta\varepsilon_0$ value of 0.96739 is in the ascending branch but is very close to the peak stress of $\zeta f_c'$.

Substituting $\sigma_d = 0.4492f_c'$ and $f_s = f_y$ into Eq. (7-111) we obtain the total percentage of steel for the balanced condition:

$$\rho_l + \rho_t = \frac{-0.4492f_c'}{f_y}\left(\frac{m_l + 2m_{lt} + m_t}{2m_{lt}}\right) \tag{7-114}$$

In the case of Example 7.2, where $m_l = 0.5$, $m_t = -0.5$, $m_{lt} = 0.866$, $f_c' = 4000$ psi, and $f_y = 60,000$ psi, the total percentage of steel ($\rho_l + \rho_t$) is

$$\rho_l + \rho_t = \frac{-0.4492(-4000)}{60,000}\left(\frac{0.5 + 2(0.866) - 0.5}{2(0.866)}\right) = 0.02995 \text{ or } 2.995\%$$

Solving the preceding equation with the equal strain condition of $\rho_l/\rho_t = 3.73$, the percentages of the steel in the longitudinal and the transverse directions for the balanced condition are

$$\rho_l = \frac{3.73}{1 + 3.73}(0.02995) = 2.362\%$$

$$\rho_t = \frac{1}{1 + 3.73}(0.02995) = 0.633\%$$

7.5.3 FAILURE REGIONS AND BOUNDARIES

The proportional loading given in Example 7.2 has a pair of principal stress variables of $S = -1$ and $\alpha_2 = 60°$. The normalized Mohr circle is plotted in

TABLE 7.3 Finding the Peak of σ_d/f_c' ($\alpha = 45°$)

ε_d	ζ Eq. (7-113)	$\varepsilon_d/\zeta\varepsilon_0$	σ_d/f_c' Eq. 7a	σ_d/f_c' Eq. 7b
-0.00050	0.46266	$0.54035 < 1$	0.36493	—
-0.00060	0.45904	$0.65354 < 1$	0.40392	—
-0.00070	0.45550	$0.76839 < 1$	0.43106	—
-0.00080	0.45204	$0.88488 < 1$	0.44605	—
-0.00086	0.45000	$0.95556 < 1$	0.44911	—
-0.00087	0.44966	$0.96739 < 1$	0.44918 peak	—
-0.00088	0.44933	$0.97924 < 1$	0.44914	—
-0.00090	0.44866	$1.00298 > 1$	—	0.44866
-0.00100	0.44535	$1.12271 > 1$	—	0.44480

Figure 7.25 with the m coefficients indicated as $m_l = 0.5$, $m_t = -0.5$, and $m_{lt} = 0.866$. The failure modes diagram for this proportional loading is shown in Figure 7.27. Such diagrams were first constructed by Han and Mau (1988).

In the failure modes diagram the percentage of longitudinal steel, ρ_l, and the percentage of transverse steel, ρ_t, are represented by the horizontal axis and the vertical axis, respectively. The inclined dotted line OA represents the equal strain condition with a ρ_l/ρ_t ratio of 3.73. The point indicated by the letter B represents the balanced condition. This balanced point has a coordinate of $\rho_l = 2.362\%$ and $\rho_t = 0.633\%$ as calculated previously.

The failure modes diagram is divided into four regions by the four solid curves radiating from the balanced point. These four regions give the four modes of failure as indicated in the diagram. They are (1) underreinforced, (2) partially underreinforced in the l direction, (3) partially underreinforced in the t direction, and (4) overreinforced. These four regions are defined as follows:

(1) Underreinforced　　　　　　　　　　　$\varepsilon_t > \varepsilon_y$, $\varepsilon_l > \varepsilon_y$, and σ_d/f_c' peaks
(2) Partially underreinforced in l direction　$\varepsilon_t < \varepsilon_y$, $\varepsilon_l > \varepsilon_y$, and σ_d/f_c' peaks
(3) Partially underreinforced in t direction　$\varepsilon_t > \varepsilon_y$, $\varepsilon_l < \varepsilon_y$, and σ_d/f_c' peaks
(4) Overreinforced　　　　　　　　　　　　$\varepsilon_t < \varepsilon_y$, $\varepsilon_l < \varepsilon_y$, and σ_d/f_c' peaks

The boundaries between the regions are defined as follows:

Inner boundary I_l:　$\varepsilon_t = \varepsilon_y$, $\varepsilon_l > \varepsilon_y$, and σ_d/f_c' peaks　Between regions (1) and (2)
Inner boundary I_t:　$\varepsilon_t > \varepsilon_y$, $\varepsilon_l = \varepsilon_y$, and σ_d/f_c' peaks　Between regions (1) and (3)
Outer boundary O_l:　$\varepsilon_t < \varepsilon_y$, $\varepsilon_l = \varepsilon_y$, and σ_d/f_c' peaks　Between regions (2) and (4)
Outer boundary O_t:　$\varepsilon_t = \varepsilon_y$, $\varepsilon_l < \varepsilon_y$, and σ_d/f_c' peaks　Between regions (3) and (4)

The boundaries of the four failure modes can be constructed by the same procedures used in Sections 7.5.1 and 7.5.2, except that the angle α is no longer 45° as required by the equal strain condition. Therefore, the equilibrium and compatibility equations will have to be generalized to include the angle α.

In the case of non-prestressed elements ($\rho_{lp} = \rho_{tp} = 0$) and neglecting the tensile stress of concrete ($\sigma_r = 0$), the three equilibrium equations for proportional loadings

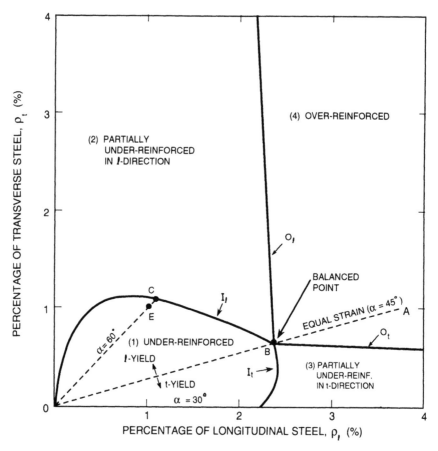

Figure 7.27 Failure modes diagram for $S = -1$, $\alpha_2 = 60°$ (for the proportional loading in Example Problem 7.2).

[Eqs. (7-88) to (7-90)] are reduced to

$$m_l \sigma_1 = \sigma_d \cos^2 \alpha + \rho_l f_l \tag{7-115}$$

$$m_t \sigma_1 = \sigma_d \sin^2 \alpha + \rho_t f_t \tag{7-116}$$

$$m_{lt} \sigma_1 = -\sigma_d \sin \alpha \cos \alpha \tag{7-117}$$

Multiply Eq. (7-115) by m_{lt} and Eq. (7-117) by m_l. Then we can subtract the latter from the former to eliminate the variable σ_1 and obtain

$$\rho_l = \frac{-\sigma_d}{f_l} \left(\frac{m_l}{m_{lt}} \sin \alpha \cos \alpha + \cos^2 \alpha \right) \tag{7-118}$$

Similarly, multiplying Eq. (7-116) by m_{lt} and Eq. (7-117) by m_t, and then subtracting the latter from the former give

$$\rho_t = \frac{-\sigma_d}{f_t} \left(\frac{m_t}{m_{lt}} \sin \alpha \cos \alpha + \sin^2 \alpha \right) \tag{7-119}$$

The relationship between the concrete stress σ_d and the steel stresses f_l and f_t in

Eqs. (7-118) and (7-119) will have to be determined from the strain compatibility conditions and the stress–strain relationships of concrete and steel.

In the case of $\varepsilon_l = \varepsilon_y$ for mild steel, the compatibility equation for ε_l (Eq. $\boxed{4}$) becomes

$$\varepsilon_r = -\varepsilon_d \cot^2 \alpha + \varepsilon_y \frac{1}{\sin^2 \alpha} \tag{7-120}$$

Similarly, in the case of $\varepsilon_t = \varepsilon_y$, the compatibility equation for ε_t (Eq. $\boxed{5}$) gives

$$\varepsilon_r = -\varepsilon_d \tan^2 \alpha + \varepsilon_y \frac{1}{\cos^2 \alpha} \tag{7-121}$$

The tensile strain ε_r in Eqs. (7-120) or (7-121) can be used in conjunction with Eq. $\boxed{8a}$ to calculate the softening coefficient of concrete ζ. Then the proportionally softened stress–strain relationship of Eqs. $\boxed{7a}$ and $\boxed{7b}$ will be used to trace the peaking of the concrete stress.

The procedures to plot the inner boundary curves, I_l and I_t, are summarized as follows:

Step 1: Select an angle α and insert it into Eqs. (7-118) to (7-121).

Step 2: Increase the strain ε_d incrementally and calculate the softening coefficient ζ by Eq. $\boxed{8a}$ in conjunction with the strain ε_r from Eqs. (7-120) or (7-121). Then we can trace and locate the peak of the normalized concrete stress $(\sigma_d)_{\text{peak}}/f_c'$ by the stress–strain relationship of Eqs. $\boxed{7a}$ and $\boxed{7b}$, similar to the process in Table 7.3.

Step 3: Substitute $\sigma_d = (\sigma_d)_{\text{peak}}$ and $f_l = f_t = f_y$ into Eqs. (7-118) and (7-119) to obtain the percentages of the longitudinal and the transverse steel ρ_l and ρ_t to locate one point on the inner boundary curves for the selected angle α.

Step 4: Vary the angle α incrementally and repeat Steps 1 to 3 to allow generation of a series of points to plot the inner boundary curves.

Take $\alpha = 60°$, for example, and locate a point on the inner boundary curve I_l, where $\varepsilon_t = \varepsilon_y$ and $\varepsilon_l > \varepsilon_y$:

Eq. (7-121) $\varepsilon_r = -\varepsilon_d \tan^2 \alpha + \varepsilon_y \dfrac{1}{\cos^2 \alpha} = -\varepsilon_d(3) + 0.00207(4)$

$$= 3(-\varepsilon_d) + 0.00828$$

Now we can find the peak of the concrete stress as shown in Table 7.4.

The longitudinal steel and the transverse steel can be calculated from Eqs. (7-118) and (7-119):

$$\rho_l = \frac{-0.3364 f_c'}{f_y} \left[\frac{m_l}{m_{lt}} \sin 60° \cos 60° + \cos^2 60° \right]$$

$$= \frac{-0.3364(4000)}{60,000} \left[\frac{0.5}{0.866}(0.866)(0.5) + (0.5)^2 \right] = 0.01121$$

TABLE 7.4   Finding the Peak of σ_d/f_c' ($\alpha = 60°$ and $\varepsilon_t = \varepsilon_y$)

ε_d	ε_r Eq. (7-121)	ζ Eq. [8a]	$\varepsilon_d/\zeta\varepsilon_0$	σ_d/f_c' Eq. [7a]	σ_d/f_c' Eq. [7b]
−0.00050	0.00978	0.34342	0.72797 < 1	0.31801	—
−0.00060	0.01008	0.33901	0.88493 < 1	0.33452	—
−0.00064	0.01020	0.33729	0.94874 < 1	0.33640	—
−0.00065	0.01023	0.33686	0.96478 < 1	0.33644 peak	—
−0.00066	0.01026	0.33644	0.98086 < 1	0.33632	—
−0.00070	0.01038	0.33476	1.04552 > 1	—	0.33473
−0.00080	0.01068	0.33067	1.20966 > 1	—	0.33007

$$\rho_t = \frac{-0.3364 f_c'}{f_y}\left[\frac{m_t}{m_{lt}}\sin 60° \cos 60° + \sin^2 60°\right]$$

$$= \frac{-0.3364(4000)}{60,000}\left[\frac{-0.5}{0.866}(0.866)(0.5) + (0.866)^2\right] = 0.01121$$

The coordinate of $\rho_l = \rho_t = 0.01121$ is given as point C in Figure 7.27. It is a point on the inner boundary I_l with $\alpha = 60°$. By varying the angle α from 45 to 90° and repeating the same procedures we have the whole inner boundary curve I_l. Similarly, the inner boundary curve I_t can be obtained using Eq. (7-120) and varying the angle α from 45° to 30°. On this boundary curve $\varepsilon_l = \varepsilon_y$ and $\varepsilon_t > \varepsilon_y$.

It is interesting to point out that the design of reinforcement using the plasticity truss model in Example 4.1 resulted in the steel percentage of $\rho_l = \rho_t = 0.01033$. This pair of ρ_l and ρ_t is plotted as point E in Figure 7.27 for a concrete strength of $f_c' = 4000$ psi. Point E is located in region (1) just within the inner boundary I_l, meaning that the 4000-psi concrete did not crush when both the longitudinal and the transverse steel yielded. If a lower strength of concrete is used, point E might be located in region (2) outside the inner boundary I_l. This situation means that the concrete would crush before the yielding of the transverse steel, and the assumption of the plasticity truss model ($f_l = f_t = f_y$) could not be ensured.

The outer-boundary curves, O_l and O_t, in Figure 7.27 can be obtained by the preceding four-step procedures with one exception in step 3. In the case of O_l, the longitudinal steel stress will yield, $f_l = f_y$, but the transverse steel stress will not, $f_t < f_y$. Hence, before the calculation of transverse steel percentage ρ_t by Eq. (7-119) the unyielded transverse steel stress f_t has to be calculated from the transverse steel strain ε_t by the strain compatibility condition, $\varepsilon_t = \varepsilon_r - \varepsilon_l + \varepsilon_d$. The strain ε_l is, of course, equal to the yield stress ε_y, and the strains ε_d and ε_r are obtained in step 2 after the peak of the normalized concrete stress is located.

8

Softened Truss Model for Torsion

8.1 Equilibrium Equations

8.1.1 SHEAR ELEMENTS IN SHEAR FLOW ZONE

In Chapter 7 we applied the softened truss model to membrane elements subjected to shear and normal stresses. In this chapter we will extend the softened truss model to members subjected to torsion (Hsu, 1984, 1988, 1990, 1991b; Hsu and Mo, 1985a, 1985b, 1985c). These two chapters, therefore, are closely related.

A reinforced concrete prismatic member is subjected to an external torque T as shown in Figure 8.1a. This external torque is resisted by an internal torque formed by the circulatory shear flow q along the periphery of the cross section. This shear flow q occupies a zone, called the shear flow zone, which has a thickness denoted t_d. This thickness t_d is a variable determined from the equilibrium and compatibility conditions. It is not the same as the given wall thickness h of a hollow member.

Element A in the shear flow zone (Figure 8.1a) is subjected to a shear stress $\tau_{lt} = q/t_d$ as shown in Figure 8.1b. The in-plane equilibrium of this element should satisfy the three equations, Eqs. (7-1) to (7-3), in Chapter 7 [see also Eqs. (4-59) to (4-61) in Chapter 4 for non-prestressed elements]:

$$\sigma_l = \sigma_d \cos^2 \alpha + \sigma_r \sin^2 \alpha + \rho_l f_l + \rho_{lp} f_{lp} \qquad \text{(8-1) or ①}$$

$$\sigma_t = \sigma_d \sin^2 \alpha + \sigma_r \cos^2 \alpha + \rho_t f_t + \rho_{tp} f_{tp} \qquad \text{(8-2) or ②}$$

$$\tau_{lt} = (-\sigma_d + \sigma_r)\sin \alpha \cos \alpha \qquad \text{(8-3) or ③}$$

In Eqs. ① to ③ the steel ratios, ρ_l, ρ_t, ρ_{lp}, and ρ_{tp}, should be taken with respect to

257

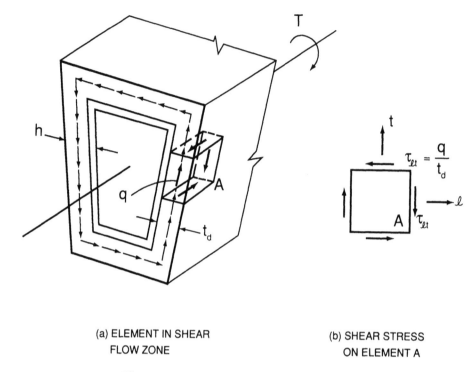

(a) ELEMENT IN SHEAR
FLOW ZONE

(b) SHEAR STRESS
ON ELEMENT A

Figure 8.1 Hollow box subjected to torsion.

the thickness of the shear flow zone t_d as

$$\rho_l = \frac{A_l}{p_0 t_d} \tag{8-4}$$

$$\rho_t = \frac{A_t}{s t_d} \tag{8-5}$$

$$\rho_{lp} = \frac{A_{lp}}{p_0 t_d} \tag{8-6}$$

$$\rho_{tp} = \frac{A_{tp}}{s t_d} \tag{8-7}$$

where

A_l, A_{lp} = total cross-sectional areas of longitudinal mild steel and longitudinal prestressing steel, respectively

A_t, A_{tp} = cross-sectional areas of one transverse mild steel bar and one transverse prestressing steel strand, respectively

Because Eqs. ① and ② involve the steel ratios ρ_l, ρ_t, ρ_{lp}, and ρ_{tp}, these two equilibrium equations are coupled to the compatibility equations through the variable

t_d. The thickness t_d is a geometric variable, which has to be determined not only by the equilibrium conditions, but also by the compatibility conditions. This is similar to the determination of the neutral axis in the bending theory (see Chapter 2), which requires the plane section compatibility condition.

8.1.2   BREDT'S EQUILIBRIUM EQUATION

In Section 8.1.1 three equations, Eqs. ① to ③, were derived from the equilibrium of a membrane element in the shear flow zone. To maintain equilibrium of the whole cross section, however, a fourth equation must be satisfied. The derivation of this additional equilibrium equation was given in Section 3.1.4.1. The resulting Eq. (3-46) is expressed in terms of the shear flow q. For a shear flow zone thickness of t_d, the shear stress τ_{lt} is

$$\tau_{lt} = \frac{T}{2 A_0 t_d} \qquad \text{(8-8) or ④}$$

This additional equation, Eq. ④, also introduces an additional variable, the torque T.

The thickness of the shear flow zone t_d is strongly involved in Eq. ④ not only explicitly through t_d, but also implicitly through A_0, which is a function of t_d. Consequently, the three equilibrium equations [Eqs. ①, ②, and ④] are now coupled to the compatibility equations through the variable t_d. These compatibility equations will be derived in the next section.

8.2   Compatibility Equations

8.2.1   SHEAR ELEMENTS

As shown in Figure 8.1a and b, the element A in the shear flow zone is subjected to a shear stress. The in-plane deformation of this element should satisfy the three compatibility equations, Eqs. (7-4) to (7-6), in Chapter 7 [see also Eqs. (5-24) to (5-26) in Chapter 5]:

$$\varepsilon_l = \varepsilon_d \cos^2 \alpha + \varepsilon_r \sin^2 \alpha \qquad \text{(8-9) or ⑤}$$

$$\varepsilon_t = \varepsilon_d \sin^2 \alpha + \varepsilon_r \cos^2 \alpha \qquad \text{(8-10) or ⑥}$$

$$\frac{\gamma_{lt}}{2} = (-\varepsilon_d + \varepsilon_r)\sin \alpha \cos \alpha \qquad \text{(8-11) or ⑦}$$

This element A will also be subjected to out-of-plane deformation. To study this out-of-plane deformation, we will first relate the shear strain γ_{lt} in Eq. ⑦ to the angle of twist θ by the geometric relationship presented in the next section.

8.2.2   SHEAR STRAIN DUE TO TWISTING

When a tube is subjected to torsion, the relationship between the shear strain γ_{lt} in the wall of the tube and the angle of twist θ of the member, can be derived from

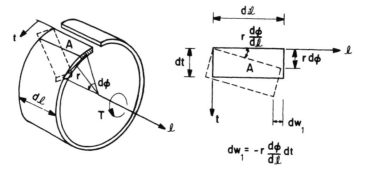

**(a) WARPING DISPLACEMENT (IN l-DIRECTION)
DUE TO ANGLE OF TWIST**

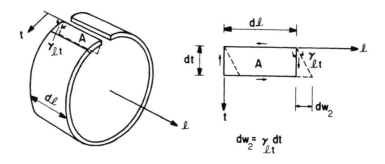

**(b) WARPING DISPLACEMENT (IN l-DIRECTION)
DUE TO SHEAR DEFORMATION**

Figure 8.2 Warping displacement in a tube.

the compatibility condition of warping deformation. In Figure 8.2, a longitudinal cut is made in an infinitesimal length d*l* of a tube. Progressing from one side of the cut along the circumference to the other side of the cut, the differential warping displacement (in the *l* direction) must be zero when integrating throughout the whole perimeter.

The warping displacement of a differential element *A*, shown in Figure 8.2, is composed of two parts. The first part is induced by the rigid rotation dϕ as shown in Figure 8.2a. The second part is caused by the shear deformation γ_{lt} as shown in Figure 8.2b. Under torsion, the element *A*, d*l* by d*t*, in Figure 8.2a rotates an angle $r\,d\phi/dl$ in the *l-t* plane. The symbol *r* is the distance from the center of twist to the center line of the element. The differential warping displacement dw_1 is therefore

$$dw_1 = -r\frac{d\phi}{dl}\,dt = -r\theta\,dt \tag{8-12}$$

The differential warping displacement dw_2 of element *A* due to shear deformation in Figure 8.2b is

$$dw_2 = \gamma_{lt}\,dt \tag{8-13}$$

Adding Eqs. (8-12) and (8-13) gives the total differential warping displacement dw due to both the rotation and the shear deformation:

$$dw = dw_1 + dw_2 = -r\theta \, dt + \gamma_{lt} \, dt \qquad (8\text{-}14)$$

For a closed section, the total differential warping displacement integrating around the whole perimeter must be equal to zero, giving

$$\oint dw = -\theta \oint r \, dt + \oint \gamma_{lt} \, dt = 0 \qquad (8\text{-}15)$$

Recalling $\oint r \, dt = 2A_0$ from Eq. (3-45), the preceding equation becomes

$$\oint \gamma_{lt} \, dt = 2A_0\theta \qquad (8\text{-}16)$$

When the wall thickness of the tube is uniform, the shear stress τ_{lt} is a constant, resulting in a uniform shear strain of γ_{lt}. Then γ_{lt} could be taken out of the integral in Eq. (8-16), giving

$$\gamma_{lt} \oint dt = 2A_0\theta \qquad (8\text{-}17)$$

Because $\oint dt$ is the perimeter of the centerline of the shear flow, p_0, we have

$$\theta = \frac{p_0}{2A_0}\gamma_{lt} \qquad (8\text{-}18) \text{ or } \boxed{8}$$

It is clear from Eq. $\boxed{8}$ that the angle of twist θ will produce a shear strain γ_{lt} in the elements of the shear flow zone. This shear strain γ_{lt} will induce the steel strains ε_l and ε_t in the l-t direction and the concrete strains ε_d and ε_r in the d-r direction. The relationships between the strains in the l-t direction (ε_l, ε_t, and γ_{lt}) and the strains in the d-r direction (ε_d and ε_r) are described by the three transformation equations, Eqs. $\boxed{5}$ to $\boxed{7}$.

8.2.3   BENDING OF DIAGONAL CONCRETE STRUTS

In a torsional member, the angle of twist θ also produces warping in the wall of the member, which, in turn, causes bending in the concrete struts. In other words, the concrete struts are not only subjected to compression due to the circulatory shear, but also subjected to bending due to the warping of the wall. The relationship between the angle of twist θ and the bending curvature of concrete struts ψ will now be studied.

A box member with four walls of thickness h and subjected to a torsional moment T is shown in Figure 8.3a. Each wall contains a shear flow zone with a thickness of t_d. The perimeter of the center line of shear flow q has a width of l_q along the top wall. The length of the member is taken to be $l_q \cot\alpha$, so that the diagonal line in the center plane of the shear flow in the top wall $OABC$ has an angle of inclination α. When this member receives an angle of twist θ, this center plane $OABC$ will become a hyperbolic paraboloid surface $OADC$ as shown in Figure 8.3b. The edge CB of the plane rotates to the position CD through an angle $\theta l_q \cot\alpha$. The curve OD will have a curvature of ψ to be determined.

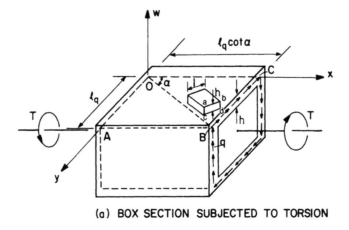

(a) BOX SECTION SUBJECTED TO TORSION

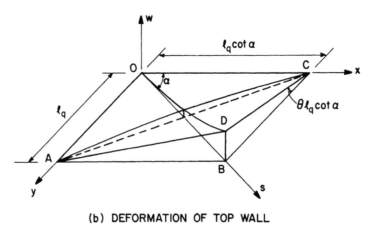

(b) DEFORMATION OF TOP WALL

Figure 8.3 Bending of a concrete strut in the wall of a box section subjected to torsion.

A coordinate system with axes x, y, and w is imposed along the edges of the center plane OC, OA, and their normal, respectively, as shown in Figure 8.3b. The hyperbolic paraboloid surface $OADC$ can then be expressed by

$$w = \theta xy \tag{8-19}$$

where w is the displacement perpendicular to the x-y plane.

Imposing an axis s through OB along the direction of the diagonal concrete struts, the slope of the curve OD can be obtained by differentiating w with respect to s. Utilizing the chain rule, the slope dw/ds is

$$\frac{dw}{ds} = \frac{\partial w}{\partial x}\frac{dx}{ds} + \frac{\partial w}{\partial y}\frac{dy}{ds} = (\theta y)\cos\alpha + (\theta x)\sin\alpha \tag{8-20}$$

The curvature of the concrete struts ψ is the second derivative of w with respect to s, resulting in

$$\frac{d^2 w}{d^2 s} = \frac{\partial(dw/ds)}{\partial x}\frac{dx}{ds} + \frac{\partial(dw/ds)}{\partial y}\frac{dy}{ds} = (\theta\sin\alpha)\cos\alpha + (\theta\cos\alpha)\sin\alpha \tag{8-21}$$

Hence,

$$\psi = \theta \sin 2\alpha \qquad \text{(8-22) or } ⑨$$

The derivation of Eq. ⑨ is illustrated by a rectangular box section in Figure 8.3, because the imposed curvature is easy to visualize in such a section. In actuality, this equation is applicable to any arbitrary bulky sections with multiple walls.

8.2.4   STRAIN DISTRIBUTION IN CONCRETE STRUTS

The curvature ψ derived in Eq. ⑨ produces a nonuniform strain distribution in the concrete struts. Figure 8.4a shows a unit width of a concrete strut in a hollow section with a wall thickness of h. The tension area in the inner portion of the cross section is neglected. The area in the outer portion that is in compression is considered to be effective to resist the shear flow. The depth of the compression zone from the neutral axis (N.A.) to the extreme compression fiber is defined as the thickness of the shear flow zone, t_d. Within this thickness, t_d, the strain distribution is assumed to be linear as shown in Figure 8.4b. This assumption is identical to Bernoulli's plane section hypothesis used in the bending theory of Chapter 2. The thickness t_d can, therefore, be related to the curvature ψ and the maximum strain at the surface ε_{ds} by the simple relationship

$$t_d = \frac{\varepsilon_{ds}}{\psi} \qquad \text{(8-23) or } ⑩$$

The average strain ε_d can be simply related to the maximum strain ε_{ds}, Figure 8.4d, by

$$\varepsilon_d = \frac{\varepsilon_{ds}}{2} \qquad \text{(8-24) or } ⑪$$

Equations ⑧, ⑨, ⑩, and ⑪ are the four additional compatibility equations for torsion. They introduce four additional variables, θ, ψ, t_d, and ε_{ds}.

8.3   Constitutive Relationships for Concrete

8.3.1   SOFTENED COMPRESSION STRESS BLOCK

When the strain distribution in the concrete struts is assumed linear, the stress distribution is represented by a curve as shown in Figure 8.4c. This softened compression stress block has a peak stress σ_p defined by

$$\sigma_p = \zeta f_c' \qquad \text{(8-25)}$$

where the softening of the concrete struts is taken into account by the coefficient ζ. The average stress of the concrete struts σ_d is defined as

$$\sigma_d = k_1 \sigma_p = k_1 \zeta f_c' \qquad \text{(8-26) or } ⑫$$

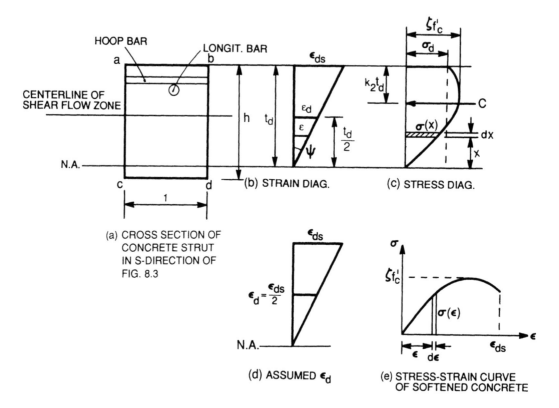

Figure 8.4   Strains and stresses in concrete struts.

where the nondimensional coefficient k_1 is defined as the ratio of the average stress to the peak stress. The symbol σ_d in Eq. ⑫ has been generalized to represent the average compression stress in a concrete strut subjected to bending and compression, rather than the compression stress of axially loaded concrete struts in a membrane element as defined for the symbol σ_d in the equilibrium equations, Eqs. ① to ③. The generalization of the symbol σ_d implies two assumptions. First, the relationship between the average stress σ_d defined by Eq. ⑫ and average strain ε_d defined by Eq. ⑪ is identical to the stress–strain relationship of an axially loaded concrete strut. Second, the softening coefficient ζ, which has been determined from the tests of membrane elements with concrete struts under axial compression, is assumed to be applicable to the concrete struts under combined bending and compression.

The resultant C of the softened compression stress block has a magnitude of

$$C = \sigma_d t_d = k_1 \sigma_p t_d = k_1 \zeta f'_c t_d \tag{8-27}$$

This resultant C is located at a distance $k_2 t_d$ from the extreme compression fiber (Figure 8.4c). The nondimensional coefficient k_2 is defined as the ratio of the distance between the resultant C and the extreme compression fiber to the depth of the compression zone t_d. The compression stress block is statically defined, when the two coefficients k_1 and k_2 are determined.

The coefficient k_1 can be determined from the equilibrium of forces by integrating the compression stress block. Designating the stress at a distance x from the neutral axis as $\sigma(x)$ (Figure 8.4c) gives

$$C = k_1 \sigma_p t_d = \int_0^{t_d} \sigma(x)\,dx \tag{8-28}$$

The geometric shape of the compression stress block in Figure 8.4c will be identical to that of the stress–strain curve of softened concrete (Figure 8.4e) if two assumptions are made. First, the strain distribution in Figure 8.4b is assumed to be linear, and the strain ε at a distance x from the neutral axis is related to x by similar triangles as

$$x = \frac{t_d}{\varepsilon_{ds}}\varepsilon \tag{8-29}$$

or

$$dx = \frac{t_d}{\varepsilon_{ds}}\,d\varepsilon \tag{8-30}$$

Second, the strain gradient in the stress block is assumed to have no effect on the stress–strain curve. Substituting dx from Eq. (8-30) into Eq. (8-28) and changing the integration limit from the distance t_d to the strain ε_{ds}, we have

$$C = k_1 \sigma_p t_d = \frac{t_d}{\varepsilon_{ds}}\int_0^{\varepsilon_{ds}}\sigma(\varepsilon)\,d\varepsilon \tag{8-31}$$

and

$$k_1 = \frac{1}{\sigma_p \varepsilon_{ds}}\int_0^{\varepsilon_{ds}}\sigma(\varepsilon)\,d\varepsilon \tag{8-32}$$

The coefficient k_2 can be determined by taking the equilibrium of moments about the neutral axis:

$$C(1 - k_2)t_d = \int_0^{t_d}\sigma(x)x\,dx \tag{8-33}$$

Substituting x from Eqs. (8-29) and dx from Eq. (8-30) into Eq. (8-33) and changing the integration limit, we obtain

$$C(1 - k_2)t_d = \frac{t_d^2}{\varepsilon_{ds}^2}\int_0^{\varepsilon_{ds}}\sigma(\varepsilon)\varepsilon\,d\varepsilon \tag{8-34}$$

Substituting C from Eq. (8-31) into Eq. (8-34) gives

$$k_2 = 1 - \frac{1}{\varepsilon_{ds}}\frac{\displaystyle\int_0^{\varepsilon_{ds}}\sigma(\varepsilon)\varepsilon\,d\varepsilon}{\displaystyle\int_0^{\varepsilon_{ds}}\sigma(\varepsilon)\,d\varepsilon} \tag{8-35}$$

The coefficients k_1 and k_2 can be calculated from Eqs. (8-32) and (8-35), respectively, if the stress–strain curve σ vs. ε is mathematically given.

8.3.2   COEFFICIENT k_1 FOR AVERAGE COMPRESSION STRESS

Coefficient k_1 can be obtained from Eq. (8-32) using the softened stress–strain curve of Eqs. 7a and 7b in Sections 7.2.1.2 or 7.3.1. These two curves are sketched in Figure 8.5a and b. Because two equations are given to describe the stress–strain curve, the coefficient k_1 will have two expressions, one for the ascending branch and another for the descending branch of the stress–strain curve.

Ascending Branch ($\varepsilon_{ds}/\varepsilon_p \le 1$ and referring to Figure 8.5a)

Inserting Eq. 7a into Eq. (8-32) gives

$$k_1 = \frac{1}{\sigma_p \varepsilon_{ds}} \int_0^{\varepsilon_{ds}} \sigma_p \left[2\left(\frac{\varepsilon}{\varepsilon_p}\right) - \left(\frac{\varepsilon}{\varepsilon_p}\right)^2 \right] d\varepsilon \qquad (8\text{-}36)$$

For the convenience of integration, the symbols of peak stress and peak strain, σ_p and ε_p, are used in Eq. (8-36), rather than $\zeta f_c'$ and $\zeta \varepsilon_0$. The dummy variable is written as ε, rather than ε_d, to avoid confusion with the symbol for the average strain. Integrating Eq. (8-36) results in

$$k_1 = \frac{\varepsilon_{ds}}{\varepsilon_p}\left(1 - \frac{1}{3}\frac{\varepsilon_{ds}}{\varepsilon_p} \right) \qquad (8\text{-}37)$$

Letting $\varepsilon_p = \zeta\varepsilon_0$ and expressing k_1 in terms of ζ, Eq. (8-37) becomes:

$$\text{When } \varepsilon_{ds}/\zeta\varepsilon_0 \le 1 \qquad k_1 = \frac{\varepsilon_{ds}}{\zeta\varepsilon_0}\left(1 - \frac{1}{3}\frac{\varepsilon_{ds}}{\zeta\varepsilon_0} \right) \qquad (8\text{-}38) \text{ or } \textcircled{13a}$$

Descending Branch ($\varepsilon_{ds}/\varepsilon_p > 1$ and referring to Figure 8.5b)

$$k_1 = \frac{1}{\sigma_p \varepsilon_{ds}} \int_0^{\varepsilon_p} \sigma_p \left[2\left(\frac{\varepsilon}{\varepsilon_p}\right) - \left(\frac{\varepsilon}{\varepsilon_p}\right)^2 \right] d\varepsilon + \frac{1}{\sigma_p \varepsilon_{ds}} \int_{\varepsilon_p}^{\varepsilon_{ds}} \sigma_p \left[1 - \left(\frac{\varepsilon - \varepsilon_p}{2\varepsilon_0 - \varepsilon_p}\right)^2 \right] d\varepsilon \quad (8\text{-}39)$$

Integration and simplification of Eq. (8-39) result in

$$k_1 = \left[1 - \frac{\varepsilon_p^2}{(2\varepsilon_0 - \varepsilon_p)^2} \right]\left(1 - \frac{1}{3}\frac{\varepsilon_p}{\varepsilon_{ds}} \right) + \frac{\varepsilon_p^2}{(2\varepsilon_0 - \varepsilon_p)^2}\frac{\varepsilon_{ds}}{\varepsilon_p}\left(1 - \frac{1}{3}\frac{\varepsilon_{ds}}{\varepsilon_p} \right) \qquad (8\text{-}40)$$

Expressing k_1 in terms of ζ by letting $\varepsilon_p = \zeta\varepsilon_0$ gives:

$$\text{When } \varepsilon_{ds}/\zeta\varepsilon_0 > 1 \qquad k_1 = \left[1 - \frac{\zeta^2}{(2 - \zeta)^2} \right]\left(1 - \frac{1}{3}\frac{\zeta\varepsilon_0}{\varepsilon_{ds}} \right)$$

$$+ \frac{\zeta^2}{(2 - \zeta)^2}\frac{\varepsilon_{ds}}{\zeta\varepsilon_0}\left(1 - \frac{1}{3}\frac{\varepsilon_{ds}}{\zeta\varepsilon_0} \right) \qquad (8\text{-}41) \text{ or } \textcircled{13b}$$

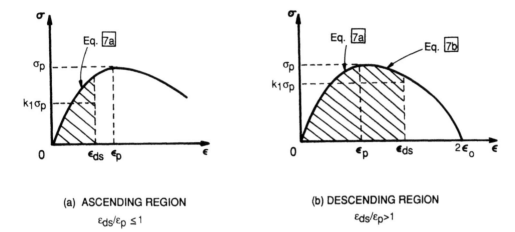

(a) ASCENDING REGION
$\varepsilon_{ds}/\varepsilon_p \leq 1$

(b) DESCENDING REGION
$\varepsilon_{ds}/\varepsilon_p > 1$

Figure 8.5 Integration of softened stress–strain curve of concrete to determine k_1.

TABLE 8.1 k_1 as a Function of ζ and ε_{ds} for Softened Concrete ($\varepsilon_0 = 0.002$)

ζ	ε_{ds}						
	0.0005	0.001	0.0015	0.002	0.0025	0.003	0.0035
0.10	0.8654	0.9215	0.9218	0.8994	0.8610	0.8089	0.7439
0.20	0.7333	0.8611	0.8883	0.8806	0.8513	0.8048	0.7429
0.30	0.6018	0.7980	0.8526	0.8604	0.8409	0.8005	0.7419
0.40	0.4948	0.7331	0.8147	0.8385	0.8294	0.7956	0.7407
0.50	0.4167	0.6667	0.7747	0.8148	0.8167	0.7901	0.7394
0.60	0.3588	0.6019	0.7325	0.7891	0.8026	0.7840	0.7379
0.70	0.3146	0.5442	0.6889	0.7613	0.7870	0.7771	0.7362
0.80	0.2799	0.4948	0.6445	0.7314	0.7698	0.7693	0.7342
0.90	0.2521	0.4527	0.6018	0.6997	0.7506	0.7603	0.7319
1.00	0.2292	0.4167	0.5625	0.6667	0.7292	0.7500	0.7292

The coefficient k_1 as expressed by Eqs. (13a) and (13b) are tabulated in Table 8.1 as a function of ε_{ds} and ζ, and ε_0 is taken as 0.002.

8.3.3   LOCATION OF CENTER LINE OF SHEAR FLOW

Similar to the calculation of coefficient k_1, coefficient k_2 for the location of resultant C can be obtained from Eq. (8-35) using the softened stress–strain curve of Eqs. 7a and 7b in Sections 7.2.1.2 or 7.3.1. These calculations show that k_2 varies generally in the range of 0.40 to 0.45. The resultant C, therefore, should lie approximately in the range of $0.40t_d$ to $0.45t_d$ from the extreme compression fiber.

The location of the center line of the shear flow is a trickier problem to determine. Considering the concrete struts, it should be located theoretically at the position of the resultant C, which is a distance of $0.40t_d$ to $0.45t_d$ from the extreme compression fiber. However, because the shear flow is constituted from the truss action of both the concrete and the steel, the location of the steel bars (or the thickness of the concrete cover) should also have an effect on the location of the center line of shear flow. Fortunately, tests (Hsu and Mo, 1985b) have shown that this effect of the steel bar location is small.

It could be argued that the average strain ε_d should be defined at the location of the resultant C, rather than at the middepth of the thickness t_d as shown in Figure 8.4b. Such treatment, of course, will complicate the calculation considerably without any convincing theoretical and experimental justification.

A simple solution to this tricky problem is to assume that the center line of shear flow lies at the middepth of the thickness t_d, the same location where the average strain ε_d has been defined. In other words, the center line of the shear flow is located at a distance of $0.5t_d$ from the extreme compression fiber. This assumption has three advantages:

1. By using $k_2 = 0.5$, a constant, the tedious process of integration to calculate the coefficient k_2 is avoided. The center line of shear flow also coincides with the location of the average strain ε_d (Figure 8.4d).

2. The calculations of A_0 and p_0 are simplified considerably (see Section 8.3.4). This simplification is slightly on the conservative side, because the constant 0.5 is somewhat greater than the actual value of the coefficient k_2 (about 0.4 to 0.45).

3. In calculating the torque $T = q(2A_0)$, the slight conservatism in A_0 is actually desirable to counteract the unconservatism involved in the use of the expression $q = (A_t/s)f_t \cot \alpha$. In the expression for q, the ratio A_t/s has been used unconservatively to describe the transverse steel areas per unit length [The unconservatism of A_t/s is caused by the discrete spacing of the transverse steel bars, explained in detail in Section 4.4.3.1 of the book by Hsu (1984)]. Tests have shown (Hsu and Mo, 1985a) that this simplifying assumption of $k_2 = 0.5$ results in good agreement between theory and tests.

8.3.4   FORMULAS FOR A_0 AND p_0

Now that the center line of the shear flow is assumed to coincide with the center line of the shear flow zone, the formula for calculating A_0 and p_0 can be derived. The formulas must be sufficiently accurate for thick tube, because the thickness of the shear flow zone, t_d, is usually quite large with respect to the overall dimensions of the cross section when the softening of concrete is taken into account. For a rectangular cross section of height a and width b as shown in Figure 8.6a, A_0 and p_0 are

$$A_0 = (a - t_d)(b - t_d) = ab - (a + b)t_d + t_d^2 = A_c - \tfrac{1}{2}p_c t_d + t_d^2 \quad (8\text{-}42)$$

$$p_0 = p_c - 4t_d \quad (8\text{-}43)$$

where

A_c = cross-sectional area bounded by the outer perimeter of the concrete

p_c = perimeter of the outer concrete cross section.

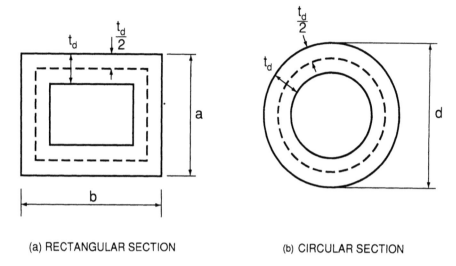

(a) RECTANGULAR SECTION   (b) CIRCULAR SECTION

Figure 8.6   Calculation of A_0 and p_0 (dotted curves represent center lines of shear flow).

For a circular cross section of diameter d as shown in Figure 8.6b, A_0 and p_0 are

$$A_0 = \frac{\pi}{4}(d - t_d)^2 = \frac{\pi}{4}d^2 - \frac{\pi}{2}dt_d + \frac{\pi}{4}t_d^2 = A_c - \frac{1}{2}p_c t_d + \frac{\pi}{4}t_d^2 \qquad (8\text{-}44)$$

$$p_0 = p_c - \pi t_d \qquad (8\text{-}45)$$

Comparing Eqs. (8-42) and (8-44) for A_0, it can be seen that the two formulas are the same, except the last term t_d^2, which has a constant 1 for rectangular section and a constant $\pi/4$ for circular section. Similarly, a comparison of Eqs. (8-43) and (8-45) for p_0 shows that the two formulas are the same, except the last term t_d, which has a constant 4 for rectangular section and a constant π for circular section. Therefore, Eqs. (8-42) to (8-45) can be summarized as follows:

$$A_0 = A_c - \tfrac{1}{2}p_c t_d + \xi t_d^2 \qquad (8\text{-}46)$$

$$p_0 = p_c - 4\xi t_d \qquad (8\text{-}47)$$

where $\xi = 1$ for rectangular section and $\xi = \pi/4$ for circular section. In other words, ξ varies from about 0.8 to 1.

Notice that the last terms in both Eqs. (8-46) and (8-47) are considerably smaller than the other terms in the formulas. A slight adjustment of these last terms will not produce a significant loss of accuracy for the formulas. Hence, a pair of formulas for A_0 and p_0 is proposed here for arbitrary bulky cross sections assuming $\xi = 1$:

$$A_0 = A_c - \tfrac{1}{2}p_c t_d + t_d^2 \qquad (8\text{-}48)$$

$$p_0 = p_c - 4t_d \qquad (8\text{-}49)$$

These two formulas are exact for rectangular sections, and should be quite accurate for any arbitrary bulky cross sections.

8.4   Governing Equations for Torsion

8.4.1   SUMMARY OF EQUATIONS

8.4.1.1   *Governing Equations*

The governing equations for equilibrium condition and compatibility condition were introduced in Sections 8.1 and 8.2. The softened constitutive equations for concrete struts in bending and compression were derived in Section 8.3. In addition, the constitutive laws for mild steel and prestressed steel were developed in Sections 7.2.3 and 7.2.4, respectively, in Chapter 7. All these equations will now be listed in this section so that a strategy for their solution can be developed.

Equilibrium Equations

$$\sigma_l = \sigma_d \cos^2 \alpha + \sigma_r \sin^2 \alpha + \rho_l f_l + \rho_{lp} f_{lp} \qquad \text{①}$$

$$\sigma_t = \sigma_d \sin^2 \alpha + \sigma_r \cos^2 \alpha + \rho_t f_t + \rho_{tp} f_{tp} \qquad \text{②}$$

$$\tau_{lt} = (-\sigma_d + \sigma_r)\sin \alpha \cos \alpha \qquad \text{③}$$

$$T = \tau_{lt}(2 A_0 t_d) \qquad \text{④}$$

Compatibility Equations

$$\varepsilon_l = \varepsilon_d \cos^2 \alpha + \varepsilon_r \sin^2 \alpha \qquad \text{⑤}$$

$$\varepsilon_t = \varepsilon_d \sin^2 \alpha + \varepsilon_r \cos^2 \alpha \qquad \text{⑥}$$

$$\frac{\gamma_{lt}}{2} = (-\varepsilon_d + \varepsilon_r)\sin \alpha \cos \alpha \qquad \text{⑦}$$

$$\theta = \frac{p_0}{2 A_0} \gamma_{lt} \qquad \text{⑧}$$

$$\psi = \theta \sin 2\alpha \qquad \text{⑨}$$

$$t_d = \frac{\varepsilon_{ds}}{\psi} \qquad \text{⑩}$$

$$\varepsilon_d = \frac{\varepsilon_{ds}}{2} \qquad \text{⑪}$$

Constitutive Laws of Materials

Concrete Struts

$$\sigma_d = k_1 \zeta f_c' \tag{12}$$

$$k_1 = f_k(\varepsilon_{ds}, \zeta) \tag{13}$$

$$\zeta = f_2(\varepsilon_d, \varepsilon_r) \tag{14}$$

$$\sigma_r = 0$$

Mild Steel

$$f_l = f_4(\varepsilon_l) \tag{15}$$

$$f_t = f_5(\varepsilon_t) \tag{16}$$

Prestressing Steel

$$f_{lp} = f_6(\varepsilon_{dec} + \varepsilon_l) \tag{17}$$

$$f_{tp} = f_7(\varepsilon_{dec} + \varepsilon_t) \tag{18}$$

Notice that the tensile stress of concrete is assumed to be zero ($\sigma_r = 0$). Thus we have a total of 18 equations, rather than 19. The six functions f_k, f_2, f_4 to f_7 in Eqs. (13) to (18) are to be selected for the desired accuracy.

8.4.1.2 Selected Constitutive Equations

For the treatment of torsion, the constitutive equations, Eqs. (13) to (18), will be selected in the following text. The simple elastic–perfectly plastic stress–strain relationship of bare mild-steel bars was assumed, because the tensile stress of concrete has been neglected.

Concrete Struts

$$k_1 = \frac{\varepsilon_{ds}}{\zeta \varepsilon_0}\left(1 - \frac{1}{3}\frac{\varepsilon_{ds}}{\zeta \varepsilon_0}\right) \qquad \frac{\varepsilon_{ds}}{\varepsilon_p} \le 1 \tag{13a}$$

$$k_1 = \left[1 - \frac{\zeta^2}{(2-\zeta)^2}\right]\left(1 - \frac{1}{3}\frac{\zeta \varepsilon_0}{\varepsilon_{ds}}\right)$$

$$+ \frac{\zeta^2}{(2-\zeta)^2}\frac{\varepsilon_{ds}}{\zeta \varepsilon_0}\left(1 - \frac{1}{3}\frac{\varepsilon_{ds}}{\zeta \varepsilon_0}\right) \qquad \frac{\varepsilon_{ds}}{\varepsilon_p} \le 1 \tag{13b}$$

$$\zeta = \frac{0.9}{\sqrt{1 + 600\varepsilon_r}} \tag{14a}$$

Mild Steel

$$f_l = E_s \varepsilon_l \qquad \varepsilon_l < \varepsilon_{ly} \qquad (15a)$$

$$f_l = f_{ly} \qquad \varepsilon_l \geq \varepsilon_{ly} \qquad (15b)$$

$$f_t = E_s \varepsilon_t \qquad \varepsilon_t < \varepsilon_{ty} \qquad (16a)$$

$$f_t = f_{ty} \qquad \varepsilon_t \geq \varepsilon_{ty} \qquad (16b)$$

Prestressing Steel

$$f_p = E_{ps}(\varepsilon_{dec} + \varepsilon_s) \qquad\qquad f_p \leq 0.7 f_{pu} \quad (17a) \text{ or } (18a)$$

$$f_p = \frac{E'_{ps}(\varepsilon_{dec} + \varepsilon_s)}{\left[1 + \left\{(E'_{ps}(\varepsilon_{dec} + \varepsilon_s))/f_{pu}\right\}^m\right]^{1/m}} \qquad f_p > 0.7 f_{pu} \quad (17b) \text{ or } (18b)$$

where

f_p = stress in prestressing steel; f_p becomes f_{lp} or f_{tp} when applied to the longitudinal and transverse steel, respectively

ε_s = strain in the mild steel; ε_s becomes ε_l or ε_t, when applied to the longitudinal and transverse steel, respectively

8.4.2 METHOD OF SOLUTION

The 18 governing equations for a torsional member, Eqs. ① to ⑱, are listed in Table 8.2 in 3 categories: the 4 equilibrium equations, the 7 compatibility equations, and the 7 constitutive equations. These 18 equations contain 21 unknown variables, which are also divided into three categories in Table 8.2: The 9 stress or force variables include σ_l, σ_t, τ_{lt}, σ_d, f_l, f_t, f_{lp}, f_{tp}, and T; the 10 strain or geometry variables include ε_l, ε_t, γ_{lt}, ε_d, ε_r, α, θ, ψ, t_d, and ε_{ds}; the 2 material coefficients are ζ and k_1. If 3 unknown variables are given, then the remaining 18 unknown variables can be solved using the 18 equations.

Table 8.2 also lists the 13 governing equation, Eqs. $\boxed{1}$ to $\boxed{13}$, for membrane elements. The similarity between this set of equations for membrane elements and the set of equations for torsion members stems from the fact that an element in the shear flow zone is subjected to the in-plane truss action of a membrane element. The additional compatibility equations and the modifications of constitutive relationships required for torsion are caused by the warping of the element in the shear flow zone.

For a member subjected to pure torsion the normal stresses σ_l and σ_t acting on an element in the shear flow zone are equal to zero, i.e., $\sigma_l = \sigma_t = 0$. If ε_d is selected as the third variable because it varies monotonically from zero to maximum, then the

TABLE 8.2 Summary of Variables and Equations for Torsion

Category	Variables			Equations		
	Stresses or Forces	Strains or Geometry	Matl.	Equil.	Compat.	Matl.
Shear Element	σ_l	ε_l	ζ	[1] ①	[4] ⑤	[7]↗ ⑬
	σ_t	ε_t		[2] ②	[5] ⑥	[8] ⑭
	τ_{lt}	γ_{lt}		[3] ③	[6] ⑦	[9]↗
	σ_d	ε_d				
	$\sigma_r \to 0$	ε_r				
	f_l	α				[10] ⑮
	f_t					[11] ⑯
	f_{lp}					[12] ⑰
	f_{tp}					[13] ⑱
Additional for Torsion	T	θ	k_1	④	⑧	⑫
		ψ			⑨	
		t_d			⑩	
		ε_{ds}			⑪	
Total for Torsion	9	10	2	4	7	7
		21			18	

remaining 18 unknown variables can be solved (Hsu, 1991b). The series of solutions for various ε_d values allows us to trace the loading history.

8.4.2.1 Characteristics of Equations

An efficient algorithm for solving the 18 equations was derived based on a careful observation of the 6 characteristics of the equations:

1. The three equilibrium equations [Eqs. ① to ③] and the three compatibility equations [Eqs. ⑤ to ⑦] are transformation equations. In other words, the stresses and strains in the l-t coordinate (σ_l, σ_t, τ_{lt}, ε_l, ε_t, and γ_{lt}) are each expressed in terms of stresses and strains in the d-r direction (σ_d, σ_r, ε_d, and ε_r).

2. Equations ⑫, ⑬, and ⑭ for the constitutive laws of concrete involve only four unknown variables in the d-r coordinate (σ_d, ε_{ds}, ε_d, and ε_r). If the strains ε_d and ε_r are given, then the stresses σ_d can be calculated from these three equations.

3. Equations ③ and ④ are independent from all other equations, because they contain two variables, τ_{lt} and T, which are not involved in any other equations.

4. The longitudinal steel stresses f_l and f_{lp} in equilibrium equation, Eq. ①, are coupled to the longitudinal steel strain ε_l in compatibility equation, Eq. ⑤, through the longitudinal steel stress–strain relationships of Eqs. ⑮ and ⑰. Similarly, the transverse steel stresses f_t and f_{tp} in equilibrium equation, Eq. ②, are coupled to the transverse steel strain ε_t in compatibility equation, Eq. ⑥, through the transverse steel stress–strain relationships of Eqs. ⑯ and ⑱.

5. Equations ⑦ to ⑩ sequentially relate the four variables γ_{lt}, θ, ψ, and ε_{ds}. Hence, these four equations can easily be combined into one equation.

6. The variable t_d is involved in Eqs. ①, ②, ⑧, and ⑩ through the terms A_0, p_0, ρ_l, and ρ_t. Therefore, these four equations are coupled.

8.4.2.2 *Thickness t_d as a Function of Strains*

The thickness of shear flow zone, t_d, can be expressed in terms of strains using the compatibility equations, Eqs. ⑤ to ⑩. To do this we will first combine the four compatibility equations, Eqs. ⑦ to ⑩, into one equation according to characteristic 5. Insert γ_{lt} from Eq. ⑦ into Eq. ⑧ to give

$$\theta = \frac{p_0}{A_0}(\varepsilon_r - \varepsilon_d)\sin\alpha\cos\alpha \tag{8-50}$$

Insert θ from Eq. (8-50) into Eq. ⑨:

$$\psi = \frac{p_0}{A_0}(\varepsilon_r - \varepsilon_d)2\sin^2\alpha\cos^2\alpha \tag{8-51}$$

Insert ψ from Eq. (8-51) into Eq. ⑩:

$$\sin^2\alpha\cos^2\alpha = \frac{A_0}{2p_0t_d}\frac{(-\varepsilon_{ds})}{(\varepsilon_r - \varepsilon_d)} \tag{8-52}$$

Equation (8-52) is the basic compatibility equation describing the warping of the shear flow zone in a member subjected to torsion. To eliminate α in Eq. (8-52) we utilize the first type of compatibility equations [Eqs. (5-30) and (5-31) in Chapter 5] to express α in terms of strains:

$$\sin^2\alpha = \frac{\varepsilon_l - \varepsilon_d}{\varepsilon_r - \varepsilon_d} \tag{8-53}$$

$$\cos^2\alpha = \frac{\varepsilon_t - \varepsilon_d}{\varepsilon_r - \varepsilon_d} \tag{8-54}$$

Substituting $\sin^2 \alpha$ and $\cos^2 \alpha$ from Eqs. (8-53) and (8-54) into Eq. (8-52) and taking $\varepsilon_{ds}/2 = \varepsilon_d$ from Eq. ⑪ results in

$$t_d = \frac{A_0}{p_0}\left[\frac{(-\varepsilon_d)(\varepsilon_r - \varepsilon_d)}{(\varepsilon_l - \varepsilon_d)(\varepsilon_t - \varepsilon_d)}\right] \qquad \text{(8-55) or ⑲}$$

It should be noted that the variable t_d is expressed in terms of strains in all d, r, l, and t directions. The variable t_d is also involved in Eqs. ①, ②, ⑧, and ⑩ through the terms, A_0, p_0, ρ_l, and ρ_t. Hence, the variable t_d must first be assumed and then checked by Eq. ⑲.

8.4.2.3 ε_l as a Function of f_l and f_{lp}

The strain ε_l can be related to the stresses f_l and f_{lp} by eliminating the angle α from the equilibrium equation, Eq. ①, and the compatibility equation, Eq. (8-52). Substituting $\sin^2 \alpha$ from Eq. (8-53) into Eq. (8-52) gives the compatibility equation:

$$\cos^2 \alpha = \frac{A_0}{2p_0 t_d}\frac{(-\varepsilon_{ds})}{(\varepsilon_l - \varepsilon_d)} \qquad \text{(8-56)}$$

From the equilibrium equation, Eq. ①, and the relationship $\sin^2 \alpha + \cos^2 \alpha = 1$, we can write

$$\cos^2 \alpha = \frac{-(\sigma_l - \sigma_r) + (\rho_l f_l + \rho_{lp} f_{lp})}{(-\sigma_d + \sigma_r)} \qquad \text{(8-57)}$$

Equating the equilibrium equation, Eq. (8-57), to the compatibility equation, Eq. (8-56), and utilizing the definitions of $\rho_l = A_l/p_0 t_d$ and $\rho_{lp} = A_{lp}/p_0 t_d$ results in

$$\varepsilon_l = \varepsilon_d + \frac{A_0(-\varepsilon_{ds})(-\sigma_d + \sigma_r)}{2\left[-p_0 t_d(\sigma_l - \sigma_r) + (A_l f_l + A_{lp} f_{lp})\right]} \qquad \text{(8-58)}$$

For pure torsion, $\sigma_l = 0$. Also inserting $\sigma_r = 0$ and $\varepsilon_{ds}/2 = \varepsilon_d$ from Eq. ⑪, Eq. (8-58) becomes

$$\varepsilon_l = \varepsilon_d + \frac{A_0(-\varepsilon_d)(-\sigma_d)}{(A_l f_l + A_{lp} f_{lp})} \qquad \text{(8-59) or ⑳}$$

In Eq. ⑳, ε_l, f_l, and f_{lp} can be solved simultaneously with the stress–strain relationships of Eqs. ⑮ and ⑰ for longitudinal steel.

8.4.2.4 ε_t as a Function of f_t and f_{tp}

Similarly, the strain ε_t can be related to the stresses f_t and f_{tp} by eliminating the angle α from equilibrium equation, Eq. ②, and the compatibility equation, Eq.

(8-52). Substituting $\cos^2 \alpha$ from Eq. (8-54) into Eq. (8-52) gives the compatibility equation

$$\sin^2 \alpha = \frac{A_0}{2 p_0 t_d} \frac{(-\varepsilon_{ds})}{(\varepsilon_t - \varepsilon_d)} \tag{8-60}$$

The equilibrium equation, Eq. ②, can also be written as

$$\sin^2 \alpha = \frac{-(\sigma_t - \sigma_r) + (\rho_t f_t + \rho_{tp} f_{tp})}{(-\sigma_d + \sigma_r)} \tag{8-61}$$

Then equating Eqs. (8-60) and (8-61), and utilizing the definitions of $\rho_t = A_t / s t_d$ and $\rho_{tp} = A_{tp} / s t_d$, result in

$$\varepsilon_t = \varepsilon_d + \frac{A_0 s(-\varepsilon_{ds})(-\sigma_d + \sigma_r)}{2 p_0 \left[-s t_d (\sigma_t - \sigma_r) + (A_t f_t + A_{tp} f_{tp}) \right]} \tag{8-62}$$

For pure torsion, $\sigma_t = 0$. Also inserting $\sigma_r = 0$ and $\varepsilon_{ds}/2 = \varepsilon_d$, Eq. (8-62) becomes

$$\varepsilon_t = \varepsilon_d + \frac{A_0 s(-\varepsilon_d)(-\sigma_d)}{p_0 (A_t f_t + A_{tp} f_{tp})} \tag{8-63 or ㉑}$$

In Eq. ㉑, ε_t, f_t, and f_{tp} can be solved simultaneously with the stress–strain relationships of Eqs. ⑯ and ⑱ for transverse steel.

8.4.2.5 Additional Equations

A_0 and p_0 have been expressed as functions of t_d by Eqs. (8-48) and (8-49):

$$A_0 = A_c - \tfrac{1}{2} p_c t_d + t_d^2 \tag{8-64 or ㉒}$$

$$p_0 = p_c - 4 t_d \tag{8-65 or ㉓}$$

and from Eqs. ⑯ and ⑰ in Chapter 7, ε_r and α are expressed in terms of strains ε_l, ε_t, and ε_d by

$$\varepsilon_r = \varepsilon_l + \varepsilon_t - \varepsilon_d \tag{8-66 or ㉔}$$

$$\tan^2 \alpha = \frac{\varepsilon_l - \varepsilon_d}{\varepsilon_t - \varepsilon_d} \tag{8-67 or ㉕}$$

Equations ㉒, ㉓, ㉔, and ㉕, are also needed in the solution procedure.

8.4.2.6 Solution Procedures

A set of solution procedures is proposed as shown in the flow chart of Figure 8.7. The procedures are described as follows:

Step 1: Select a value of strain in the d direction, ε_d.

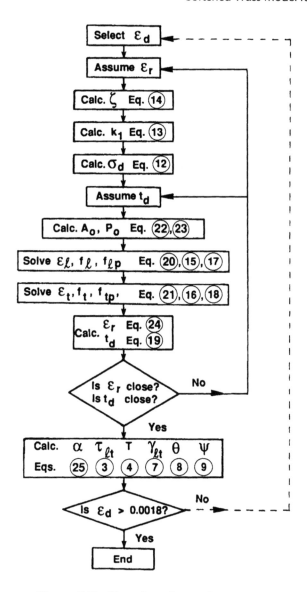

Figure 8.7 Flow chart for torsion analysis.

Step 2: Assume a value of strain in the r direction, ε_r.

Step 3: Calculate the coefficients ζ and k_1, and the average concrete stresses σ_d from Eqs. (14), (13), and (12), respectively.

Step 4: Assume a value of the thickness of shear flow zone, t_d, and calculate the cross-sectional properties A_0 and p_0 by Eqs. (22) and (23).

Step 5: Solve the strains and stresses in the longitudinal steel (ε_l, f_l, and f_{lp}) from Eqs. (20), (15), and (17), and those in the transverse steel (ε_t, f_t, and f_{tp}) from Eqs. (21), (16), and (18).

Step 6: Calculate the strain $\varepsilon_r = \varepsilon_l + \varepsilon_t - \varepsilon_d$ from Eq. (24) and t_d from Eq. (19). If ε_r or t_d are the same as assumed, the values obtained for all the strains are

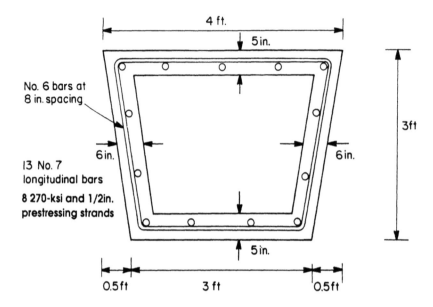

Figure 8.8   Box section for Example Problem 8.1.

correct. If t_d is not the same as assumed, then another value of t_d is assumed and Steps 4 to 6 are repeated. If ε_r is not the same as assumed, then another value of ε_r is assumed and Steps 3 to 6 are repeated.

Step 7: Calculate the angle α, the shear stress τ_{lt}, the torque T, the shear strain γ_{lt}, the angle of twist θ, and the curvature of the concrete struts ψ from Eqs. (25), (3), (4), (7), (8), and (9), respectively. This will provide one point on the T vs. θ curve or on the τ_{lt} vs. γ_{lt} curve.

Step 8: Select another value of ε_d and repeat Steps 2 to 7. Calculation for a series of ε_d values will provide the whole T vs. θ and τ_{lt} vs. γ_{lt} curves.

The preceding solution procedures have two distinct advantages. First, the variable angle α does not appear in the iteration process from Steps 2 to 6. Second, the calculation of ε_l and ε_t in Step 5 can easily accommodate the nonlinear stress–strain relationships of reinforcing steel, including those for prestressing strands. These advantages were derived from an understanding of the 6 characteristics of the 18 governing equations. Steps 1 to 3 are proposed because of characteristics 2, Step 4 is required due to characteristics 6, Steps 5 and 6 are the results of characteristics 1 and 4, and Step 7 is possible based on characteristics 1, 3, and 5.

EXAMPLE PROBLEM 8.1

Analyze the torque-twist behavior of the hollow box girder with trapezoidal cross section as shown in Figure 8.8. The girder is reinforced with 13 No. 7 longitudinal bars and No. 6 transverse hoop bars at 8-in. spacing. Both sizes of mild-steel bars have a yield strength of 60,000 psi. It is also prestressed longitudinally by eight 270-ksi strands of $\frac{1}{2}$ in. diameter. The strain due to decompression, ε_{dec}, is taken as 0.005 and the stress–strain curve can be represented by the Ramberg–Osgood formula, Eqs. (17a) and (17b). The cylinder compressive strength of concrete is 6000 psi. The softening coefficient of concrete ζ is specified by Eq. (14a), and the averaging coefficient k_1 by Eqs. (13a) and (13b). The tensile stress of concrete is neglected ($\sigma_r = 0$), and the net concrete cover is 1.5 in.

To illustrate the method of calculation, two values of strain ε_d are selected for examples. The first value of $\varepsilon_d = -0.0005$ is chosen to demonstrate the case at first yield of the transverse steel while the longitudinal steel and the prestressed steel are still in their elastic ranges. In the second case of $\varepsilon_d = -0.0015$, both the mild steel and the prestressed steel have yielded and the torsional moment obtained is close to the ultimate strength.

SOLUTION

(1) Select $\varepsilon_d = -0.0005$

$$\varepsilon_{ds} = 2(-0.0005) = -0.001$$

Assume $\varepsilon_r = 0.00414$ after several cycles of trial-and-error process:

Eq. (14a)
$$\zeta = \frac{0.9}{\sqrt{1 + 600\varepsilon_r}} = \frac{0.9}{\sqrt{1 + 600(0.00414)}} = 0.482$$

$$\frac{\varepsilon_{ds}}{\zeta\varepsilon_0} = \frac{-0.001}{0.482(-0.002)} = 1.037 > 1 \quad \text{descending branch}$$

$$\frac{\zeta^2}{(2-\zeta)^2} = \frac{0.482^2}{(2-0.482)^2} = 0.1008$$

Eq. (13b) $k_1 = \left[1 - \frac{\zeta^2}{(2-\zeta)^2}\right]\left(1 - \frac{1}{3}\frac{\zeta\varepsilon_0}{\varepsilon_{ds}}\right) + \frac{\zeta^2}{(2-\zeta)^2}\frac{\varepsilon_{ds}}{\zeta\varepsilon_0}\left(1 - \frac{1}{3}\frac{\varepsilon_{ds}}{\zeta\varepsilon_0}\right)$

$$= [1 - 0.1008]\left(1 - \frac{1}{3}\frac{1}{(1.037)}\right) + 0.1008(1.037)\left(1 - \frac{1}{3}1.037\right)$$

$$= 0.6102 + 0.0684 = 0.6786$$

k_1, of course, can also be obtained from Table 8.1.

Eq. (12) $\sigma_d = k_1\zeta f_c' = 0.6786(0.482)(-6000) = -1963$ psi

Assume $t_d = 3.79$ in. after two or three trials

$$A_c = \frac{3(3+4)}{2}(12)^2 = 1512 \text{ in.}^2$$

$$p_c = \left(3 + 4 + 2\sqrt{3^2 + 0.5^2}\right)(12) = 157 \text{ in.}$$

Eq. ㉒ $A_0 = A_c - \frac{1}{2}p_c t_d + t_d^2 = 1512 - \frac{1}{2}(157)(3.79) + (3.79)^2$

$= 1229$ in.^{2}

Eq. ㉓ $p_0 = p_c - 4t_d = 157 - 4(3.79) = 141.8$ in.

$A_l = 13(0.60) = 7.8$ in.^{2}

$A_{lp} = 8(0.153) = 1.224$ in.^{2}

Solve ε_l:

Eq. ⑳ $\varepsilon_l = \varepsilon_d + \dfrac{A_0(-\varepsilon_d)(-\sigma_d)}{(A_l f_l + A_{lp} f_{lp})}$

$= -0.0005 + \dfrac{1229(0.0005)(1963)}{(7.8 f_l + 1.224 f_{lp})}$

Assume elastic range

Eq. ⑮a $f_l = 29 \times 10^6 \; \varepsilon_l$ before yielding of mild steel

Eq. ⑰a $f_{lp} = 29 \times 10^6(0.005 + \varepsilon_l)$ before elastic limit of prestressed steel

Then

$(\varepsilon_l + 0.0005)(\varepsilon_l + 0.000678) = 4.609 \times 10^{-6}$

$\varepsilon_l = 0.00156$

$\varepsilon_y = 0.00207 > 0.00156$ **O.K** for mild steel

ε_{ps} at $0.7f_{pu} - \varepsilon_{dec} = 0.00652 - 0.005 = 0.00152 \approx 0.00156$

Say **O.K** for prestressed steel

Solve ε_t:

Eq. ㉑ $\varepsilon_t = \varepsilon_d + \dfrac{A_0 s(-\varepsilon_d)(-\sigma_d)}{p_0(A_t f_t + A_{tp} f_{tp})}$

$= -0.0005 + \dfrac{1229(8)(0.0005)(1963)}{141.8(0.44 f_t + 0)}$

$\varepsilon_t + 0.0005 = \dfrac{154.7}{f_t}$

Assume elastic range

Eq. (16a) $f_t = 29 \times 10^6 \, \varepsilon_t$ before yielding

Then

$$\varepsilon_t^2 + 0.0005 \, \varepsilon_t - 5.333 \times 10^{-6} = 0$$

$$\varepsilon_t = 0.00207 = \varepsilon_y \quad \text{(transverse steel just yielded)}$$

Check ε_r:

Eq. (24) $\varepsilon_r = \varepsilon_l + \varepsilon_t - \varepsilon_d = 0.00156 + 0.00207 + 0.0005$

$$= 0.00413 \approx 0.00414 \text{ (assumed)} \quad \textbf{O.K.}$$

Check t_d:

Eq. (19) $t_d = \dfrac{A_0}{p_0} \left[\dfrac{(-\varepsilon_d)(\varepsilon_r - \varepsilon_d)}{(\varepsilon_l - \varepsilon_d)(\varepsilon_t - \varepsilon_d)} \right]$

$$= \frac{1229}{141.8} \left[\frac{(0.0005)(0.00413 + 0.0005)}{(0.00156 + 0.0005)(0.00207 + 0.0005)} \right]$$

$$= 3.79 \text{ in.} \approx 3.79 \text{ in. (assumed)} \quad \textbf{O.K.}$$

Calculate α, τ_{lt}, T, γ_{lt}, and θ:

Eq. (25) $\tan^2 \alpha = \dfrac{\varepsilon_l - \varepsilon_d}{\varepsilon_t - \varepsilon_d} = \dfrac{0.00156 + 0.0005}{0.00207 + 0.0005} = 0.8016$

$$\alpha = 41.84° \quad \sin \alpha = 0.667 \quad \text{and} \quad \cos \alpha = 0.745$$

Eq. (3) $\tau_{lt} = (-\sigma_d + \sigma_r)\sin \alpha \cos \alpha = (1963 + 0)(0.667)(0.745)$

$$= 976 \text{ psi}$$

Eq. (4) $T = \tau_{lt}(2 A_0 t_d) = 976(2)(1229)(3.79)$

$$= 9{,}092{,}000 \text{ in.-lb}$$

$$= 9092 \text{ in.-kip (587 kN-m)}$$

Eq. (7) $\gamma_{lt} = 2(-\varepsilon_d + \varepsilon_r)\sin \alpha \cos \alpha$

$$= 2(0.0005 + 0.00413)(0.667)(0.745)$$

$$= 0.00460$$

Eq. ⑧
$$\theta = \gamma_{lt}\left(\frac{p_0}{2A_0}\right) = 0.00460\frac{141.8}{2(1229)}$$

$$= 0.000265 \text{ rad/in.}$$

Note that the curvature of the concrete struts ψ was not calculated because it was not required.

(2) Select $\varepsilon_d = -0.0015$

$$\varepsilon_{ds} = 2(-0.0015) = -0.0030$$

Assume $\varepsilon_r = 0.00903$ after several cycles of trial-and-error process:

Eq. ⑭ₐ
$$\zeta = \frac{0.9}{\sqrt{1 + 600(0.00903)}} = 0.355$$

$$\frac{\varepsilon_{ds}}{\zeta\varepsilon_0} = \frac{-0.0030}{0.355(-0.002)} = 4.23 > 1 \quad \text{descending branch}$$

Eq. ⑬ᵦ
$$k_1 = 0.798 \quad \text{(from Table 8.1)}$$

Eq. ⑫
$$\sigma_d = 0.798\,(0.355)\,(-6000) = -1700 \text{ psi}$$

Assume $t_d = 5.00$ in. after two or three trials:

Eq. ㉒
$$A_0 = 1512 - \tfrac{1}{2}(157)\,(5.00) + (5.00)^2 = 1145 \text{ in.}^2$$

Eq. ㉓
$$p_0 = 157 - 4(5.00) = 137.0 \text{ in.}$$

Solve ε_l:

Eq. ⑳
$$\varepsilon_l = -0.0015 + \frac{1145\,(0.0015)\,(1700)}{(7.8f_l + 1.224f_{lp})}$$

Assume $\varepsilon_l = 0.00257 > \varepsilon_y$ after a couple of trials:

Eq. ⑮ᵦ $f_l = 60,000$ psi

$$E_p'(\varepsilon_{dec} + \varepsilon_l) = 30,060\,(0.005 + 0.00257) = 227.6 \text{ ksi}$$

Eq. ⑰ᵦ
$$f_{lp} = \frac{E_p'(\varepsilon_{dec} + \varepsilon_l)}{\left[1 + \{(E_p'(\varepsilon_{dec} + \varepsilon_l))/f_{pu}\}^4\right]^{1/4}} = \frac{227.6}{\left[1 + (227.6/270)^4\right]^{1/4}}$$

$$= 205.5 \text{ ksi}$$

Then

$$\varepsilon_l = -0.0015 + \frac{1145(0.0015)(1700)}{7.8(60,000) + 1.224(205,500)} = -0.0015 + 0.00406$$

$$= 0.00256 = 0.00257 \text{ (assumed)} \quad \textbf{O.K.}$$

Solve ε_t:

Eq. ㉑ $$\varepsilon_t = -0.0015 + \frac{1145\ (8)\ (0.0015)\ (1700)}{137.0\ (0.44f_t + 0)}$$

Assume yielding

Eq. ⑯ⓑ $$f_t = 60,000 \text{ psi}$$

Then

$$\varepsilon_t = -0.0015 + 0.00646 = 0.00496 > \varepsilon_y \quad \textbf{O.K.}$$

Check ε_r:

Eq. ㉔ $$\varepsilon_r = \varepsilon_l + \varepsilon_t - \varepsilon_d = 0.00257 + 0.00496 + 0.0015$$

$$= 0.00903 \approx 0.00903 \text{ (assumed)} \quad \textbf{O.K.}$$

Check t_d:

Eq. ⑲ $$t_d = \frac{1145}{137.0}\left[\frac{(0.0015)\ (0.00903 + 0.0015)}{(0.00257 + 0.0015)\ (0.00496 + 0.0015)}\right]$$

$$= 8.358\ (0.601) = 5.02 \text{ in.} \approx 5.00 \text{ in. (assumed)} \quad \textbf{O.K.}$$

Calculate α, τ_{lt}, T, γ_{lt}, and θ:

Eq. ㉕ $$\tan^2\alpha = \frac{0.00257 + 0.0015}{0.00496 + 0.0015} = 0.6300$$

$$\alpha = 38.44° \quad \sin\alpha = 0.622 \quad \text{and} \quad \cos\alpha = 0.783$$

Eq. ③ $$\tau_{lt} = (1700 + 0)\ (0.622)(0.783) = 828 \text{ psi}$$

Eq. ④ $$T = 828\ (2)\ (1145)\ (5.00) = 9,481,000 \text{ in.-lb}$$

$$= 9481 \text{ in.-kip}$$

Eq. ⑦ $$\gamma_{lt} = 2\ (0.0015 + 0.00903)\ (0.622)(0.783) = 0.01025$$

Eq. ⑧ $$\theta = 0.01025\frac{137.0}{2\ (1145)} = 0.000613 \text{ rad/in.}$$

TABLE 8.3 Results of Calculations for Example Problem 8.1
($\varepsilon_d = -0.25, -0.5, -1.0, -1.5,$ and -1.75×10^{-3})

Variables	Eqs.	Calculated Values				
ε_d (10^{-3}) selected	—	−0.250	−0.500	−1.000	−1.500	−1.750
ε_{ds} (10^{-3})	(11)	−0.500	−1.000	−2.000	−3.000	−3.500
ε_r (10^{-3}) last assumed	—	2.25	4.14	7.34	9.03	9.29
ζ	(14a)	0.587	0.482	0.387	0.355	0.351
$\varepsilon_{ds} / \zeta\varepsilon_0$	—	0.426	1.037	2.59	4.23	4.99
k_1	(13)	0.365	0.679	0.842	0.798	0.741
σ_d (psi)	(12)	−1287	−1963	−1955	−1700	−1561
t_d (in.) last assumed	—	3.58	3.79	4.28	5.00	5.50
A_0 (in.^{2})	(22)	1244	1229	1194	1145	1111
p_0 (in.)	(23)	142.7	141.8	139.9	137.0	135.0
ε_l (10^{-3})	(20)	0.791	1.56	2.28	2.57	2.48
ε_t (10^{-3})	(21)	1.207	2.07	4.06	4.96	5.06
ε_r (10^{-3}) checked	(24)	2.25	4.13	7.34	9.03	9.29
t_d (in.) checked	(19)	3.59	3.79	4.29	5.02	5.52
α (deg)	(25)	40.21	41.84	38.84	38.44	38.24
τ_{lt} (psi)	(3)	634	976	955	828	759
T (in.-kip)	(4)	5647	9092	9761	9481	9276
γ_{lt} (10^{-3})	(7)	2.47	4.60	8.15	10.25	10.73
θ (rad / in. $\times 10^{-3}$)	(8)	0.142	0.265	0.477	0.613	0.652
f_l (ksi)	(15)	22.9	45.2	60.0	60.0	60.0
f_t (ksi)	(16)	35.0	60.0	60.0	60.0	60.0
f_{lp} (ksi)	(17)	167.9	190.2	200.0	205.5	203.8
f_{tp} (ksi)	(18)	0	0	0	0	0

The results of the preceding calculations for $\varepsilon_d = -0.0005$ and -0.0015 are summarized in Table 8.3, together with three additional cases of $\varepsilon_d = -0.00025$, 0.001, and -0.00175. Comparison of these five cases illustrates clearly the trends of all the variables. The relationship between the torsional moment T and the angle of twist θ is shown in Figure 8.9.

Table 8.3 clearly shows that $\varepsilon_d = -0.0005$ closely represents the point of first yield of the transverse mild steel. At $\varepsilon_d = -0.001$ both the longitudinal and transverse mild steels are yielding and the torque resisted reaches a maximum shortly thereafter. When ε_d is increased further, the torsional resistance decreases. At $\varepsilon_d = -0.0015$ the calculated thickness t_d is equal to the actual wall thickness of 5.0 in. When ε_d is

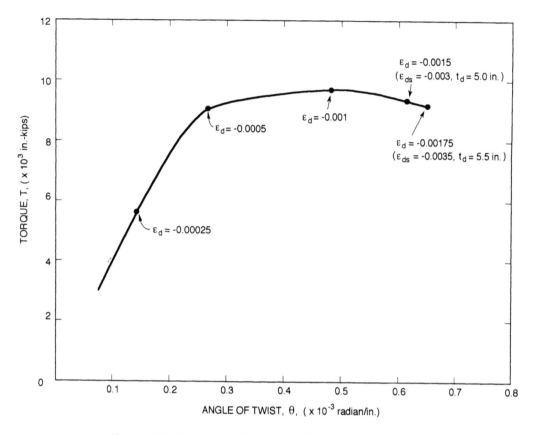

Figure 8.9 Torque-twist curve for Example Problem 8.1.

taken as -0.00175, the calculated thickness $t_d = 5.5$ in., i.e., greater than the actual thickness. This situation indicates that the calculation is actually invalid and that the torque-twist curve would drop off more quickly for $\varepsilon_d > -0.0015$ than is shown in Figure 8.9.

The angle of twist at first yield, θ_y, is 0.000265 rad/in. as given in Figure 8.9 or Table 8.3. The maximum calculated angle of twist, θ_u, for the given wall thickness of 5.0 in. is 0.000613 rad/in. Let us define the torsional ductility μ_t as the ratio θ_u/θ_y. Then the torsional ductility of this box girder is

$$\mu_t = \frac{\theta_u}{\theta_y} = \frac{0.000613}{0.000265} = 2.31$$

8.5   Design for Torsion

8.5.1   ANALOGY BETWEEN TORSION AND BENDING

A prismatic member of arbitrary cross section subjected to a torsional moment T was shown in Figure 3.6 (see Section 3.1.4). This external torsional moment T is resisted by an internal moment that is the product of the shear flow q in the shear flow zone and twice the lever arm area A_0:

$$T = q(2A_0) \tag{8-68}$$

In a reinforced concrete member after cracking of concrete and yielding of steel, the shear flow q can be expressed by the properties of the transverse steel:

$$q = \frac{A_t f_{ty}}{s} \cot \alpha \tag{8-69}$$

Inserting q from Eq. (8-69) into Eq. (8-68) and noticing $T = T_n$ we have

$$T_n = \frac{A_t f_{ty} \cot \alpha}{s}(2A_0) \tag{8-70}$$

Equation (8-70) is the fundamental equation for torsion in the truss model. Because A_0 is defined by the center line of shear flow, the crucial problem is to determine the thickness of the shear flow zone t_d.

The preceding analysis of torsion is analogous to the analysis of bending in a prismatic member discussed in Chapter 2. In Figure 2.1, A rectangular cross section is subjected to a bending moment M. This external bending moment M is resisted by an internal bending moment that is the product of the resultant of the compressive stresses C in the compression zone of depth c and the lever arm jd:

$$M = C(jd) \tag{8-71}$$

After cracking of the flexural member and the yielding of steel, the equilibrium of the compressive force of concrete C and the tensile forces of longitudinal steel $A_s f_y$ give

$$C = A_s f_y \tag{8-72}$$

Inserting C from Eq. (8-72) into Eq. (8-71) and noticing $M = M_n$ we have

$$M_n = A_s f_y(jd) \tag{8-73}$$

Equation (8-73) is the fundamental equation for bending. Because jd is defined by the position of the resultant of the compression stress block, the crucial problem is to determine the depth of the compression zone c.

Equation (8-73) shows that the bending moment capacity M_n is equal to the longitudinal steel force $A_s f_y$ times the resultant lever arm jd. Similarly, in Eq (8-70) the torsional moment capacity T_n is equal to a certain stirrup force per unit length, $(A_t f_{ty}/s)\cot \alpha$, times twice the lever arm area $2A_0$. In other words, the term of twice the lever arm area $2A_0$ in torsion is equivalent to the resultant lever arm jd in bending; the shear flow q is similar to the resultant of compressive stresses C.

In bending, an increase of the nominal bending strength M_n due to increasing reinforcement results in an increase of the depth of the compression zone c, and a reduction of the resultant lever arm jd. The relationships among M_n, c, and jd can be derived from equilibrium, compatibility, and constitutive relationships of materials. Similarly in torsion, an increase of the nominal torsional strength T_n due to increasing reinforcement results in an increase of the thickness of shear flow zone t_d and a reduction of the lever arm area A_0. The relationships among T_n, t_d, and A_0 can also be derived from the equilibrium, compatibility and constitutive relationships of materials. The crucial problem in torsion of reinforced concrete is to find the thickness of the shear flow zone t_d and the lever arm area A_0. This is analogous to finding the depth of the compression zone c and the lever arm jd in bending.

8.5.2   VARIOUS DEFINITIONS OF LEVER ARM AREA A_0

When Rausch derived the basic torsion equation [Eq. (8-70) (with $\alpha = 45°$)] in 1929, a reinforced concrete member after cracking was idealized as a space truss with linear, one-dimensional members, Figure 1.3. Each diagonal concrete strut is idealized as a straight line lying in the center surface of the hoop bars. Hence, the lever arm area A_0 is defined by the area within the center surface of the hoop bars. This area is commonly denoted as A_1. It has been adopted by the German Standard (1958) and others. Using the bending analogy, this definition is equivalent to assuming that the resultant lever arm jd is defined as the distance between the centroid of the tension bars and the center line of the stirrups in the compression zone. In terms of torsional strength, this assumption is acceptable near the lower limit of the total steel percentage of about 1%, but becomes increasingly unconservative with an increasing amount of steel (Figure 8.10). For a large steel percentage of 2.5–3% near the upper limit of underreinforcement (both the longitudinal steel and stirrups reach yielding), the overprediction of torsional strength by Rausch's equation using A_1 exceeds 30%. This large error has two causes. First, the thickness of the shear flow zone t_d may be very large, in the order of $1/4$ of the outer cross-sectional dimension, due to the softening of concrete. Second, in contrast to the bending strength M_n, which is linearly proportional to the resultant lever arm jd, the torsional strength T_n is proportional to the lever arm area A_0, which, in turn, is proportional to the square of the lever arm r (see Figure 3.6).

To reduce the unconservatism of using A_1 in Rausch's equation, Lampert and Thurlimann (1968) have proposed that A_0 be defined as the area within the polygon connecting the centers of the corner longitudinal bars. This area is commonly denoted as A_2 and has been adopted by the CEB-FIP Code (1978). In terms of the bending analogy, this definition is equivalent to assuming that the resultant lever arm jd is defined as the distance between the centroid of the tension bars and the centroid of the longitudinal compression bars. The introduction of A_2 has reduced the unconservatism of Rausch's equation for high steel percentages. However, the assumption of a constant lever arm area (not a function of the thickness of shear flow zone) remains unsatisfactory.

Another way of modifying Rausch's equation was suggested by the author (Hsu, 1968a, 1968b) and adopted by the ACI Building Code (1971):

$$T_n = T_c + \frac{A_t f_{ty}}{s}(\alpha_t A_1) \tag{8-74}$$

where

$\alpha_t = 0.66 + 0.33\, y_1/x_1 \le 1.5$

x_1 = shorter center-to-center dimension of a closed stirrup

y_1 = longer center-to-center dimension of a closed stirrup

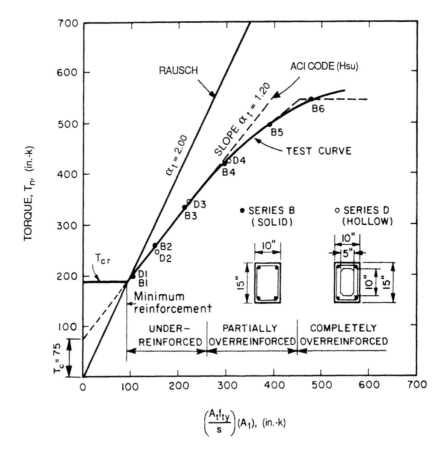

Figure 8.10   Comparison of Rausch's formula and ACI Code formula with tests (1 in. = 25.4 mm; 1 in.-kip = 113 N-m).

T_c = nominal torsional strength contributed by concrete

$\quad = 0.8x^2y\sqrt{f'_c}$ where x and y are the shorter and longer sides, respectively, of a rectangular section

Two modifications of Rausch's equation are made in Eq. (8-74) based on tests. First, a smaller lever arm area $(\alpha_t/2)A_1$ is specified, where α_t varies from 1 to 1.5. Second, a new term T_c is added. This term represents the vertical intercept of a straight line in the T_n vs. $(A_t f_{ty}/s)(A_1)$ diagram (Figure 8.10).

The preceding definitions of the lever arm areas A_1, A_2, or $(\alpha_t/2)A_1$ all have a common weakness. They are not related to the thickness of the shear flow zone or the applied torque. A logical way to define A_0 must start with the determination of the thickness of shear flow zone.

8.5.3   THICKNESS OF SHEAR FLOW ZONE FOR DESIGN

The thickness of the shear flow zone t_d given in Eq. (8-55) is suitable for the analysis of torsional strength. It is, however, not useful for the design of torsional members. In design, the thickness t_d should be expressed in terms of the torsional strength, T_n. This approach will now be introduced.

The stress in the diagonal concrete struts, σ_d, can be related to the thickness t_d and the shear flow q by inserting $\sigma_r = 0$ and $\tau_{lt} = q/t_d$ into the equilibrium equation, Eq. (8-3):

$$\sigma_d = \frac{q}{t_d \sin\alpha \cos\alpha} \tag{8-75}$$

At failure, σ_d in Eq. (8-75) reaches the maximum $\sigma_{d,\,max}$, whereas the torsional moment reaches the nominal capacity T_n. Substituting $q = T_n/2A_0$ at failure into Eq. (8-75) gives

$$t_d = \frac{T_n}{2A_0\,\sigma_{d,\,max}\,\sin\alpha\cos\alpha} \tag{8-76}$$

If t_d is assumed to be small, the last term ξt_d^2 in Eq. (8-46) is neglected and A_0 can be expressed by the thin-tube approximation:

$$A_0 = A_c - \frac{t_d}{2}p_c \tag{8-77}$$

Substituting A_0 from Eq. (8-77) into (8-76) and multiplying all the terms by $2p_c/A_c^2$ result in:

$$\left(\frac{p_c}{A_c}t_d\right)^2 - 2\left(\frac{p_c}{A_c}t_d\right) + \frac{T_n p_c}{A_c^2}\frac{1}{\sigma_{d,\,max}\,\sin\alpha\cos\alpha} = 0 \tag{8-78}$$

Let

$$t_{d0} = \text{thickness defined by } A_c/p_c$$

$$\tau_n = \text{nominal torsional stress defined by } T_n p_c/A_c^2$$

$$\tau_{n,\,max} = \text{maximum torsional stress defined by } \sigma_{d,\,max}\,\sin\alpha\cos\alpha$$

Equation (8-78) becomes

$$\left(\frac{t_d}{t_{d0}}\right)^2 - 2\left(\frac{t_d}{t_{d0}}\right) + \frac{\tau_n}{\tau_{n,\,max}} = 0 \tag{8-79}$$

When t_d/t_{d0} is plotted against $\tau_n/\tau_{n,\,max}$ in Figure 8.11, Eq. (8-79) represents a parabolic curve. Solving t_d from Eq. (8-79) gives

$$t_d = t_{d0}\left[1 - \sqrt{1 - \frac{\tau_n}{\tau_{n,\,max}}}\right] \tag{8-80}$$

This approach of determining the thickness of the shear flow zone was first proposed

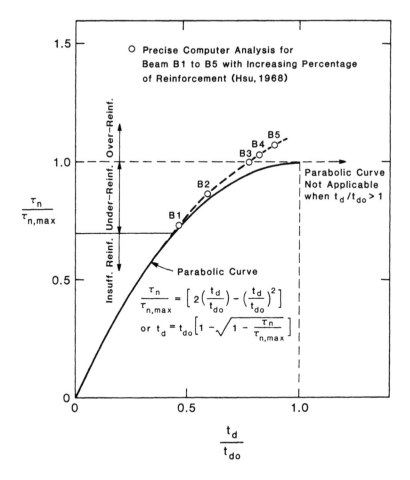

Figure 8.11 Graphical presentation of Eqs. (8-79) or (8-80).

by Collins and Mitchell (1980). Equation (8-80) was later adopted by the Canadian Code (Canadian Standard Association, 1984), in the form

$$t_d = \frac{A_1}{p_1}\left[1 - \sqrt{1 - \frac{T_n p_1}{0.7\,\phi_c f'_c A_1^2}\left(\tan\alpha + \frac{1}{\tan\alpha}\right)}\right] \qquad (8\text{-}81)$$

In Eq. (8-81), A_c and p_c are replaced by A_1 and p_1, respectively, because the concrete cover is considered to be ineffective. $\sigma_{d,\,max}$ is assumed to be $0.7\,\phi_c f'_c$, in which the material reduction factor ϕ_c can be taken as 0.6.

Equations (8-81) and (8-80) clearly show that the thickness ratio t_d/t_{d0} is primarily a function of the nominal shear stress ratio τ_n/f'_c. The thickness ratio t_d/t_{d0} is also a function of the crack angle α, but is not sensitive when α varies in the vicinity of 45°.

In Eq. (8-80), $\tau_n \leq \tau_{n,\,max}$ represents the case of underreinforcement and $\tau_n > \tau_{n,\,max}$ indicates overreinforcement. The case of overreinforcement cannot be expressed by Eq. (8-80), because it gives a complex number ($\sqrt{-1}$). Figure 8.11 shows that Eq. (8-80) is applicable when τ_n is less than about 0.9 $\tau_{n,\,max}$. However, when τ_n exceeds 0.9 $\tau_{n,\,max}$, t_d is increasing unreasonably fast. This problem reflects the difficulty in using the thin-tube approximation for A_0 [Eq. (8-77)] to find t_d. When t_d

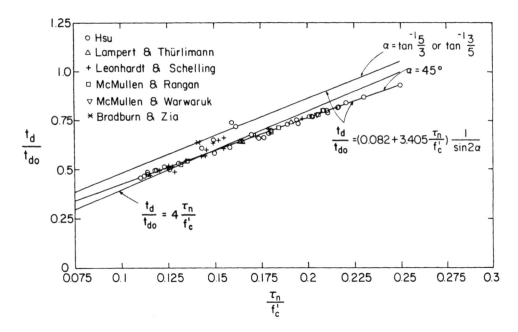

Figure 8.12 Thickness ratio t_d/t_{d0} as straight line functions of nominal shear stress ratio τ_n/f_c'.

exceeds about 0.7 t_{d0}, the tube becomes so thick that the term ξt_d^2 cannot be neglected.

To avoid this weakness, a different approach is adopted. Using the method presented in Section 8.4, a computer program was written to analyze the torsional behavior of reinforced concrete members and to calculate the thickness t_d (Hsu and Mo, 1985a). This computer program was used to analyze the 61 eligible torsional members available in literature. The thicknesses t_d of these test beams are calculated and a linear regression analysis of the thickness ratios t_d/t_{d0} is made as a function of τ_n/f_c'. This analysis provides the expression (Hsu and Mo, 1985b)

$$t_d = \frac{A_c}{p_c}\left(0.082 + 3.405\frac{\tau_n}{f_c'}\right)\frac{1}{\sin 2\alpha} \tag{8-82}$$

Equation (8-82) is plotted in Figure 8.12 for the cases of $\alpha = 45°$ and $\alpha = \tan^{-1}(5/3)$ or $\tan^{-1}(3/5)$, which are the limits adopted by the CEB-FIP Code (1978). The 61 test points are also included and the correlation is shown to be excellent. The t_d/t_{d0} values calculated from Eq. (8-82) for Hsu's Series B (1968a) are recorded in Table 8.4. When compared to the t_d/t_{d0} values obtained from the computer program, the correlation is again excellent. The ten beams in Series B were chosen because they have total reinforcement ratios varying from a low of 1.07% to a high of 5.28%, and a volume ratio of longitudinal steel to stirrups varying from 0.205 to 4.97. The wise range of application of Eq. (8-82) is evident. It is not only applicable to underreinforced members, but also to overreinforced members.

Although Eq. (8-82) is found to be excellent, it is considered somewhat unwieldly for practical design. In the next section a simplified expression for t_d is proposed. The simplicity is obtained with a small sacrifice in accuracy.

TABLE 8.4 Comparison of Thickness of Shear Flow Zone.

Beam	ρ_l (%)	ρ_t (%)	T_n (in.-kip)	α (deg)	ζ	t_d (in.)	t_d/t_{do}	Hsu/Mo Eq. (8-82) t_d/t_{do}	Eq. (8-85) t_d/t_{do}
B1	0.534	0.537	202	46.5	0.372	1.41	0.470	0.464	0.449
B2	0.827	0.823	287	44.8	0.340	1.78	0.593	0.606	0.615
B3	1.17	1.17	370	44.5	0.510	2.32	0.770	0.770	0.808
B4	1.60	1.61	431	44.6	0.531	2.46	0.820	0.818	0.865
B5	2.11	2.13	446	44.5	0.560	2.65	0.883	0.884	0.942
B8	0.534	2.61	278	56.7	0.440	2.12	0.707	0.681	0.637
B10	2.67	0.537	280	33.2	0.444	2.16	0.720	0.691	0.648

Note: Cross section 10 in. × 15 in.; $f_y = 47,000$ psi; $f_c' = 4000$ psi.

8.5.4 SIMPLIFIED DESIGN FORMULAS

A simple expression for the thickness of shear flow zone t_d can be obtained directly from Eq. (8-8), noting that $T = T_n$ at the maximum load:

$$t_d = \frac{T_n}{2A_0\tau_{lt}} \tag{8-83}$$

Assuming that $A_0 = m_1 A_c$ and $\tau_{lt} = m_2 f_c'$, where m_1 and m_2 are nondimensional coefficients, substituting them into Eq. (8-83) gives

$$t_d = C\frac{T_n}{A_c f_c'} \tag{8-84}$$

where $C = 1/2m_1 m_2$. For underreinforced members, m_1 varies from 0.55 to 0.85 and m_2 varies from 0.13 to 0.22. These values are obtained from the Appendix of Hsu and Mo's report (1983). The low values of m_2 are due to the softening of concrete. For an increasing amount of reinforcement, m_2 increases and m_1 decreases. Therefore, the product $m_1 m_2$ can be taken approximately as a constant 0.125, making C into a constant of 4. Then

$$t_d = \frac{4T_n}{A_c f_c'} \tag{8-85}$$

Equation (8-85) is also plotted in Figure 8.12. Comparison with Eq. (8-82) shows the difference to be small. Actually, Eq. (8-85) can be considered as a simplification of Eq. (8-82) by neglecting the small first term with constant 0.082 and increasing the constant in the second term from 3.405 to 4. The small effect of α is also neglected by taking $\sin 2\alpha = 1$, which is the exact value when $\alpha = 45°$. The t_d/t_{do} ratios calculated from Eq. (8-85) for Series B (Hsu, 1968a) are also recorded in Table 8.4. A comparison with the computer values also shows the correlation to be reasonable.

Inserting Eq. (8-85) into the thin-tube expression of A_0 in Eq. (8-77) and p_0 in

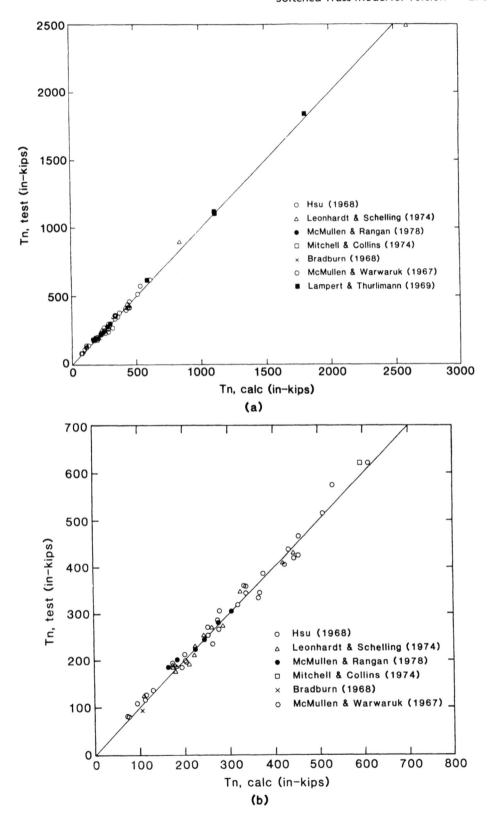

Figure 8.13 (a) Comparison of test strengths with calculated strengths using proposed t_d [Eq. (8-85)]. (b) Comparison of test strengths with calculated strengths using proposed t_d [Eq. (8-85)]. Expanded scale for lower portion of a.

Eq. (8-49) gives

$$A_0 = A_c - \frac{t_d}{2} p_c = A_c - \frac{2T_n p_c}{A_c f_c'} \tag{8-86}$$

$$p_0 = p_c - 4t_d = p_c - \frac{16T_n}{A_c f_c'} \tag{8-87}$$

A_0 in Eq. (8-86) is used in conjunction with Eq. (8-70) to calculate the torsional strength T_n for the 61 beams available in literature, (Hsu, 1990). The calculated values are plotted in Figure 8.13a and b and compared to the test values. The average $T_{n,\text{test}}/T_{n,\text{calc}}$ value is 1.013 and the standard deviation is 0.055.

The maximum thickness of the shear flow zone, $t_{d,\text{max}}$, to ensure that the beam remains underreinforced can be obtained by examining the test results. A typical series of beams with increasing percentage of reinforcement is given by series B in Figure 8.10. It can be seen that the maximum torsional resistance lies in between those of beams B3 and B4. The thickness ratios t_d/t_{d0} for B3 and B4 are calculated from the computer program and are given in Table 8.4 to be 0.77 and 0.82, respectively. Taking a close average of 0.8 for these two beams gives

$$t_{d,\text{max}} = 0.8t_{d0} = 0.8 \frac{A_c}{p_c} \tag{8-88}$$

This maximum thickness is slightly conservative when used in conjunction with the simplified Eq. (8-85) as shown in the last column of Table 8.4. The validity of Eq. (8-88) can also be observed in Figure 8.11 by the trend of the test points of Series B. The dotted curve crosses from the underreinforced region into the overreinforced region at about $0.8t_{d0}$. Series B is quite representative of the other series of test points.

Substituting $t_{d,\text{max}}$ from Eq. (8-88) into Eq. (8-85), the maximum torsional resistance $T_{n,\text{max}}$ of a member becomes

$$T_{n,\text{max}} = 0.2f_c' \frac{A_c^2}{p_c} \tag{8-89}$$

This maximum torsional resistance can be used to check the minimum size of cross section in the design of torsional beams.

8.5.5   DESIGN LIMITATIONS

Three limitations are required in practical design:

1   *Minimum Torsional Reinforcement*

The torsional reinforcement in a beam must be designed to exceed a minimum amount in order to avoid a brittle failure. This minimum torsional reinforcement should provide a post-cracking strength T_n equal to or greater than the cracking torque T_{cr}:

$$T_n \geq T_{\text{cr}} \tag{8-90}$$

Based on the PCA tests (Hsu, 1968a), the cracking torque of solid sections subjected to pure torsion T_{cr} can be predicted by the formula

$$T_{cr} = 5\sqrt{f_c'}\,\frac{A_c^2}{p_c} \tag{8-91}$$

where f_c' and $\sqrt{f_c'}$ are in pounds per square inch (psi). In the case where torsion is combined with shear and torsion, the cracking torque may be greatly reduced.

Dividing Eq. (8-91) by Eq. (8-89), the ratio of cracking torque to maximum torque $T_{cr}/T_{n,max}$ for solid sections is

$$\frac{T_{cr}}{T_{n,max}} = \frac{25}{\sqrt{f_c'}} \tag{8-92}$$

where f_c' is expressed in pounds per square inch (psi) and $\sqrt{f_c'}$ is dimensionless. Equation (8-92) shows that the ratio $T_{cr}/T_{n,max}$ is a function of the concrete strength f_c'.

In the case of hollow sections, the cracking torque under pure torsion is

$$T_{cr} = 5\sqrt{f_c'}\,A_c h \tag{8-93}$$

where f_c' and $\sqrt{f_c'}$ are in pounds per square inch (psi) and h is the minimum wall thickness.

2   Limitation of the Angle α to Prevent Premature Failure

If a beam is designed for its maximum torque $T_{n,max}$, the angle α must be 45° to ensure that both the longitudinal steel and the transverse steel will yield simultaneously when the concrete crushes. That is to say, the section must be designed for the balanced point represented by point B in Figure 7.27 and Sections 7.5.2 and 7.5.3. If an angle α is chosen other than 45°, then premature failure will occur because either the longitudinal steel or the transverse steel will not yield at failure.

If the beam is designed for a torque T_n less than the maximum torque $T_{n,max}$, then the angle α could be designed within a range either less than or greater than 45°. This is possible because redistribution of stresses between the longitudinal steel and the transverse steel could occur so that the steel in both directions could yield before the crushing of concrete. Assuming that the range of α increases linearly with the ratio $T_n/T_{n,max}$, then the range of α is

$$\left(\frac{T_n}{T_{n,max}}\right)45° \le \alpha \le 90° - \left(\frac{T_n}{T_{n,max}}\right)45° \tag{8-94}$$

3   Limitation of the Angle α to Prevent Excessive Cracking

When the angle α is designed too far away from 45°, that is, when the range of α is specified to be overly large, then the condition of cracking will be excessive. In order to prevent excessive cracking, the 1978 CEB-FIP code limits the range of α to

$$\tan^{-1}\left(\frac{3}{5}\right) \le \alpha \le \tan^{-1}\left(\frac{5}{3}\right) \tag{8-95}$$

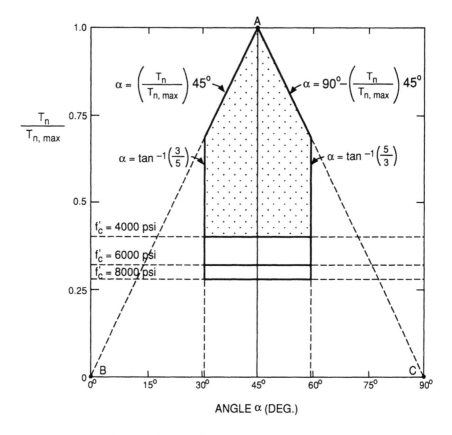

Figure 8.14 Design limitations for torsion.

The validity of this limitation has been demonstrated in Figure 5.9 and Section 5.4.

In short, the design must satisfy three conditions: (1) Condition (8-90) together with Eq. (8-91) or (8-92) for solid sections, or Eq. (8-93) for hollow sections; (2) Condition (8-94); and (3) Condition (8-95). Conditions 2 and 3 are valid for both solid and hollow sections.

A window of design opportunity that satisfies all three conditions is given by the shaded area in Figure 8.14. The window is bounded on the top by an inverted V-shaped line expressing Condition (8-94), on the sides by two vertical lines expressing Condition (8-95), and on the bottom by a horizontal line expressing Eq. (8-92). Because the ratio $T_{cr}/T_{n,\,\mathrm{max}}$ in Eq. (8-92) is a function of the concrete strength f'_c, three bottom horizontal lines are given for $f'_c = 4000$, 6000, and 8000 psi. It is also interesting to note that the sloping lines AB and AC, which express Condition (8-94), are reasonable approximations of the inverted V-shaped curves in Figure 7.30b (page 347) of the book by Hsu (1984). The line AB represents the case of $\varepsilon_l = \varepsilon_y$ and $\varepsilon_t > \varepsilon_y$, whereas the line AC gives the case of $\varepsilon_t = \varepsilon_y$ and $\varepsilon_l > \varepsilon_y$.

Condition (8-94) has the advantage that it can be easily extended to the case of combined shear and torsion. In this case, the minimum cross section can be checked by the CEB-FIP linear interaction criterion [Eq. (3-143)]:

$$\frac{V_n}{V_{n,\,\mathrm{max}}} + \frac{T_n}{T_{n,\,\mathrm{max}}} \leq 1 \tag{8-96}$$

Accordingly, in the case of combined shear and torsion, Condition (8-94) can be generalized as

$$\left(\frac{V_n}{V_{n,\max}} + \frac{T_n}{T_{n,\max}} \right) 45° \le \alpha \le 90° - \left(\frac{V_n}{V_{n,\max}} + \frac{T_n}{T_{n,\max}} \right) 45° \qquad (8\text{-}97)$$

DESIGN EXAMPLE 8.2 (RECTANGULAR SOLID SECTION)

Design a reinforced concrete rectangular beam to resist a torsional moment of 2000 in.-kip (226 kN-m). The net concrete cover is 1.5 in. (3.81 cm) and the material strength is $f_c' = 4000$ psi (27.5 MPa) and $f_y = 60,000$ psi (413 MPa).

SOLUTION

Select a section 20 in. wide by 30 in. high

$$A_c = (20)(30) = 600 \text{ in.}^2 \ (3871 \text{ cm}^2)$$

$$p_c = 2(20 + 30) = 100 \text{ in.} \ (254 \text{ cm})$$

Check maximum torque and cracking torque

Eq. (8-89) $\displaystyle T_{n,\max} = 0.2 f_c' \frac{A_c^2}{p_c} = 0.2(4000)\frac{(600)^2}{100}$

$$= 2880 \text{ in.-kip} \ (325 \text{ kN-m})$$

Eq. (8-91) $\displaystyle T_{cr} = 5\sqrt{f_c'}\,\frac{A_c^2}{p_c} = 5\sqrt{4000}\,\frac{(600)^2}{100} = 1138 \text{ in.-kip} \ (128 \text{ kN-m})$

$$T_{cr} < T_n = 2000 \text{ in.-kip} < T_{n,\max} \quad \textbf{O.K.}$$

Calculate t_d, A_0, and p_0

Eq. (8-85) $\displaystyle t_d = \frac{4T_n}{A_c f_c'} = \frac{4(2000)}{600(4)} = 3.33 \text{ in.} \ (8.45 \text{ cm})$

Eq. (8-86) $A_0 = A_c - p_c t_d/2 = 600 - 100(3.33)/2 = 434 \text{ in.}^2 \ (2800 \text{ cm}^2)$

Eq. (8-87) $p_0 = p_c - 4t_d = 100 - 4(3.33) = 86.7 \text{ in.} \ (220 \text{ cm})$

Design of stirrups

Eq. (8-70) $\displaystyle \frac{A_t}{s} = \frac{T_n \tan\alpha}{2A_0 f_{ty}} = \frac{2000 \tan\alpha}{2(434)(60)}$

$$= 0.0384(\tan\alpha) \text{ in.}^2/\text{in.} \ (0.0975 \, (\tan\alpha) \, \text{cm}^2/\text{cm})$$

The angle of crack inclination α can be chosen to suit the designer's purpose. For best crack control, α should be taken as 45°. However, for economic reasons, α can be taken to be less than 45°. The minimum α angle is

Eq. (8-94) $\alpha_{min} = \left(\dfrac{T_n}{T_{n,max}}\right)45° = \left(\dfrac{2000}{2880}\right)45° = 31.25°$ governs

Eq. (8-95) $\alpha_{min} = \tan^{-1}\left(\dfrac{3}{5}\right) = 30.9° < 31.25°$ **O.K.**

Select $\alpha = 31.25°$ for economic reason (then $\tan\alpha = 0.607$)

$$\frac{A_t}{s} = 0.0384(0.607) = 0.0233 \text{ in.}^2/\text{in. } (0.0592 \text{ cm}^2/\text{cm})$$

Select No. 4 bars $s = \dfrac{0.20}{0.0233} = 8.6$ in. (22 cm)

Check ACI stirrup spacing

$$p_1/8 = 10.75 \text{ in. } (27.3 \text{ cm}) < 12 \text{ in. } (30 \text{ cm})$$

$$s = 8.6 \text{ in. } (22 \text{ cm}) < 10.69 \text{ in. } (27.1 \text{ cm}) \quad \textbf{O.K.}$$

Use No. 4 stirrups at 8.5-in. (21.5-cm) spacing. The stirrups are designed to have a net concrete cover of 1.5 in. The center line of the steel cage, which is made up of the longitudinal steel bars and the stirrups, can be represented by the inner surface of the stirrups. The distance from the inner surface of the stirrups to the outer surface of the beam, defined as $\bar{c}$, should be less than $0.75t_d$ (Hsu and Mo, 1985b) to ensure the attainment of the calculated torsional strength:

$$0.75t_d = 0.75(3.33) = 2.50 \text{ in. } (6.35 \text{ cm})$$

$$\bar{c} = 1.5 + 0.5 = 2.0 \text{ in. } (5.08 \text{ cm}) < 2.5 \text{ in. } (6.35 \text{ cm}) \quad \textbf{O.K.}$$

Design of longitudinal steel

$$A_l = \frac{T_n p_0}{2A_0 f_{ly}\tan\alpha} = \frac{2000(86.7)}{2(434)(60)(0.607)} = 5.49 \text{ in.}^2 (35.4 \text{ cm}^2)$$

Select ten No. 7 longitudinal bars so that spacing will be less than 12 in. (30 cm) specified by ACI Code:

$$\text{Actual } A_l = 10(0.60) = 6.0 \text{ in.}^2 (38.7 \text{ cm}^2) > 5.49 \text{ in.}^2 (35.4 \text{ cm}^2) \quad \textbf{O.K.}$$

DESIGN EXAMPLE 8.3 (TRAPEZOID HOLLOW SECTION)

Design the reinforcement for the hollow box beam with the trapezoidal cross section as shown in Figure 8.8. The beam should be able to resist a torsional moment of 7400 in.-kip (836 kN-m). The net concrete cover is 1.5 in. (3.81 cm) and the material strengths are $f_c' = 4000$ psi (27.6 MPa) and $f_y = 60,000$ psi (413 MPa).

SOLUTION

For the given outer cross-sectional dimensional shown in Figure 8.8,

$$A_c = \frac{(3+4)(3)(12)^2}{2} = 1512 \text{ in.}^2 \ (9755 \text{ cm}^2)$$

$$p_c = \left(3 + 4 + 2\sqrt{3^2 + 0.5^2}\right)(12) = 157 \text{ in. } (399 \text{ cm})$$

Check maximum torque and cracking torque

Eq. (8-89) $T_{n,\max} = 0.2 f_c' \dfrac{A_c^2}{p_c} = 0.2(4000)\dfrac{(1512)^2}{157}$

$$= 11,650 \text{ in.-kip } (1316 \text{ kN-m})$$

Eq. (8-93) $T_{cr} = 5\sqrt{f_c'}\, A_c h = 5\sqrt{4000}\,(1512)(5) = 2391 \text{ in.-kip } (270 \text{ kN-m})$

$$T_{cr} < T_n = 7400 \text{ in.-kip} < T_{n,\max} \quad \textbf{O.K.}$$

Calculate t_d, A_0, and p_0

Eq. (8-85) $t_d = \dfrac{4T_n}{A_c f_c'} = \dfrac{4(7400)}{1512(4)} = 4.89 \text{ in. } (12.4 \text{ cm})$

$$< 5 \text{ in. } (12.7 \text{ cm}) \text{ wall thickness } \textbf{O.K.}$$

Eq. (8-86) $A_0 = A_c - p_c t_d/2 = 1512 - 157(4.89)/2 = 1128 \text{ in.}^2 \ (7277 \text{ cm}^2)$

Eq. (8-87) $p_0 = p_c - 4t_d = 157 - 4(4.89) = 137.4 \text{ in. } (349 \text{ cm})$

Design of stirrups

Eq. (8-70) $\dfrac{A_t}{s} = \dfrac{T_n \tan \alpha}{2 A_0 f_{ty}} = \dfrac{7400 \tan \alpha}{2(1128)(60)}$

$$= 0.0547(\tan \alpha) \text{ in.}^2/\text{in. } (0.139\,(\tan \alpha)\ \text{cm}^2/\text{cm})$$

Select $\alpha = 45°$ for best crack control

$$\frac{A_t}{s} = 0.0547(1) = 0.0547 \text{ in.}^2/\text{in. } (0.139 \text{ cm}^2/\text{cm})$$

Select No. 6 bars

$$s = \frac{0.44}{0.0547} = 8.04 \text{ in. } (20.4 \text{ cm})$$

Check stirrup spacing

$$\frac{p_1}{8} = \frac{157 - 4(2)(1.5 + 0.75/2)}{8} = 17.75 \text{ in. } (45.1 \text{ cm})$$

$$s = 8.04 < 12 \text{ in. } (30 \text{ cm}) < 17.75 \text{ in. } (45.1 \text{ cm}) \quad \textbf{O.K.}$$

Use No. 6 transverse hoop bars at 8-in. (20.3-cm) spacing.

Design of longitudinal steel

$$A_l = \frac{T_n p_0}{2 A_0 f_{ly} \tan \alpha} = \frac{7400(137.4)}{2(1128)(60)(1)} = 7.51 \text{ in.}^2 (48.4 \text{ cm}^2)$$

Select 13 No. 7 longitudinal bars so that spacing will be less than 12 in. (30 cm)

Actual $A_l = 13(0.60) = 7.80 \text{ in.}^2 (50.3 \text{ cm}^2) > 7.51 \text{ in.}^2 (48.4 \text{ cm}^2)$ **O.K.**

The arrangement of the mild steel rebars are shown in Figure 8.8.

References

ACI Committee 318 (1971), *Building Code Requirements for Reinforced Concrete*, ACI 318-71, American Concrete Institute, Detroit, MI, 78 pp.

ACI Committee 340 (1973), *Design Handbook in Accordance with Strength Design Method of ACI 318-71*, Volume 1, Special Publication SP-17(73), American Concrete Institute, Detroit, MI, 403 pp.

ACI Committee 318 (1983), *Commentary on Building Code Requirements for Reinforced Concrete*, ACI 318-83, American Concrete Institute, Detroit, MI, 155 pp.

ACI Committee 318 (1989), *Building Code Requirements for Reinforced Concrete*, ACI 318-89, American Concrete Institute, Detroit, MI, 353 pp.

Bai, S. L. et al. (1991), Experimental investigation of the static and seismic behavior of knee joints in in-situ reinforced concrete frames, Research Report, Chongching Architectural and Civil Engineering Institute, Chongching, Szechuang, and Central Design Institute of Metallurgy, Beijing, China, 203 pp. (in Chinese).

Belarbi, A. and Hsu, T. T. C. (1990), Stirrup stresses in reinforced concrete beams, *Structural Journal of the American Concrete Institute*, Vol. 87, No. 5, Sept.–Oct., pp. 350–358.

Belarbi, A. and Hsu, T. T. C. (1991), Constitutive laws of reinforced concrete in biaxial tension-compression, Research Report UHCEE 91-2, Dept. of Civil and Environmental Engineering, University of Houston, Houston, TX, July, 155 pp.

Branson, D. E. (1965), Instantaneous and time-dependent deflections on simple and continuous reinforced concrete beams, HPR Report No. 7, Part 1, Alabama Highway Department, Bureau of Public Roads, Birmingham, Alabama, August, pp. 1–78.

Branson, D. E. (1977), *Deformation of Concrete Structures*, McGraw-Hill Book Co., New York, 546 pp.

Bredt, R. (1896), Kritische Bemerkungen zur Drehungselastizitat, *Zeitschrift des Vereines Deutscher Ingenieure*, Band 40, No. 28, July 11, pp. 785–790; No. 29, July 18, pp. 813–817 (in German).

Canadian Standard Association (1984), Design of concrete structures and buildings, CAN3-A23.3-M84, Canadian Standard Association, Rexdale (Toronto), Ontario, 281 pp.

CEB-FIP (1978), *Model Code for Concrete Structures, CEB-FIP International Recommendation*, Third Edition, Comite Euro-International du Beton (CEB), 348 pp.

Collins, M. P. (1973), Torque-twist characteristics of reinforced concrete beams, *Inelasticity and Non-Linearity in Structural Concrete*, Study No. 8, University of Waterloo Press, Waterloo, Ontario, Canada, pp. 211–232.

Collins, M. P. and Mitchell, D. (1980), Shear and torsion design of prestressed and non-prestressed concrete beams, *Journal of the Prestressed Concrete Institute*, Vol. 25, No. 5, Sept.–Oct., pp. 32–100.

Elfgren, L. (1972), Reinforced concrete beam loaded in torsion, bending and shear, Publications 71:3, Division of Concrete Structures, Chalmers University of Technology, Goteborg, Sweden, 249 pp.

Fialkow, M. N. (1985), Design and capacity evaluation of reinforced concrete shell membranes, *Journal of the American Concrete Institute*, Proceedings, Vol. 82, No. 6, Nov.–Dec. pp. 844–852.

German Standard DIN 4334 (1958), *Bemessung im Stahlbetonbau (Design of Reinforced Concrete)*, Wilhelm Ernst and Sohn, Berlin, Germany, 57 pp.

Gupta, A. K. (1981), Membrane reinforcement in shells, *Journal of the Structural Division, ASCE*, Vol. 107, No. ST1, January, pp. 41–56.

Gupta, A. K. (1984), Membrane reinforcement in concrete shells: A review, *Nuclear Engineering and Design*, Vol. 82, October, pp. 63–75.

Han, K. J. and Mau, S. T. (1988), Membrane behavior of R/C shell elements and limits on the reinforcement, *Journal of Structural Engineering, ASCE*, Vol. 114, No. 2, Feb., pp. 425–444.

Hsu, T. T. C. (1968a), Torsion of structural concrete—Behavior of reinforced concrete rectangular members, *Torsion of Structural Concrete*, Special Publication SP-18, American Concrete Institute, Detroit, MI, pp. 261–306.

Hsu, T. T. C. (1968b), Ultimate torque of reinforced rectangular beams, *Journal of the Structural Division, ASCE*, Vol. 94, No. ST2, Feb., pp. 485–510.

Hsu, T. T. C. (1982), Is the staggering concept of shear design safe?, *Journal of the American Concrete Institute*, Proceedings Vol. 79, No. 6, Nov.–Dec., pp. 435–443.

Hsu, T. T. C. (1983), Author's closure to the paper, Is the "staggering concept" of shear design safe?, *Journal of the American Concrete Institute*, Proceedings Vol. 80, No. 5, Sept.–Oct., pp. 450–454.

Hsu, T. T. C. (1984), *Torsion of Reinforced Concrete*, Van Nostrand Reinhold, Inc., New York, 544 pp.

Hsu, T. T. C. (1988), Softening truss model theory for shear and torsion, *Structural Journal of the American Concrete Institute*, Vol. 85, No. 6, Nov.–Dec., pp. 624–635.

Hsu, T. T. C. (1990), Shear flow zone in torsion of reinforced concrete, *Journal of the Structural Division, ASCE*, Vol. 116, No. 11, Nov., pp. 3205–3225.

Hsu, T. T. C. (1991a), Nonlinear analysis of concrete membrane elements, *Structural Journal of the American Concrete Institute*, Vol. 88, No. 5, Sept.–Oct., pp. 552–561.

Hsu, T. T. C. (1991b), Nonlinear analysis of concrete torsional members, *Structural Journal of the American Concrete Institute*, Vol. 88, No. 6, Nov.–Dec., pp. 674–682.

Hsu, T. T. C., Mau, S. T., and Chen, B. (1987), A theory on shear transfer strength of reinforced concrete, *Structural Journal of the American Concrete Institute*, Vol. 84, No. 2, Mar.–Apr., pp. 149–160.

Hsu, T. T. C. and Mau, S. T., Eds. (1992), *Concrete Shear in Earthquake*, Elsevier Science Publishers, Inc., New York, 518 pp.

Hsu, T. T. C. and Mo, Y. L. (1983), Softening of concrete in torsional members, Research Report No. *ST-TH-001-83*, Department of Civil Engineering, University of Houston, Houston, Texas, 107 pp.

Hsu, T. T. C. and Mo, Y. L. (1985a), Softening of concrete in torsional members—Theory and tests, *Journal of the American Concrete Institute*, Proceedings Vol. 82, No. 3, May–June, pp. 290–303.

Hsu, T. T. C. and Mo, Y. L. (1985b), Softening of concrete in torsional members—Design recommendations, *Journal of the American Concrete Institute*, Proceedings Vol. 82, No. 4, July–Aug., pp. 443–452.

Hsu, T. T. C. and Mo, Y. L. (1985c), Softening of concrete in torsional members—Prestressed concrete, *Journal of the American Concrete Institute*, Proceedings Vol. 82, No. 5, Sept.–Oct., pp. 603–615.

Hsu, T. T. C. and Mo, Y. L. (1985d), Softening of concrete in low-rise shear walls, *Journal of the American Concrete Institute*, Proceedings Vol. 82, No. 6, Nov.–Dec., pp. 883–889.

Lampert, P. and Thurlimann, B. (1968, 1969), Torsionsversuch an Stahlbetonbalken (torsion tests of reinforced concrete beams), Bericht Nr. 6506-2, 101 pp.; Torsion-Biege-Versuche an Stahlbetonbalken (Torsion-bending tests on reinforced concrete beams), Bericht Nr. 6506-3, Institute of Baustatik, ETH, Zurich, Switzerland (in German).

Mau, S. T. and Hsu, T. T. C. (1986), Shear design and analysis of low-rise structural walls, *Journal of American Concrete Institute*, Proceedings Vol. 83, No. 2, Mar.–Apr., pp. 306–315.

Mau, S. T. and Hsu, T. T. C. (1987a), Shear behavior of reinforced concrete framed wall panels with vertical loads, *Journal of American Concrete Institute*, Proceedings Vol. 84, No. 3, May–June, pp. 228–234.

Mau, S. T. and Hsu, T. T. C. (1987b), Shear strength prediction for deep beams with web reinforcement, *Journal of American Concrete Institute*, Proceedings Vol. 84, No. 6, Nov.–Dec., pp. 513–523.

Mau, S. T. and Hsu, T. T. C. (1989), A rational formula for the shear design of deep beams, *Structural Journal of the American Concrete Institute*, Vol. 86, No. 5, Sept.–Oct., pp. 516–523.

Morsch, E. (1902), *Der Eisenbetonbau, seine Anwendung und Theorie*, 1st edition, Wayss and Freytag, A. G., Im Selbstverlag der Firma, Neustadt, A. D. Haardt, 118 pp.; 2nd edition, Verlag Von Konrad Wittmer, Stuttgart, 1906, 252 pp.; 3rd edition (English Transl. by E. P. Goodrich), McGraw-Hill Book Company, New York, 1909, 368 pp.

Nielsen, M. P. (1967), Om forskydningsarmering i jernbetonbjaelker (On shear reinforcement in reinforced concrete beams), *Bygningsstatiske Meddelelser*, Vol. 38, No. 2, pp. 33–58.

Nielsen, M. P. and Braestrup, M. W. (1975), Plastic shear strength of reinforced concrete beams, *Bygningsstatiske Meddelelser*, Vol. 46, No. 3, pp. 61–99 (Copenhagen, Denmark).

Pang, X. B. and Hsu, T. T. C. (1992), Constitutive laws of reinforced concrete in shear, Research Report UHCEE 92-1, Dept. of Civil and Environmental Engineering, University of Houston, Houston, TX.

Rausch, E. (1929), *Berechnung des Eisenbetons gegen Verdrehung* (*Design of Reinforced Concrete in Torsion*), Technische Hochschule, Berlin, Germany, 53 pp. (in German). A second edition was published in 1938. The third edition was titled *Drillung (Torsion) Schub und Scheren in Stahlbetonbau*, Deutscher Ingenieur-Verlag GmbH, Dusseldorf, Germany, 1953, 168 pp.

Ritter, W. (1899), Die Bauweise Hennebique, *Schweizerische Bauzeitung*, Vol. 33, No. 7, pp. 59–61 (Zurich, Switzerland).

Robinson, J. R. and Demorieux, J. M. (1972), Essais de traction-compression sur modeles d'ame de poutre en Beton Arme, IRABA Report, Institut de Recherches Appliquees du Beton de l'Ame, Part 1, June 1968, 44 pp.; Resistance Ultimate du Beton de l'ame de poutres en Double te en Beton Arme, Part 2, May 1972, 53 pp.

Schlaich, J., Schafer, K., and Jennewein, M. (1987), Toward a consistant design of structural concrete, *Prestressed Concrete Institute Journal*, Vol. 32, No. 3, May–June, pp. 74–150.

Swann, R. A. (1969), Flexural strength of corners of reinforced concrete portal frames, Technical Report TRA 434, Cement and Concrete Association, London, England, 14 pp.

Tamai, S., Shima, H., Izumo, J., and Okamura, H. (1987), Average stress–strain relationship in post yield range of steel bar in concrete, *Concrete Library of JSCE*, No. 11, June 1988, pp. 117–129. (Translation from *Proceedings of JSCE*, No. 378/V-6, Feb. 1987.)

Thurlimann, B. (1979), Shear strength of reinforced and prestressed concrete—CEB approach, *ACI–CEB–PCI–FIP Symposium, Concrete Design: U.S. and European Practices*, ACI Special Publication SP-59, American Concrete Institute, Detroit, MI, pp. 93–115.

Vecchio, F. and Collins, M. P. (1981), Stress–strain characteristic of reinforced concrete in pure shear, *IABSE Colloquium, Advanced Mechanics of Reinforced Concrete*, Delft, Final Report, International Association of Bridge and Structural Engineering, Zurich, Switzerland, pp. 221–225.

Vecchio, F. and Collins, M. P. (1982), The response of reinforced concrete to in-plane shear and normal stresses, Publication No. 82-03 (ISBN 0-7727-7029-8), Department of Civil Engineering, University of Toronto, Toronto, Canada, 332 pp.

Vecchio, F. J. and Collins, M. P. (1986), The modified compression field theory for reinforced concrete elements subjected to shear, *Journal of American Concrete Institute*, Proceedings Vol. 83, No. 2, Mar.–Apr., pp. 219–231.

Index

Bernoulli compatibility condition, 2, 5, 11, 12, 19
  compression rebar stresses in, 33
  compression zone depth in, 75
  curvature and, 51
  moment-axial-load-curvature relationship and, 74
Bernoulli compatibility truss model, 11, 21–74
  applications of, 6
  bending and, see Bending
  bending deflections and, 37–40
  bending ductility and, 61–62
  bending rigidities and, see Bending rigidities
  compression failure in, 69–72
  doubly reinforced beams and, 31–33, 51–55
  flanged beams and, 33–34, 36
  general theory of bending in, 21–24
  history of, 7–8
  linear bending theory in, see Linear bending theory
  mild-steel reinforced beams and, 40–41
  moment-curvature relationships in, 59–62
  nonlinear bending theory in, see Nonlinear bending theory
  overreinforced beams and, 45–46
  principles of, 6, 24–25
  singly reinforced beams and, 30–31, 41–46
  tension failure and, 67–69
  transformed area for rebars and, 29–30
  underreinforced beams and, 46–51
Bernoulli hypothesis, 3, 7, 21, 24, 25, 33
Bernoulli linear strains, 1
Biaxial constitutive laws, 167
Biaxial stresses, 250
Boundary conditions, 187, 189
Boundary stresses, 5, 187, 189
Bredt's formula, 86–87, 259
B region, 5, see Main region

C

CEB-FIP Model Code, 110–121
  basic principles of, 112–116
  beam shear in, 110–112, 114
  contribution of concrete and, 116–119
  practical considerations in, 120–121
  reinforcement design and, 119–120
  shear in, 112–116, 119–120, 121
  shear-torsion interactions and, 119–120
  torsion in, 115–116, 118–120, 121
Centroidal axis, 30
Concentric load, 63
Chain rule, 262
Closing moment, 15–17, 18
Columns, 63–74, see Axial load and Bending
Compatibility conditions, 5, 9, 12, 23, 255, 256, see also specific types
  Bernoulli, see Bernoulli compatibility condition
  equations for, see Compatibility equations
  Mohr, 2, 5, 9, 11, 12, 167–168
  two-dimensional, 3

Compatibility equations, 10, 11, 68, 157–160, 190
  for membrane elements, 157–160, 168, 219
  in softened truss model, 194, 259–263, 270–271
  for torsion, 259–263, 270–271
Compression failure, 69–72
Compression rebar stresses, 33
Compression steel, 52, 53–54, 55
  strains on, 68, 69
  stresses on, 53, 55, 68
  stress-strain relationships of, 67
Compression strains, 9, 21, 41, 68, 69
Compression strength, 6
Compression stresses, 9, 21, 41, 53, 54, 75, 141
  average, 266–267
  principal, 133
  resultant of, 286
  softened, 263–265
  of stirrup forces, 182
  transverse, 144
Compression zone, 73, 75, 76, 286, 287
Concrete contribution, see Contribution of concrete
Constitutive equations, 271–272
Constitutive laws, 26, see also specific types
  biaxial, 167
  of concrete, 1, 9, 40, 194
    equations for, 219
    nonlinear, 41–43
  Hooke's, see Hooke's law
  of materials, 5, 25, 194–196, 219
  of mild steel, 40, 43, 194, 219
  nonlinear, 40, 41–43
  of reinforcement, 24, 40
  in softened truss model, 193, 194–196, 263–270
  of steel, 1, 44, 198
  uniaxial, 1, 167
Contribution of concrete, 8, 9–10, 116–117
Cracking
  bending rigidities and, 30–34
  control of, 160–163, 168, 298, 299
  of doubly reinforced beams, 31–33
  excessive, 295–297
  of flanged beams, 33–34
  of membrane elements, 170–175
  patterns of, 13
  severity of, 202
  of singly reinforced beams, 30–31
  strain at, 204
  strength of concrete, 209
  stress at, 204
  torque of, 294, 295, 297, 299
  width of, 207
  at yielding of steel, 164–166
Curvature, 51, 73–74

D

Deflections, 37–40
Deformations, 51, 221, 260, see also specific types

I

Inclined web reinforcement, 110–112
Inner boundary, 253–256
Internal couple concept, 24, 25, 75
Internal resisting moment, 23–24, 34
Internal torsion moment, 8

J

Joint design, 5

K

Knee joints, 13–18

L

Lateral loads, 38
Lever arm area, 87–88, 268–270, 286, 287–288
Linear analysis, 11, 193, see also specific types
Linear bending theory, 25–40
  analysis problem solutions in, 27–29
  bending deflections and, 37–40
  bending rigidities and, 30–36
  transformed area for rebars and, 29–30
Linear membrane element theory, 11
Linear regression analysis, 291
Load, see also specific types
  axial, see Axial load
  concentric, 63–66
  eccentric, 63–66
  lateral, 38
  midspan concentrated, 103–106, 181
  proportional, 234–237, 238, 239, 242, 249
  service, 5, 6, 34, 167
  static, 221
  ultimate, 40–41, 53, 167
  uniformly distributed, see Uniformly distributed load
Load-deformation behavior, 1
Local regions, 3–5, 10, 19
Longitudinal steel, 82, 86, 102, 183–185, 227
  in beam shear, 117–118
  design of, 298, 300
  in torsion, 119
  yielding of, 164–166, 247

M

Main regions, 3–5, 18–19
Mechanics principles, 2, 5, 24
Member design, 5

Membrane elements, 1, 10
  balanced conditions in, 251–252
  basic equations for, 193–196
  compatibility equations for, 157–160, 168
  cracked, 170–175
  equal strain condition in, 250–251, 252
  equations for, 219–222
  equilibrium of, 136–143
  equilibrium truss model for, 143–147
  failure modes of, 249–256
  failure regions in, 252–256
  linear theory of, 11
  in Mohr compatibility truss model, 167–178
    cracked, 170–175
    example problem in, 172–175
    Hooke's law and, 167, 168–170
  nonlinear theory of, 12
  non-prestressed, 239, 250
  overreinforced, 291, 294
  prestressed, 194
  proportional load in, 234–237, 238, 239, 242, 249
  reinforcement in, 161–163
  shear, 257–259
  softened truss model for, see Softened truss model
  strains in, 11, 149–166
    crack control and, 160–163
    principal, 153–154
    RC sign conventions for, 152–153
    transformation of, 149–154
      equations for, 155–156
      in terms of principal strains, 155–160
  strength of, 237–239
  stresses in, 1, 2, 11, 123–147
    equilibrium and, 136–143
    equilibrium truss model for, 143–147
    example problem in, 145–147
    principal, 130–131
    transformation of, 123–131
      equations for, 131–133
      in terms of principal stresses, 131–136
  uncracked, 176–178
  underreinforced, 81, 249, 251, 287, 291, 292, 294
Midspan concentrated load, 103–106, 181
Mild steel, 63
  constitutive equations for, 219, 220
  constitutive laws of, 40, 43, 194, 219
  rebars of, 300
  reinforced beams of, 40–41
  strains in, 218
  stress-strain relationships of, 40, 43, 205–217
    bare bars and, 205–206
    steel bars embedded in concrete and, 206–217
Minimum reinforcement, 84
Models, 1, 5–11, see also specific types
  applications of, 5–6
  Bernoulli compatibility truss, see Bernoulli compatibility truss model
  equilibrium truss, see Equilibrium truss model
  future development of, 9–11
  historical development of, 6–9

9 780367 450137